THE SUSTAINABILITY OF AIR T

To my family

The Sustainability of Air Transportation
A Quantitative Analysis and Assessment

MILAN JANIĆ
Delft University of Technology, The Netherlands

Routledge
Taylor & Francis Group

LONDON AND NEW YORK

First published 2007 by Ashgate Publishing

2 Park Square, Milton Park, Abingdon, Oxon OX14 4RN
711 Third Avenue, New York, NY 10017, USA

Routledge is an imprint of the Taylor & Francis Group, an informa business

First issued in paperback 2016

British Library Cataloguing in Publication Data
Janić, Milan
  The sustainability of air transportation: a quantitative
  analysis and assessment
  1. Aeronautics, Commercial - Environmental aspects
  2. Aeronautics, Commercial - Management 3. Aeronautics,
  Commercial - Environmental aspects - Mathematical models
  4. Sustainable development 5. Environmental indicators
  I. Title
  387.7

Library of Congress Cataloging-in-Publication Data
Janić, Milan.
  The sustainability of air transportation: a quantitative analysis and assessment /
by Milan Janić.
      p. cm.
  Includes bibliographical references and index.
  ISBN 978-0-7546-4967-0
  1. Air travel. 2. Aeronautics, Commercial. 3. Sustainable development I. Title.

  HE9776.J367 2007
  387.7--dc22

ISBN 978-0-7546-4967-0 (hbk)   2006103136
ISBN 978-1-138-26273-7 (pbk)

Transferred to Digital Printing in 2012

Contents

List of Figures

List of Tables

Preface

Transport has always played an important role in the economic development of almost all countries worldwide. In general, growth of transport volumes (both passengers and freight) over time has significantly contributed to the overall economic growth, and vice versa. However, at the same time, it has created increasingly noticeable and worrying negative impacts on society and the environment, in terms of air pollution of both a local and global character, local noise and vibration, waste, land use, congestion, and traffic accidents. Such developments have given rise to the question of proper balancing between the positive effects (benefits) and unavoidable negative impacts (cost of damages) of transport operations over the short, medium, and long term. The concept of sustainable transport has emerged as the potential solution and as a major and rather popular policy framework. It was expected that this would be acceptable for almost all actors directly or indirectly involved in transport system issues. These have been users – passengers and freight shippers, transport infrastructure providers, service operators, manufacturers of transport vehicles, facilities and equipment, local communities, and authorities and other market and system regulatory bodies at different – local, national and international – levels.

The air transport system is a component of the transport system. However, the term 'air transportation' in the title of this book has been often used instead of the term 'air transport system'. Nevertheless, as the author I have in mind the same contents of both terms, and therefore I prefer to keep the term 'system', aiming to point out the complexity of the air transport system structure and the internal relations between its particular components.

The air transport system consists of several major components – subsystems – such as the users-passengers and air cargo shipments, airlines, airports, air traffic control (management), aerospace manufacturers, and dedicated national and international regulatory authorities (bodies). Commonly, the airports and air traffic control are regarded as the fixed system infrastructure providing the space, facilities and equipment for servicing the main users – passengers and freight, and airline aircraft. In this context, users-passengers, freight (cargo), and airlines (aircraft) represent the demand for the 'fixed infrastructure'. Regarding the relationship with users (passengers and freight shippers), the airlines provide the aircraft as the 'mobile infrastructure' to serve them. The airlines, airports, and air traffic control use facilities and equipment supplied by the aerospace manufacturers according to given institutional (national and international) regulations aiming to guarantee safe, efficient and effective operations.

Nowadays, the air transport system is recognised as one of the fastest growing sectors within the transport sector as well as the general regional economies. Such development has already created, and will certainly continue to create both positive and negative effects on society and the environment similar to the entire transport system. Generally, the positive effects, in addition to its contribution to globalisation of the economy and people's everyday life, embraces employment within and around the system, which in turn stimulates local (regional) and global (national and international) economies and welfare. The negative impacts include the users and airlines; losses of time due to the air traffic congestion (delays), and damages to the environment and people's health caused by depletion of the non-renewable fuel resources, air pollution from burning fuels at both local – airport – and global – airspace – level, aircraft noise around airports, land take for building up the fixed infrastructure, contamination of soil and drinking water, air traffic incidents and accidents, and waste. Expressed in monetary terms, these damages are regarded as externalities, which still need market recognition, i.e., internalising. The specific and often neglected 'externality' also appears to be disruptions of planned operations caused by bad weather, terrorist attacks (and/or threats), epidemic diseases, and other unavoidable causes. This 'externality' is usually expressed in the losses for almost all actors involved.

The positive and negative effects (impacts) interrelate with each other and are permanently in dynamic interaction. Recognising such dependability raises the question of operationalisation of the strategies for driving development of the air transport system in the short and long-term along the trajectory permanently (re-) balancing positive effects and negative impacts. Similarly, as with the entire transport system, the concept of sustainable development has emerged as an optional strategy. In general, the core of the concept is to find an appropriate balance between the benefits-revenues and impacts-costs while respecting the long-term equality and fairness of treatment of the interests and preferences of the particular actors involved.

This book elaborates the concept of sustainable air transport system using a quantitative approach. This implies applying quantitative methods/tools to analyse and assess the performances of the main system components relevant to its sustainability – airlines, airports and air traffic control. In my opinion, such an approach seems convenient regarding an inherent complexity and diversity of the system components, performances and their (inter)-relationships. Consequently, aiming at fulfilling the expectations of the readers, the book contains: i) a description of the system, components and their performances; ii) representative real-life cases making particular topics easier to understand; and iii) adequate cases providing a relatively clear picture of the current and prospective (forthcoming) achievements and problems in dealing with sustainability of the air transport system.

The book consists of five chapters. Chapter 1 introduces the concept of sustainable development, sustainable entities, sustainable transport, and a sustainable air transport system. Chapter 2 describes the technical/technological, operational, economic, social, environmental and institutional performances of the main components of air transport system – airlines, airports and air traffic control. The performances of users – passengers and freight shippers – as well as of the aerospace manufacturers are implicitly taken into account. Chapter 3 deals with the framework, principles, and prospective methodologies for analysis and assessment of the effects and impacts of particular components of the air transport system on society and the environment.

Chapter 4 contains the selected cases of modelling sustainability of particular components of the air transport system. The material is mainly taken from my already published research work. Some, however, are still awaiting publication. In particular this chapter aims to illustrate the complexity of dealing with particular sustainability issues, the necessity for an integral approach, and the potential of modelling while appropriately addressing and resolving specific sustainability problems. The final chapter (Chapter 5) summarises the contents of the previous chapters and addresses some future issues with respect to the sustainability of the air transport system.

To my knowledge, this present book appears as a unique example since it tries to elaborate causes and consequences through an explicit consideration of the air transport system technical/technological and operational performances as the main driving forces of other – economic, social, environmental, and institutional – performances, and vice versa. Therefore, the book is of interest to a relatively wide auditorium of readers, including those involved in (re)-solving the problems of the development of the (air) transport system – professionals within the system, researchers and consultants, transport and spatial policy makers at regional (community), national, and international level. In addition, advanced undergraduates and postgraduates in the field of air transport planning, technology, applied operational research, (air) transport economics and policy could find the book of value for their courses.

Even though the current air transport system is changing relatively rapidly, the book focuses on the generic and principal rather than case-specific issues and will therefore have a relatively lasting value. The book provides a relatively solid and comprehensive framework for the analysis and assessment of changes, identification of problems, and looking for satisfactory solutions for the future sustainable development of the air transport system. In some sense, regarding the above-mentioned contents, the book represents a complement to my previous book: *Analysis and Modelling of Air Transport Systems: Capacity, Quality of Services and Economics* published in the year 2001.

This is an important time for both the air transport industry, and the academic, research and policy community. First, the system has recovered from the crisis of the terrorist attacks of September 11, 2001 and is again on a relatively stable trajectory of growth despite the flattering (rising) fuel prices threatening to again slow down the expected growth. According to many forecast exercises this growth will continue at an average rate of 5 per cent in passenger and 6 per cent in freight transport demand over the next two decades. It will primarily be driven by the overall economic growth (Gross Domestic Product – GDP), further globalisation of the regional and world's economy, and even further decreasing of the airfares achieved among other factors also by the growth of low-cost carriers. The growth rates in the international markets are expected to be about twice as high as those in the domestic markets, as well as higher in the developing than in the developed countries. Despite temporal concerns, continuous flattering (rising) of the fuel prices is not expected to significantly affect the above-mentioned long-term trends. The system infrastructure – airports and air traffic control/management – is expected to support such growth safely, efficiently and effectively. In this context, new technologies will provide great support. Nevertheless, the scale and scope of the impacts of such growth on society and the environment will continue to be increasingly a matter of concern occupying the number of actors involved. Second, in addition to the industry, the academic and research community is seemingly strengthening its efforts to educate staff able to deal with solutions for policy makers. The two local efforts in Europe close to me are illustrative: i) setting up the Delft Centre for Aviation Research (DCA) at Delft University of Technology (Delft, the Netherlands) in spring 2006; and ii) starting the international MSc studies in aviation as the joint project of the University of Belgrade (Belgrade, Serbia), University of Trieste (Trieste, Italy), University of Graz (Graz, Austria), and EUROCONTROL (Budapest, Hungary) in the autumn of 2007. Finally, at the global scale, Air Transport Research Society (ATRS) as part of the World Conference on Transport Research (WCTR) continues to strengthen its activities by increasing the number and quality of papers presented at its regular annual conferences. The same is happening at the regular annual U.S. Transportation Research Board (TRB) meetings in Washington DC (U.S.). Therefore, the book is expected to be a complement to the above-mentioned academic efforts.

In writing the book I am particularly grateful to the OTB Research Institute of Delft University of Technology (Delft, the Netherlands), which has permanently offered good working conditions, an inspiring scientific atmosphere and a positively energised environment.

I am also grateful to my family – spouse, Vesna, and son, Miodrag (student of medicine) – for their considerable and permanent support, inspiration and very high level of tolerance and understanding of my devotion to this project instead of sometimes to them. Finally, I would like to express my particular gratitude to my brother Predrag (a famous architect) whose professional approach, and continuous insistence on the highest quality and perfection of his work has always strongly driven and inspired me.

Milan Janić

List of Abbreviations

ACI	Airport Council International
AEA	Association of European Airlines
ATAG	Air Transport Action Group
ATC/ATM	Air Traffic Control/Management
BAA	British Airport Authority
BTS	Bureau of Transport Statistics (U.S.)
CAA	Civil Aviation Authorities
CEC	Commission of European Communities
CTP	Common Transport Policy
DETR	Department of Transport and Regions (UK)
EC	European Commission
EEC	EUROCONTROL Experimental Centre
EU	European Union
EPA	Environmental Protection Agency
ERA	European Regional Airline Association
FAA	Federal Aviation Administration
GAO	Government Accountancy Office (U.S.)
GDP	Gross Domestic Product
GNP	Gross National Product
IATA	International Air Traffic Association
ICAO	International Civil Aviation Organisation
IPCC	Intergovernmental Panel of Climate Change
NASA	National Astronautics and Space Administration
OECD	Organisation for Economic Cooperation and Development
RCEP	Royal Commission of Environmental Pollution
SFC	Specific Fuel Consumption
TRB	Transportation Research Board (U.S.)
UIC	International Union of Railways
UN	United Nations
WTO	World Trade Organisation

Chapter 1

The Science and Policy of Sustainability

1.1 Background

The current production and consumption systems across the world have different impacts on the natural systems that sustain life on Earth. Mankind is a part of such constantly evolving systems. The science dealing with the systems sustaining life is ecology, which investigates ecosystems whose basic parts are plants and organisms mostly formed of water and organic materials. These ecosystems permanently change trying to survive using their reproductive and survival strategies. Except in specific situations, the human population is not considered a part of these ecosystems frequently called the life-sustaining systems. Nevertheless, it also fights for survival in the same way as ecosystems by adapting to natural and self-induced impacts appropriately. Under such circumstances, there might be some interrelations between the people-made and natural causes of impacts on particular ecosystems, which have frequently been referred to as environmental problems.

In general, people have changed their natural environment and consequently affected the life sustaining systems by almost all their activities. Historically, there are four categories of environmental problems that have been identified:

- in the life sustaining systems threatening to people's health;
- in the life sustaining systems threatening the cultivation of spices;
- in human society affecting the economic benefits obtained from natural processes; and
- In the life sustaining systems and human society affecting the benefits obtained from natural processes of future generations and/or developing nations.

The first category of impacts refers to contamination of the environment by man-made waste, excessive emissions of air pollutants from the industrial processes and noise (all usually with a strong postponing effect).

The second category of impacts originates from the attitude that Nature has an intrinsic value, which has to be saved at any price. At a global scale this attitude appears to be in conflict with the accelerating industrial development of particularly the 20th and the beginning of twenty-first century.

The third category of impacts relates to the growth of the industrial society through the over-exploitation of nature. In other words, natural resources to support such growth have become scarcer at both a local and global scale. In many cases it seems that the speed and intensity of harvesting these still-available resources has been higher than the speed and intensity of their natural recovery, which in turn has caused their depletion and irreversible damage. Under such circumstances, the need to control the harvesting processes has emerged.

The last category of impacts relates to the rights to use the natural (constrained) resources. In other words, as the amount of these resources is limited and generally decreasing, the question is whose right is it to use them? For example, is it right to leave future generations without crude oil? Will they need some quantities or at least an adequate alternative? Are today's generations obliged to provide such adequate alternatives? Or, is it justified that currently 20 per cent of the world's population uses about 80 per cent of the world's resources?

The above-mentioned impacts materialised in different forms and were recognised relatively early, at the end of nineteenth century, but have become particularly relevant during the last two decades of the twentieth century and the beginning of the twenty-first century. Consequently, the concepts of 'sustainable development' and 'sustainable society' have been promoted both at the national and international (global) level.

1.2 The Concept of Sustainable Development

The social-economic development of the twentieth century showed that a liberal conventional (capitalist) economy, as a counterpart to the central planning economy, has seemingly been unable to provide sustainable development for the human population. Pure economic growth as an exclusive objective of the overall social-economic development has not, until recently, appropriately considered the side effects in terms of environmental, intergenerational and livability effects. Most businesses have not been held accountable for the costs of damages made to the society and environment, and eventually their restoration (Griethuysen, 2002; Mikesell, 1992). The damages and restoration costs mainly included those of noise and air pollution. In general, the approach aiming to prove these statements uses the following equation to quantify the environmental degradation caused by economic development (Hooper and Gibbs, 1995):

$$TED = P * PCI * ED \qquad (1.1)$$

where
   TED is the total environmental degradation (billion $US);
   P is the population (number of people);
   PCI is the Gross National Product (GNP) per capita ($US per inhabitant); and
   ED is the environmental degradation per unit of GNP ($US per $US/ inhabitant).

According to Equation 1.1, for example, if the population doubles, the environmental degradation per unit of GNP per capita will have to half in order to keep the overall degradation at the current level. Similarly, if GNP per capita increases at an average annual rate of about 2.5–5 per cent, the required reduction of the environmental degradation would have to be by about 5–10 times, respectively. Nevertheless, this 'one-sided' equation has a limited value since it does not explicitly reflect a balance between the environmental degradation and other social and economic benefits to be obtained by the economic development.

Efforts to improve this rather rough and simplistic approach have resulted in the launching of the concept of 'sustainable development' (Kelly, 1998). Generally, this concept is defined as the 'development that meets the present needs without compromising the ability of the future generations to meet their needs'. This embraces: 'i) prioritizing particular needs, and ii) respecting limitations of the environment to meet the present and future societal needs' or 'maximization of welfare and provision of a sound, economic, social and environmental base for both present and future generations' (Brundtland, 1987; OECD, 1998). Daly (1991) identified three requirements for sustainable development as follows: i) the rates of use of renewable resources should not exceed the rates of their regeneration; ii) the rates of use of non-renewable resources should not exceed the rates of development of their substitutes; and iii) the rates of pollution emission should not exceed the assimilative capacity of the environment. Particularly, Gale and Gordray (1994) set up four questions for sustainable development, bearing in mind the differences between people, their culture, habits, and the specific objectives as follows: 'What is sustained? What sustains it? How is sustainability measured? What are the policies?' In addition, the concept is related to individual entities as the 'capacity of the entity enabling its continuation in the future' (Ekins, 1996), the 'maintenance of the natural resource base including the waste absorptive capacity of the environment ...' (Mikesell, 1992) and/or the 'valuation of the environment including the time horizon for decision making and issues of equity' (Pearce, Markandya and Barbier, 1989).

Generally, the above-mentioned statements imply that long-term sustainable development, in addition to the positive overall socio-economic effects, should always take into account the constraints of the non-renewable resources and absorptive capacity of the environment for different types of impacts. In such a context, the sustainable entity is usually assumed to possess at least three – social, economic and environmental – dimensions. The economic dimension refers to the conventional economy and services, which increase the living standards of individuals and groups beyond the monetary income ('man-made capital'). The social dimension relates to the people's intra-personnel qualities such as skills and experiences ('human capital'). The environmental dimension considers the sum of all bio-geological processes and their associated elements ('environmental capital') (Spangenberg, 2002). However, the essence of the above-mentioned dimensions appears to be very specific to the systems with solid technical/technological and operational dimension such as transport system.

1.3 The Sustainability of the Transport System

Transport systems have always played an important role for people and society. On the one hand it has provided clear socio-economic benefits. On the other it has generated a series of socially and environmentally adverse effects. For some time the transport system has been considered as the main source of the man-made

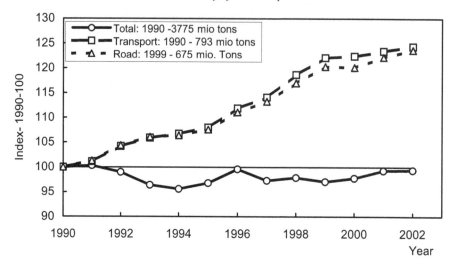

Figure 1.1   Emissions of CO$_2$ in the EU (25 Member States)
Compiled from EC (2005)

emissions of air pollutants, particularly of carbon dioxide – CO$_2$. For example, in the EU (European Union) like in other regions of the world, the transport system has permanently increased its share in the total man-made emissions of CO$_2$, from about 21 per cent in the year 1990 to about 26 per cent in the year 2002 (EC, 2005). The total emission of CO$_2$ from all man-made polluting sources such as power and heat generation, industry, transport, households, and services had been 3,775 million tons in the year 1990 and 3,750 tons in the year 2002. Figure 1.1 shows changes during the observed period.

As can be seen, while the total man-made emissions in CO$_2$ equivalents have been kept almost continuously below or around the 1990 level, emissions from the transport system have continuously increased, faster during the earlier and slower during the later part of the observed period. In the total amounts, the road transport is responsible for about 85 per cent. Therefore, the substantive research, policy and practical efforts, not only in the European Union (EU) but also in other parts of the Developed World, at the different geographical and institutional scales, have been undertaken aiming at appropriately balancing these and other effects in the long-term. Consequently, the concept of the sustainability of the transport system has been created (OECD, 2000, 2002).

As always when dealing with an innovative concept, understanding of particular terms and the definition has appeared to be of the greatest importance. Nevertheless, at present, a universally accepted definition of the sustainable transport system does not exist. Some definitions used for different purposes by both the academic community and practitioners are as follows (Litman, 2003):

> Sustainable transportation is that, which does not endanger public health or ecosystems and that meets needs for access consistent with (a) use of renewable resources that are below their rates of regeneration, and (b) use of non-renewable resources below the rates of development of renewable substitutes (OECD, 2001).

A sustainable transport system i) allows basic access and development needs of individuals, companies and society to be met safely in a manner consistent with human and ecosystem health and promotes equity between successive generations; ii) is affordable, operates fairly and efficiently, offers a choice of transport mode and supports a competitive economy as well as balanced regional development; and iii) limits emissions and waste within the planet's ability to absorb them, uses renewable resources at or below their rates of generation, and uses non-renewable resources at or below the rates of development of renewable substitutes while minimising the impacts of the use of land and generation of noise (CEC, 1999).

The goal of sustainable transport is to ensure that environment, social and economic considerations are factored into decisions affecting transportation activity (TC, 1999).

A sustainable transport system is one in which fuel consumption, vehicle emissions, safety, congestion, and social and economic access are of such levels that they can be sustained into the indefinite future without causing great or irreparable harm to future generation of people throughout the world (Richardson, 1999).

The above-mentioned definitions point out that sustainable transport systems should simultaneously and continuously contribute to increasing the social-economic welfare without depleting the natural resources, destroying the environment and harming people's health. Nevertheless, these definitions are still not operational, mainly due to the lack of the quantitative targets on the positive and negative effects including the time schedule for their implementation. Over time, some efforts for operationalisation of the sustainability definitions have been made. This has included setting up the quantitative targets on particular impacts, refining the scope of sustainability and building up the systems for monitoring developments. Some examples of such efforts have been the Kyoto Protocol at global, the EU Common Transport Policy (CTP) at regional-continental, and the numerous state policies at the national scale (CEC, 1998). In most cases, the states have agreed to set up their own long-term quantitative targets as contributions to the common more general objectives. Table 1.1 shows an example of such long-term targets for reducing the emissions of greenhouse gases from the transport sector in Sweden (SEPA, 1996).

The EU has set up its targets for the impacts of the transport system on the environment based on the OECD Environmental Sustainability Targets (EST). Table 1.2 shows the example (CEC, 2000; OECD, 2000, 2002).

Table 1.1 **An example of setting up the national targets for reduction of carbon dioxide (CO_2) in Sweden**

Transport mode/year	2005[1]	2020[1]	2050[1]
Road traffic	−10%	−20%	−75%
Aviation	+30%	0%	−20%
Railroad	0%	−20%	−20%
Shipping	0%	−20%	−20%
Total:	**−5%**	**−15%**	**−60%**

[1] Compared with the year 1990

Source: SEPA (1996)

Table 1.2   The targets for reduction of some environmental impacts in the EU until the year 2020

Type of pollutant	Long term (quantitative) targets
CO_2	should not exceed 20% of the total CO_2 emissions in 1990
VOCs	should not exceed 10% of the total VOCs emissions in 1990
NO_x	should not exceed 10% of the total transport-related NO_x emissions in 1990
Particulates	Reduction of fine particulate (PM_{10}) emissions from transport for 55–99%
Noise	should not exceed a maximum of 55–70 dB during the day and 45 dB at night and indoors
Land-Use/Land Take	compared with 1990, a smaller proportion of urban land devoted to transport infrastructure

Source: CEC (2000); OECD (2000, 2002)

These targets are set up aiming to control the impact of the EU transport system on the environment. Some estimates have shown that the cost of damages to people's health and the environment in the EU amounted to about 10 per cent of its GDP in the year 1995 (i.e., about €700 billion) with a tendency to increase by about 42 per cent by the year 2010. Road transport accounts for about 91.5 per cent, air transport 6.1 per cent, rail 1.9 per cent, and waterway transport about 0.5 per cent of the total figure. Regarding the type of impact, the cost of air pollution and climate change is about 48.5 per cent, noise 6 per cent, traffic accidents about 29.4 per cent. These figures exclude congestion externalities (EC, 2000). Nevertheless, these estimates should be considered with caution since they have been based on the estimates of the prospective damages of the transport system on people's health and the environment and the monetary values put on these damages either to prevent or recover them.

Consideration of sustainability mainly depends on the perceptions, preferences and objectives of the particular actors involved, referred to in different publications as 'Anthropocentric vs. Deep Ecology', 'Rich vs. Poor' and 'Techno-positive vs. Techno-phobic' (Litman, 2003). The last case could be applied to transport systems when opposing attitudes of different actors regarding transport sustainability are discussed. On the one hand, some of them claim that new technologies, in addition to the necessary commercial viability, can resolve all sustainability problems ('techno-positive'). These actors generally neglect new technologies thanks to their higher operational/economic efficiency, and effectiveness may also act as an internal driving force for instigating transport demand. Such developments in many cases might either diminish or completely neutralise their own efficiency gains. On the other hand some other actors might be against new technologies by a priori saying that they only create and exacerbate most problems ('techno-phobic'). In many cases these actors seem to neglect the positive economic and social contributions of these technologies. One example of such polarisation of attitudes is the discussion about the 'sustainability' of High-Speed Transport systems (Whitelegg, 1993). On the one hand, these systems have generally enabled significant savings of travel time compared with the 'conventional' systems. On the other, nevertheless, they have been considered as generally unsustainable seemingly due to: i) serving the needs of

Table 1.3   Possible adverse impacts of a given transport system on its
             sustainability

Economic impacts	Social impacts	Environmental impacts
Congestion	Inequity of impacts	Air pollution
Barriers to mobility	Mobility disadvantaged	Climate change
Accident damages	Human health impacts	Water pollution
Facility costs	Community cohesion	Noise
Users' costs	Community livability	Habitat damage and loss
Depletion of non-renewable resources	Aesthetic	Depletion of non-renewable resources

Source: Litman (2003, 2003a)

privileged individuals with higher time value and ability to pay; ii) requiring more land for an adequate infrastructure, iii) consuming more energy from the non-renewable resources in an absolute sense; and iv) encouraging more intensive mobility because of the substantial time savings within the available time budget. In addition, at the beginning, the scope of the concept of sustainable transport had a relatively narrow focus – on just a few adverse impacts of transport on the environment – but over time it has been widened to include a series of the adverse economic, social, and environmental impacts. Table 1.3 gives a summary.

The EU Common Transport Policy (CTP), as one of the best-known examples of the international policies dealing with sustainability of transport systems, was in place from 1992 (the year of the Commission's first White Paper on the future common transport policy) to 2001 (the year of the most recent Commission's White paper). According to the CTP, 'The European transport systems should achieve their full potential to promote the competitiveness of European businesses, employment and environmental sustainability' through 'liberalizing market access, ensuring integrated transport systems across Europe, fair and efficient pricing within and between transport modes, enhancing the social dimension, and making sure the agreed rules be properly implemented'. Consequently, citizens should be provided with 'safe, environmentally and consumer friendly, and quality driven transport systems and services'. These objectives are to be achieved by the development of sustainable transport by gradually 'shifting the balance between transport modes as the heart of sustainable strategy' over the next 30 years. In particular, this will include the operationalisation of a series of policy measures such as the gradual breaking up of the links (i.e., decoupling) between economic and transport growth, fair and efficient pricing, revitalising transport modes as alternatives to road mode and further investing in the trans-European networks. The target is to decrease the negative environmental impacts of transport systems to their 1990 level, starting from the year 2010. This is despite emerging 'barriers' such as the continuing economic growth, the growth of mobility after enlargement of the EU (from 15 to 25 members) and saturation of the major transport infrastructure and links, combined with an increased complexity of providing reasonably good accessibility to the remote areas and the infrastructure modernisation in new member states requiring significant investments (CEC, 1998; CEC, 2001a; CEC, 2001).

Generally, it seems that sustainable development of a given transport system in the long-term could be achieved if its overall positive contributions to the economic and social welfare continuously increase (or social negative impacts decrease) and the total negative impacts on people's health and the environment decrease to or below given absolute targets or thresholds (Janić, 2003). The above-mentioned progress towards sustainability could be assessed by comparing the absolute measures on the total volumes of positive and/or negative impacts relative to the specified targets (thresholds). In addition, some relative measures could be used, but only to eventually indicate the general directions of development.

1.4 The Sustainability of the Air Transport System

As is the case with the transport system, the generally acceptable and agreed definition of sustainability of the air transport system is still lacking. Nevertheless, over the past two and a half decades, dealing with the sustainability of the system has been consolidated as an important part of the agenda for almost all the actors involved. This has opened up the space for a wide range of interpretations. In most cases, the concept of sustainability has very often been used in the narrow sense, reflecting the eco-efficiency, which has embraced only two parameters – greenhouse gas emissions and noise. In parallel, the concept of the costs of the environmental and social damages has been elaborated mainly in the academic literature, but still with little practical application.

1.4.1 Main Actors

The main actors currently dealing with sustainability of the air transport system are as follows (EEC, 2004): i) Organisations for international cooperation; ii) International aviation organisations; iii) Air transport system operators – airports, air traffic control/management, and airlines; iv) Aerospace manufacturers – aircraft, engines, avionics, and other equipment; v) Non-Governmental organisations & lobby groups; vi) Users – passengers and freight shippers; and vii) Research and scientific organisations.

1.4.1.1 Organisations for international cooperation
The most well known organisations facilitating international cooperation and dealing with sustainability of air transport system are the United Nations (UN), Organisation for Economic Cooperation and Development (OECD), European Union (EU) and The Intergovernmental Panel on Climate Change (IPCC).

- The UN has used the very general approach to sustainability based on the above-mentioned Brundtland report (1987), which was additionally elaborated at a Conference in Rio de Janeiro (Brazil) in 1992. Dealing with sustainability of the air transport sector has been delegated to International Civil Aviation Organisation (ICAO).

- The OECD has developed a set of 10 guidelines for sustainability of the transport sector in the above-mentioned document Environmental Sustainability Targets (EST), which was presented at the meeting of the Environmental Ministers of OECD in May 2001. The guidelines pointed out the necessity for taking into account the economic and social benefits on the one hand and the environmental impacts on the other.

- The EU Commission has identified four main issues for integration of the strategic environmental concerns into air transport policy as follows: i) improving the environmental standards on noise and emissions; ii) strengthening the economic and market incentives; iii) assisting airports in dealing with environmental problems; and iv) improving technology in the long-term. These issues have been addressed in such documents as: 'Air Transport and the Environment: Towards Meeting Challenges of Sustainable Development' (COM (1999 (640)), Directive on aircraft noise (banning the so-called Stage 2 aircraft from operating at EU airports) (1992), Directive on hush kit Stage 1 and 2 aircraft (1998), and Directive on noise restrictions at Community airports (2002). In addition, the EU has launched the CTP policy document – white paper titled 'The European Transport Policy for 2010 – Time to Decide' (2001), in which it has been pointed out that transport has played an important role in the economic and social development of the Community and that balancing the impacts on the environment is of the greatest importance. For the air transport sector this actually means the more efficient and effective use of the airspace and airport capacity (the policy of the 'Single European Sky' starting from the year 2002 is supposed to make the contribution).

- The IPCC under support of the ICAO and UN has carried out investigations into the impacts of the air transport system on climate change (the study 'Aviation and Climate Change' was published in 1999 but did not tackle the problem of sustainability in the wider context).

1.4.1.2 The international aviation organisations
Some important international aviation organisations dealing with sustainability of the air transport system are ICAO, International Air Traffic Association (IATA), Air Transport Action Group (ATAG), Airport Council International (ACI), European Regional Airline Association (ERA), and EUROCONTROL.

- The ICAO has been mostly responsible for the coordinated development of the air transport system in terms of technology and safety. While dealing with sustainability, ICAO has been mainly focused on the environmental problems such as the aircraft emissions at both a local and global scale, and noise around airports.

- The IATA has mainly adopted the UN concept of dealing with sustainability of the air transport system, which has additionally been articulated through cooperation with organisations such as ICAO.

- The ATAG has dealt with sustainability of the air transport system by pointing out its importance for economic and social development, but without sacrificing other essential human and ecological values at present and in the future.

- The ACI has proposed an environmental management concept for the airports in order to enable systematic coping with their environmental problems related to operations and development.
- ERA has emphasised the role of airlines in promoting regions, contributions to external trade and GDP, i.e., the economic benefits of the air transport system.
- EUROCONTROL has dealt with sustainability of the air transport system by including the environmental issues in different programmes such as for example ATM 2000+ and EATMP programme. The main environmental issues taken into account are noise and air pollution, both closely related to the efficiency, effectiveness, and safety of services for the service users – airlines.

1.4.1.3 The air transport system operators
The air transport operators dealing with sustainability of the air transport system are airports, air traffic control/management, and airlines.

- Airports deal with their own sustainable development differently. Some of them only implement policies and measures for mitigating the environmental impacts if necessary. For example, at many airports worldwide, there are noise restrictions and at a very few restrictions to the emission of air pollutants. However, some other airports apply a stricter approach by including the environmental issues in their management model.
- The Air Traffic Control/Management (ATC/ATM) service providers have mainly focused on the safety, efficiency and effectiveness of services, and better use of the available airspace capacity. Although they do not directly address the issue of sustainability, the improved efficiency and effectiveness of services reduces the fuel consumption and associated emissions of air pollutants, noise around airports, and lost time due to delays, which indirectly contribute to the social and environmental component of the sustainability of the air transport system.
- The airlines have generally dealt with system sustainability by including the economic, social and environmental issues in their business model. The preconditions for dealing with these issues are profitability and survival as the main elements of any airline business strategy. Nevertheless, the airlines have improved their economic efficiency and in many cases replaced their fleets with aircraft that consume less fuel and are less noisy, both of which have indirectly contributed to mitigating the associated environmental impacts. In addition, many airlines consider that their social and economic contribution to sustainability is sufficient through their survival and the direct and indirect employment they provide.

1.4.1.4 The aerospace manufacturers: Aircraft, engines, avionics, and other equipment
The aircraft and engine manufacturers have been strongly focused on the performances of their products, which enable reduction of the fuel consumption, associated emissions of air pollutants, and noise. They have also considered the processes of recycling particular components (materials) both during their production and when they need to be scrapped.

1.4.1.5 The non-governmental organisations and lobby groups

The non-governmental organisations and lobby groups deal with the sustainability of the air transport system through involvement in airport development and expansion plans. Their objectives are mostly to protect the environment and the neighbouring people from increased noise expected from the growth of air traffic.

At an international level, these organisations and groups generally use the UN approach while dealing with sustainability of the air transport system. In many cases, they apply a one-sided approach by considering only the negative impacts of the system on the environment and people's health.

1.4.1.6 Users – passengers and freight shippers

Users of air transport services – passengers and freight shippers – usually consider safety, price, convenience and diversity of choice as the main features of the air transport system. They do not particularly consider the negative effects of the air transport system on the environment and society, but mostly look at the direct social and economic benefits.

1.4.1.7 The research and scientific organisations

Researchers and scientists usually deal with sustainability of the air transport system through considering simultaneously both its positive effects and negative impacts on the society and the environment. Such fair treatment could help in giving input to the creation of fair and balanced policies on the current and long-term system development.

The above-mentioned overview indicates that seemingly sustainable development of the air transport system needs to be generally based on balancing (i.e., trading-off) between the system's positive effects (benefits) and negative impacts (costs) in both the short- and long-term.

1.4.2 Scale of Dealing with Sustainability

Balancing the impacts and effects of the air transport short- and long-term operations can be carried out at a global (intercontinental), regional (continental, national) and local (community) scale, individually and/or through cooperation between particular actors involved (INFRAS, 2000). The global scale implies the entire world, particular continents or their large parts; regional scale embraces a country or its part(s); local scale relates to an individual airport(s) and the surrounding community. At each scale, sustainability can be considered regarding the specific effects and impacts from the perspective of particular actors involved.

1.4.2.1 Global scale

At the global (international) scale, a particularly important element for dealing with the effects and impacts of the air transport system might be monitoring current and expected volumes of emissions of carbon dioxide (CO_2). For example, in the EU the share of the air transport emissions of CO_2 in the total man-made emissions of CO_2 has been relatively constant – about 2 per cent – despite the fact that air transport has been increasing its relative share in the total volumes of transport output in terms of p-km (p-km – passenger kilometres) from about 4 per cent in the year 1990 to about 9 per cent in the year 2002 (see Figure 1.1) (CEC, 1999a; EC, 2004). This share of CO_2 appears to be relatively small when compared with the share of the emissions of the transport sector in the total man-made emissions (about 19–23 per cent during the period 1990–2002). However, emissions from the air transport system are very specific mainly due to their being deposited at altitudes between 9 km and 12 km (below and above tropopause) and the latitudes from 40° N to 60° N, where, together with the emissions of nitrogen oxides (NO_x), they act as greenhouse gases and thus contribute to global warming. Regarding the predicted annual growth rate of air traffic until the year 2050 of about 5 per cent, it is expected that the share of its emissions in the total man-made emissions of CO_2 equivalents will increase to about 3–3.5 per cent regardless of the improvements in the aircraft engines/airframe technology and efficiency of operations (CEC, 1999a; IPCC, 1999; RCEP, 2001). In order to prevent such development, some quantitative targets on the emissions have been set up. For example, in Europe, the Advisory Council for Aeronautical Research in Europe (ACARE) has set up the targets for improving the environmental performances of the whole system, covering the aircraft engines, airframe and operations. These targets imply reduction of the fuel consumption and CO_2 emissions (units per passenger-kilometre) by about 50 per cent by the year 2020 (about 10–20 per cent of the contributions by improving the engines and the rest by improving the airframe and operational efficiency). During the same period the emissions of nitrogen oxides NO_x are expected to reduce by 50 per cent, of which about 60–80 per cent will be by engine improvements and the rest by improvements in the airframe and operations. Noise will be reduced by 50 per cent mainly thanks to the improvements of engines and airframe (RCEP, 2001). Nevertheless, these improvements are expected to be efficient but not sufficient to offset the expected growth of air pollutants due to the growth of air traffic during the forthcoming years. Therefore, the additional economic and regulatory measures, of which some are internalising externalities through taxation of the fuel and emissions and the emission trading schemes, will need to be introduced in order to maintain the desired balance between the system effects and impacts. In the scope of such intentions, the assessment and quantification of the current and prospective impacts appears to be still fuzzy. Figure 1.2 shows an example of predicting the long-term relative and absolute emissions of CO_2 in the EU and the ambiguity of using different types of results by both supporters and opponents of sustainable development of the system. As can be seen, during the period 1992–2025, the volume of passenger kilometres is expected to increase by about 363 per cent and the total emission of CO_2 by about 190 per cent.

In this context, the supporters will use the relative measure(s) on the fuel (and corresponding CO_2 emissions) efficiency to claim that the sector is strongly moving towards or that it has already achieved the desired 'sustainability' state. On the contrary, many opponents will argue that such relative measure(s) can only be used as one of the indicators that the sector is beginning to move towards sustainability. Nevertheless, referring to some original principles of sustainability – to ensure the long-term development within given constraints (targets-thresholds) on the non-renewable resources and adverse impacts – the absolute emissions of CO_2 indicate that the sector, driven by the very strong external economic/business and social

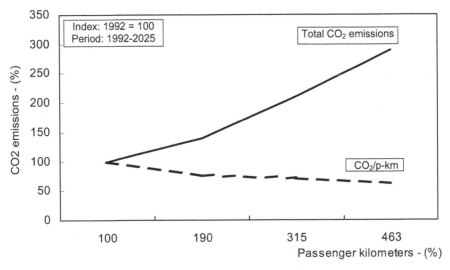

Figure 1.2   The long-term expected emissions of CO$_2$ from the air transport system in the EU

Compiled from CEC (2000)

forces, appears not to be moving towards sustainability at all. Even though it shows the tendency to move in the opposite direction due to the continuous increase in the total fuel consumption and corresponding emissions of CO$_2$, neither of which will be able to guarantee reaching the absolute future targets (thresholds) (1 kg of jet fuel ≈ 3.18 kg CO$_2$). Some figures can be used to confirm this latest statement. In the EU, in the year 1990, the total man-made emission of CO$_2$ was 3,775 million tons and that of the air transport system 83 million tons (about 2 per cent). The share of the external cost of these emissions in the total emissions of CO$_2$ made by transport is similar (EC, 2000). By the year 2050, the total man-made emissions of CO$_2$ are expected to increase by about 60 per cent as compared with the year 1990. At the same time, these emissions by the air transport system are likely to increase by about five times in absolute terms, if no improvements to the current state take place (AEA, 2005). The modified Equation 1.1 reflects some of the above-mentioned attitudes and expectations as follows:

$$TE = P * D * FC * SE \tag{1.2}$$

where
- TE is the total emission of greenhouse gases (tonnes);
- P is the number of air travellers;
- D is the average travel distance (km or miles);
- FC is the fuel consumption (tonnes per p-km); and
- SE is the specific emission (tonnes of pollutant per tonne of fuel consumed).

If the type of fuel does not change, the variable SE for particular polluting gases such as CO$_2$ and H$_2$O will remain the same. The specific fuel consumption FC could be eventually further decreased by the above-mentioned development of new aircraft engines, more efficient and effective maintenance, and significantly improved flight and air traffic flows guidance and control, respectively. Consequently, two variables, which might be affected by the economic and regulatory measures, are the number of passengers P and travel distance D. This, however, might be a very complex task. Due to the increase in GDP and the overall welfare, more people will travel, using the opportunity of the constantly diminishing airfares. In addition, the globalisation of business and tourism will probably contribute to increasing the average travel distances, which in turn, together with an increase in the number of people travelling, will make the volume of passenger kilometres (p-km), fuel consumption and the associated emissions of pollutants grow and very complex to manage.

1.4.2.2 Regional (national) scale

At the regional (national, continental) scale, particularly in the U.S. and Western Europe, growth in the demand for air transport has been driven by liberalisation of the already matured air transport markets, increased productivity, and lower prices. Such growth has been confronted with the limited capacity of the infrastructure – airports and ATC/ATM causing increased congestion, delays, and a consequent diminishing of the expected effectiveness and efficiency of the services. Such developments have given rise to the question of an appropriate trading-off between demand and capacity at the regional scale in order to prevent further deterioration in the efficiency and effectiveness of services and an increase in the external impacts on people's health and the environment. This has usually implied provision of sufficient infrastructure capacity at airports as the main locations of congestion and delays, more efficient utilisation of this capacity, and the non-significant increase or stabilising of the impacts of noise and air pollution. Consequently, the problem of balancing the effects and impacts has moved from the regional to the local scale, and vice versa, indicating the necessity for considering the interrelation between particular levels while dealing with the sustainability of the air transport system at each of them.

1.4.2.3 Local (community) scale

Dealing with the sustainability of air transport at a local scale usually implies airports. In this case the sustainability includes establishing a trade-off between the positive effects and associated negative impacts of the airport growth on the local community and environment. In the scope of solutions, the physical expansion represents one extreme and constraining growth another. Managing sustainable airport growth might be the third compromise solution, in which case the given airport, dominant airline(s), and local community all take over some of the responsibilities. The most recent (typical) example is the public enquiry on positioning a new runway at one of London's five airports (DETR, 2002). Despite the fact that the decision has been made to build the new runway at Stansted Airport, Heathrow, London's biggest airport, which hosts British Airways (BA) as the dominant airline and the leader of the Oneworld alliance, has remained under focus due to its global importance for the London region and entire UK economy. One of the main objectives of BA is to contribute to the airport's sustainable development through its own sustainable development. In such a context, the airline has to consider balancing the three sets of effects and impacts around Heathrow Airport.

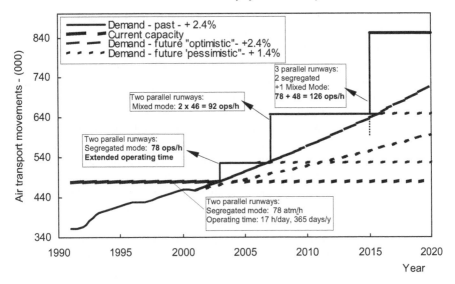

Figure 1.3   Balancing between demand and capacity at local level: London Heathrow Airport

Compiled from Janić (2004)

Firstly, the economic effects of providing air services to users flying to very diverse destinations; direct employment by the airline and related induced services, trade and investments; and indirect contributions to the local economy. Second, there are the social effects and impacts. The effects are stimulus to the dialogue of particular stakeholders, employment diversity (opportunity), community initiatives, and the promotion and stimulation of the use of public transport. The impacts are contributions to the road congestion and related delays around the airport due to airline-related activities. The last are the environmental effects and impacts. The effects include improvement of the aircraft technological and operational performances. The impacts embrace increasing noise and air pollution at the airport due to the growth of air traffic (BA, 2004).

The main objective of Heathrow Airport can be considered to be sustainable growth, which implies the handling of growing traffic with the simultaneous mitigation of the associated impacts, mostly in terms of noise, air pollution, and land use. Figure 1.3 shows possible scenarios of the prospective long-term growth of the air transport movements and solutions for provision of the adequate runway capacity at this airport (DETR, 2002; Janić, 2004).

Currently, two parallel runways are used in the segregated mode, aiming to mitigate and fairly distribute the noise burden on the local community during the day. This is combined with a restriction on the airport operating time by imposing a night ban on operations from 22.00 to 5.00. Such constraints generally compromise the runway capacity, which could be achieved if only the safety constraints were in place. Under such circumstances, the existing and predicted growth of demand will lead to airport saturation in the near future. In order to prevent the negative impacts of saturation in terms of the increased congestion, delays and emissions of air pollutants on the one hand and successfully coping with the sustainability policy of the dominant (BA) airline with respect to the users' welfare, contribution to local economy, etc., on the other, the alternative solutions for more efficient utilisation and even expansion of the existing runway capacity have been considered (BA, 2001). The solutions based on the more efficient utilisation of the runway system capacity include extending the airport operating time and changing the modality of operation from the segregated to mix mode. Both solutions would increase the runway capacity on the one hand and the noise burden and air pollution on the other. The increased noise and air pollution burden could eventually be compensated by the gradual introduction of quieter aircraft, the provision of more noise-efficient Air Traffic Control/Air Traffic Management (ATC/ATM) guidance in the vicinity of the airport, and the implementation of an appropriate pricing policy for externalities – congestion, noise and air pollution – which would regulate access and consequently maintain the impacts under prescribed limits. Implementation of both solutions requires relieving if not sacrificing some of the environmental constraints. Otherwise, the question is how the airline(s) will accept operations and further growth within the highly constrained environment at the airport. What could the domain of their contributions be? Is there a real danger of collision between the long-term objectives of a given airport, airline(s) and local community, all actually benefiting from the airport growth? The solution implying building the new third runway might be reasonable in the long-term if it could balance the effects and impacts. In this case, the effects would be the economic benefits achieved by the airport's further growth. The immediate impacts would be usage of the additional land, increased and re-distributed noise, and increased air pollution, all expected to be gradually neutralised by the innovative aircraft and ATC/ATM technologies. In addition, co-operation in terms of sharing the spare capacity with other airports in the area could be considered as an option. For Heathrow, these would be London Stansted and Luton Airport. At the European scale such cooperation is also considered. For example, Munich airport cooperates with Augsburg, Frankfurt with Hahn, and Rome Leonardo Da Vinci (Fiumicino) International Airport with Rome Ciampino Airport (INFRAS, 2000). Anyway, any solution requires assessment of its overall social viability by taking into account the overall socio-economic benefits and costs of impacts regarding the preferences and interests of particular actors involved. Such an approach brings the issue of the long-term sustainable development of a particular airport to the fore.

The discussion so far has indicated that one of the pre-assumptions for successfully dealing with sustainability of the air transport system is a rather good knowledge of the technical/technological, operational, economic, social, environmental and institutional performances of its main components – airports, air traffic control/ management, and airlines. This also implies understanding of the interrelationships between components and particular performances. Chapter 2 will describe particular performances of airlines, airports, and *ATC/ATM* regarding their relevance for the quantitative analysis and assessment of the sustainability of the air transport system.

References

AEA (2005), *European Aviation Industry Joint Position Paper on Emissions Containment Policy* (Brussels: Association of European Airlines).

BA (2001), *From the Ground Up: Social and Environmental Report 2001, British Airways PLC, Waterside, (HBBG)* Middlesex, UK.

BA (2004), *Social and Environmental Report–2004, British Airways PLC, Waterside, (HBBG)* Middlesex, UK.

Brundtland, G.H. (1987), *Our Common Future* (Oxford: Oxford University Press).

CEC (1998), *The Common Transport Policy: Sustainable Mobility – Perspectives for the Future*, COM (1998) 716 final (Brussels: Commission of the European Communities).

CEC (1999), *Integrating the Environmental Dimension: A Strategy for the Transport Sector, A Status Report* (Brussels: Commission of the European Communities).

CEC (1999a), *Air Transport and the Environment: Towards Meeting the Challenges of Sustainable Development, Commission of European Communities*, COM (1999), 640 final (Brussels: Commission of the European Communities).

CEC (2000), *Defining an Environmentally Sustainable Transport System, Commission Expert Group on Transport and Environment, Report* (Brussels: Commission of the European Communities).

CEC (2001), *A Sustainable Europe for a Better World: A European Union Strategy for Sustainable Development*, COM (2001), 264 (Luxembourg: Commission of the European Communities, Office for Official Publications of the European Communities).

CEC (2001a), *European Transport Policy for 2020: Time to Decide*, COM [(2001) 370] (Luxembourg: Commission of the European Communities, Office for Official Publications of the European Communities).

Daly, H. (1991), *Steady State Economics* (Washington, DC: Island Press).

DETR (2002), *The Future Development of Air Transport in the United Kingdom: South-East*, A National Consultation Document (London: Department for Transport and Regions).

EC (2000), *The Way to Sustainable Mobility: Cutting the External Cost of Transport* (Brussels, Belgium: Brochure of the European Commission).

EC (2005), *Energy & Transport in Figures 2004* (Brussels: European Commission Directorate-General for Energy and Transport).

ECMT (2000), *Sustainable Transport Policies* (Paris: European Conference of Ministers of Transport).

EEC (2004), *Defining Sustainability in the Aviation Sector*, Report EEC/SEE/2004/03 (Bretigny Sur Orge Cedex: EUROCONTROL Experimental Centre).

Ekins, P. (1996), 'Limits to Growth and Sustainable Development: Grappling with Ecological Realities' in *The Political Economy of Development and Underdevelopment*, Jameson, K. and Wilber, C. (eds) (New York: McGraw-Hill), pp. 53–69.

Gale, R.P. and Gordray, S.M. (1994), 'Making Sense of Sustainability: Nine Answers to What Should Be Sustained', *Rural Sociology*, **59**, 311–332.

Griethuysen, P. (2002), 'Sustainable Development: An Evolutionary Economic Approach', *Sustainable Development*, **10**, 1–11.

Hooper, P.D. and Gibbs, D. (1995), 'Cleaner Technology: A Means of an End or an End to a Means', *Greener Management International*, **3**, 28–40.

INFRAS (2000), *Sustainable Aviation*, Pre-Study, MM-PS (Zurich, Switzerland: INFRAS).

IPCC (1999), *Aviation and the Global Atmosphere, Intergovernmental Panel on Climate Change* (Cambridge: Cambridge University Press).

Janić, M. (2003), 'An Assessment of the Sustainability of Air Transport System: Quantification of Indicators,' 2003 ATRS (Air Transport Research Society) Conference, 10-12 July, Toulouse, France, p. 35.

Janić, M. (2004), 'Expansion of Airport Capacity: Case of London Heathrow Airport', *Transportation Research Record 1888*, 7–14.

Kelly, K.L. (1998), 'A System Approach to Identifying Decisive Information for Sustainable Development', *European Journal of Operational Research*, **109**, 452–464.

Litman, T. (2003), 'Issues in Sustainable Transportation,' *T.D.M. Encyclopedia* (Victoria: Victoria Transport Policy Institute). http://www.vtpi.org.

Litman, T. (2003a), *Sustainable Transportation and TDM: Planning that Balances Economics, Social and Ecological Objectives* (Victoria: Victoria Transport Policy Institute). http://www.vtpi.org.

Mikesell, R.F. (1992), *Economic Development and the Environment: A Comparison of Sustainable Development with Conventional Development Economics* (New York: Mansell Publishing Limited).

OECD (1998), *Sustainable Development: A Renewed Effort by the OECD*, Policy Brief No. 8 (Organisation for Economic Co-operation and Development). www. oecd.org.

OECD (2001), *Policy Instruments for Achieving Project Environmentally Sustainable Transport* (Organisation for Economic Co-operation and Development). www. oecd.org.

Pearce, D.W., Markandya, A. and Barbier, E. (1989), *Blueprint for a Green Economy* (London: Earthscan).

RCEP (2001), *The Environmental Effects of Civil Aircraft in Flight, Special Report* (London: Royal Commission of Environmental Pollution).

Richardson, B. (1999), 'Towards A Policy on a Sustainable Transportation System', *Transportation Research Record 1670*, 27–34.

SEPA (1996), *Towards an Environmentally Sustainable Transport System*, Final Report from the Swedish EST-Project, Report 4682 (Stockholm: The Swedish Environmental Protection Agency).

Spangenberg, J.H. (2002), 'Institutional Sustainability Indicators: An Analysis in Agenda 21 and a Draft Set of Indicators for Monitoring Their Effectivity', *Sustainable Development*, **10**, 103–115.

TC (1999), 'Towards Sustainable Transportation, Transport Canada'. www.tc.gc. ca/envaffairs/english.

Whitelegg, J. (1993), *Transport for a Sustainable Future: The Case for Europe* (London: Belhaven Press).

Chapter 2

Performances of the Air Transport System

2.1 The Concept of Performances

This chapter aims to familiarise the reader with those performances of the air transport system components – airlines, airports, and ATC/ATM – which are relevant for the analysis and assessment of sustainability. The performances of each component are grouped into six categories as such i) Technical/technological; ii) Operational; iii) Economic; iv) Social; v) Environmental; and vi) Institutional. Unlike in other relevant books and studies dealing with transport sustainability, which focus exclusively on the economic, social, and environmental performances, this approach explicitly embraces the technical/technological and operational performances as the main driving forces of other performances. This makes it easier to understand the relationships between the ultimate causes and consequences because each category of performances can be at the same time a cause and a consequence (INFRAS, 2000; Spangenberg, 2002; Janić, 2003). Such extension of the scope of approach primarily aims to address the inherent complexity and hierarchy of the interrelations between the performances of the air transport system components – from the technical/ technological to the institutional performances, and vice versa. Figure 2.1 shows a generic scheme of such interrelationships.

It should be mentioned that this scheme is sufficiently generous that it could also be applied to other transport modes.

The *technical/technological performances* relate to the main technical/ technological characteristics of the system components – vehicles (aircraft), infrastructure (airports and airspace), and the (air) traffic guidance/management systems complemented/supported by ITS (Intelligent Transport Systems) (ATC/ATM – Air Traffic Control/Management). The vehicle (aircraft) characteristics include the engine technology referring to the engine power (thrust), fuel consumption, emissions of air pollutants from burning fuel, and noise. In addition, the number of vehicles (aircraft) of particular types and engine technologies, their utilisation and proportion in total numbers, and geographical distribution could also be included at different levels, local, intermediate and global – dependent on the purpose of analysis. The infrastructure (airport) characteristics refers to the interoperability, i.e., suitability of a given infrastructure in both the airside and landside areas to be flexibly (smoothly) used regardless of the different types/categories of vehicles (aircraft) and volume and fluctuations of traffic up to a certain level. The characteristics of traffic guidance/management systems (Air Traffic Control/Management System) refer to the facilities, equipment and devices enabling (and supporting) the efficient and

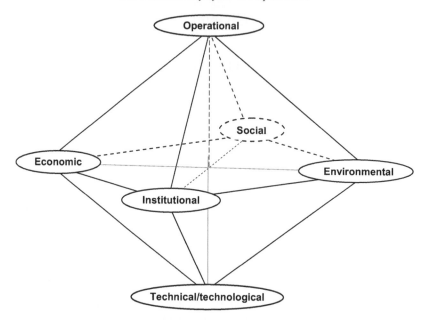

Figure 2.1   Generic scheme of performances of the air transport system for analysis and assessment of sustainability

effective transport of air passengers and air cargo between origin and destination airports and consequently diminishing (or not additionally increasing) energy consumption, related air pollution, and other adverse effects.

The *operational performances* include undertaking the specified operations and processes aiming to produce transport services by using the system components. In such a context, there is always a sound operational objective to 'optimise' the utilisation of the disposable resources (transport means – aircraft fleet, labour, and fuel consumption) under given (technological, operational-safety, economic and environmental) constraints, which in turn is expected to contribute to the higher efficiency in energy consumption and associated adverse effects such as air pollution and noise. These performances are mainly driven and influenced by the technical/ technological, economic and environmental performances (Janić, 2003).

The *economic performances* include investments for providing components of the air transport system, generally in terms of labour, capital and energy. In addition they relate to the internal costs, revenues, and productivity of particular system components. As well, they may include direct, indirect, and induced employment, contribution to GDP and equity, which can also be considered as the social performances (CEC, 2001). The economic performances mainly depend on the system's technical/technological and operational performances.

The *social performances* reflect the effects of the air transport system on society in terms of congestion and delays, accessibility of services, damages from air traffic incidents and accidents, community cohesion, liveability, and social interactions (Button and Stough, 1998; DETR, 1999, 2000). They mainly depend on the system operational performances.

The *environmental performances* reflect the physical impacts of the air transport system on people's health and the environment in terms of local and global air and water pollution, noise, generation of waste, and land use. Most of these impacts directly depend on the system's technical/technological and operational performances (Janić, 1999).

The *institutional performances* have recently been recognised as relevant because of increasing the need for establishing the institutional forms in order to be able to deal more comprehensively with the system's sustainability. These forms are supposed to bring the various external social entities and the system actors together, set up and facilitate rules of behaviour through the various institutions mediating the interests of actors involved. In general, other performances may influence these in terms of the number, structure, hierarchy and effectiveness of institutions and their objectives (Spangenberg, 2002).

2.2 Airlines

2.2.1 Background

The airline industry represents the first component of the air transport system. It consists of airlines, which according to their commercial needs operate different sizes of fleets, aircraft types, and air route network configurations. The airline technical/ technological performances mainly relate to the aircraft characteristics they use. These include the aircraft speed, carrying capacity, and technical productivity. In addition to these are the characteristics of aircraft engines such as power (thrust), fuel consumption and noise, and aircraft safety. All these characteristics are considered as derived and materialised through aircraft design.

The airline operational performances include demand, capacity, and quality of services. The economic performances embrace the airline revenues, aircraft and airline operational costs, and profits dependent on the airline economic/business model. The latest generally can be the full cost (legacy), charter, and low-cost model. The social performances mainly relate to the airline employment, usually related to the size of the airline fleet in terms of the number and size of aircraft. The environmental performances include noise, local air quality, climate change, waste and the use of resources. Finally, the institutional performances relate to the structure of airline ownership, and market regulation for the airline industry.

Particular performances influence the air transport system's sustainability directly or indirectly both positively and negatively.

2.2.2 Technical/Technological Performances

2.2.2.1 Aircraft speed, capacity and technical productivity
The main feature of aircraft that has enabled the air transport system to develop globally is its high speed compared with other transport means (vehicles). Generally, the intention to increase transport speed has been an obsession for people in recent times. One of the investigations (Ausubel and Marchetti, 1996) has shown

that humans as territorial animals have always instinctively tried to maximise territory, which has been equated with opportunities and resources. However, the contemporary life offers always limitations in terms of time and monetary budget for travelling acting as significant barriers to increase distances (i.e., 'territory'). Under such circumstances the higher travelling speed combined with lower costs have become crucial requirements for developers of new transport technologies, which over time also emerged as their goals. Evidently, the higher speed enables longer distance travelling within a given time budget, and very often is preferred as compared with the alternatively greater number of shorter trips. Consequently, increasing transport speed has significantly extended personal travel distance over the past two centuries, from about 50 km per day by horseback to about 650 km per day by car and about 650 km per hour by the contemporary air transport system − aircraft. The increase in speed has gradually enabled the market penetration and substitution of currently 'conventional' transport options, mostly with respect to time and much less respecting the monetary budget. Since most people generally have a fixed time and money budget for travel, they have always bought the maximal distance within it. With an increase in personal income, they have tended to buy faster transport services and thus to travel further. Thus, the increase in travelling speed has also been stimulated by the economic development (i.e., Per Capita Income) (Ausubel and Marchetti, 1996).

Similarly, as with other sophisticated technologies, high-speed transport technologies have been evolving over the last half of the twentieth century. Air transport emerged as the earliest high-speed alternative at the beginning of the twentieth century and has been always modernising through improving the 'aircraft capabilities,' 'airline strategy' and 'governmental regulation' (Boeing, 1998). The development of the 'aircraft capabilities such as speed, payload and take-off-weight has been of the greatest importance, since they have mostly affected other two sub-processes, and vice versa. Both the speed and payload have contributed to the enormous increase in the aircraft technical productivity, by more than 100 times during the last 40 years' (Horonjeff and McKelvey, 1994). This productivity can be roughly estimated as follows:

$$TP = PL*V \qquad\qquad\qquad\qquad\qquad\qquad (2.1)$$

where
    TP is technical productivity (t-km/h).
    PL is payload (tonnes); and
    V is the commercial (block) speed (km/h).

Development of the aircraft speed over the past 50 years has been illustrated using the 'Speed Ratio' comparing the speeds of the various commercial aircraft with the speed of the slowest aircraft − DC3 (238 km/h). Figure 2.2 shows the example.

The trend line suggests that there has been more than proportional progress of the aircraft speed over time, currently completed by the development of the supersonic aircraft − Concorde − in 1974 (Horonjeff and McKelvey, 1994; Boeing, 1998). Simultaneously, the aircraft carrying capacity has increased to include a wide

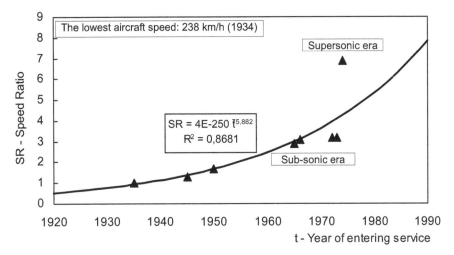

Figure 2.2   Development of aircraft speeds over time

diversity of aircraft types – from a small number (for example, 30–50) to more than 500 seats. The most recent example is the launch of the new giant Airbus A380 with a capacity of 500–600 seats.

Such a development has certainly contributed to an increase in the above-mentioned technical productivity in static terms. In addition, this productivity can also be considered as the aircraft's 'dynamic' capacity, expressed by the realised passenger (or tonne) kilometres (miles) per unit of time of a given non-stop flight. With most contemporary commercial aircraft, in order to reach the longer ranges the payload – passengers and freight – need to be reduced. At the same time, the speed generally increases (i.e., this is block speed as the average speed between the aircraft engine starting up at origin and the engine shutting down at the destination airport). Consequently the aircraft technical productivity as the product of the payload and average speed changes due to such variations as shown in Figure 2.3 (Doganis, 2002).

As can be seen, if the payload is maximal, the aircraft technical productivity will increase with increasing distance, i.e., the stage length, thanks to an increase in the average block speed. After the payload is reduced due to taking more fuel for covering the longer distance, the technical productivity diminishes despite further increasing the average block speed. In addition, substantive progress has been achieved over the past three decades in increasing the technical productivity of a new generation aircraft, i.e., the differences in productivity between the aircraft A310, B757, B767 300ER and the new B787 are obvious.

2.2.2.2 Aircraft engines
Coverage of the relatively long distances in a relatively short time and increasing the aircraft technical productivity have been particularly possible thanks to the development of the high bypass turbofan engines (Jenkinson, Simpkin and Rhodes, 1999). However, these engines powering most of the growing world's aircraft fleet have directly induced the most adverse effects of the contemporary air transport

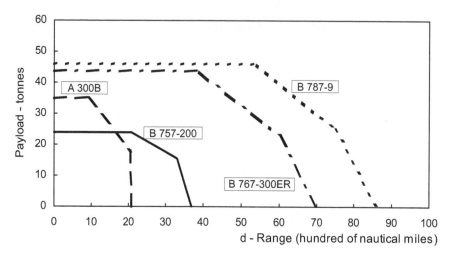

a) Payload vs. range

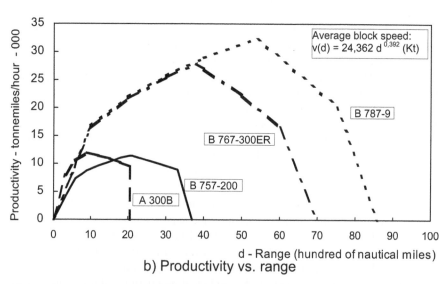

b) Productivity vs. range

Figure 2.3   Dependence of aircraft production performances on the stage length: a) Payload vs range; b) Productivity vs range (http:// www.boeing.com)

Compiled from Janić (2001)

system on the environment – increased fuel consumption, air pollution, and noise. Consequently, in addition to the number scale and scope of operations, this piece of technology seemingly holds a key for the prospective more unsustainable development of the air transport system.

The aircraft engines consume fuel (kerosene-JP1) to generate the power needed to move the aircraft from the ground – the origin airport. Then together with the (air) lift force, which counter balances the aerodynamic drag and the aircraft weight, the engine power moves the aircraft through to the destination airport. The most important parameters of performances of turbofan engines are thrust, efficiency, Specific Fuel Consumption (SFC), emissions, and noise.

Engine thrust is derived from the change in momentum of the air through the engine and the thrust occurred due to the static pressure ratio across the final (exhaust) nozzle. Analytically, this follows (Jenkinson, Simpkin and Rhodes, 1999):

$$T = m(v_1 - v_0)/g + (p - p_0)/A \qquad (2.2)$$

where

m is the air flow through the engine (kg/s);
v_1 is the velocity of exhaust jet (m/s);
v_0 is the velocity of air entering the engine (m/s);
g is the gravitational acceleration (m/s^2);
p, p_0 is the pressure at the intake and exhaust station, respectively (N/m);
A is the nozzle cross sectional area (m^2).

In Equation 2.2, thrust T is usually expressed in kiloNewton (kN) (SI units) or Libras (lb) (British units).

Engine efficiency (η_e) directly expresses the rationale of the engine fuel consumption, i.e., the higher efficiency implies lower fuel consumption per unit of the engine thrust.

Specific Fuel Consumption (SFC) expresses the amount of fuel generating one unit of thrust in a given unit of time. It is expressed in kg of fuel per kN of thrust per h. The SPC and engine efficiency η_e are interrelated as follows:

$$SFC = M/4\eta_e \qquad (2.3)$$

where

M is Mach number.

For example, the SFC for most contemporary engines with an efficiency of about $\eta_e = 35-40$ per cent operating above the tropopause at $M = 0.84$ is about 0.55–0.60 (Jenkinson, Simpkin and Rhodes, 1999). Multiplying the SFC with the thrust per engine and the number of engines per aircraft might give an estimation of the total fuel consumption per unit of time of a given flight. For example, this hourly consumption increases with increasing aircraft size (i.e., seating capacity) and the number of engines implying that the larger aircraft are provided with the higher total engine thrust. Figure 2.4 shows an example of such a relationship.

For the large commercial aircraft the fuel consumption represents a significant portion of the direct operating costs – about 30 per cent. Therefore, the fuel-efficient engines play an important role in the overall economic efficiency of the aircraft and airline. Indirectly, the lower fuel consumption per single unit (engine) implies the

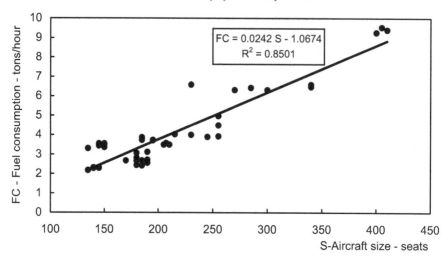

Figure 2.4   The average fuel consumption per unit of flying time vs the aircraft seat capacity

Compiled from Janić (2001)

lower consumption of non-renewable energy sources and consequently the lower emission of air pollutants.

Emissions from the aircraft engines in terms of type and quantity of air pollutants depend on the engine bypass ratio and fuel type. The bypass ratio of the contemporary turbofan engines is up to about 10. These engines consume JP-1 jet-fuel, a derivative of crude oil. The most important air pollutants are HC, CO, NO_x and SO_2. Their quantities change around airports and in the high-altitude atmosphere, depending on the phase of flight and which specific engine regime is required. In general, as emitted during given time such as for example Landing/Take-Off (LTO) cycle, they are positively correlated with the aircraft size (i.e., take-off weight, which in turn is also correlated with the number and thrust of engines) (Janić, 2003a). Nevertheless, two polluting components are directly proportional to the quantity of fuel burned – CO_2 – 3.18 and H_2O – 1.2 kg/kg of fuel burned. Thus, the longer flights consume larger quantities of fuel and consequently emit larger quantities of CO_2 and H_2O, over relatively large geographical areas (intercontinental flights are a typical example). Consequently, regarding sustainability, these aircraft/flights appear to be a matter of concern. Usually, quantitative information on the emissions of particular engine types are available from the national and international regulatory bodies such as Civil Aviation Authorities (CAA), Federal Aviation Administration (FAA) US, EUROCONTROL and ICAO, respectively.

Noise primarily comes from the aircraft engines while flying near the ground, i.e., during take-off and landing. The noise spreads in front of and behind the aircraft engine. Noise-spreading generators in front of the engine are the compressor and fan. The back noise-spreading generators are the turbine, fan, and jet-afflux. In general, larger aircraft and their more powerful engines are noisier in absolute terms. However, both closeness of flight paths to the human population and the aircraft-engine operating regime have shown to be the most important factors of noise annoyance. Therefore, noise became an issue for national and international regulation at a very early stage of the air transport system development when ICAO recommended national legislation (Annex 16 to 1944 Chicago Convention) according to the US FAR (Federal Aviation Regulation) – Part 36. Consequently, the maximum noise exposure levels for three critical operating conditions around airports – landings, taking-off, and sideline – were imposed. Consequently, noise measurement points for aircraft certification have been defined requiring the newly designed aircraft to comply with these noise certification rules. They have also become a component of the noise regulation institution JAR (Joint Airworthiness Requirements).

In general, at the noise certification points around an airport the maximum aircraft noise cannot exceed 108 Effective Perceived Noise Level in decibels (EPNdB) (this is an equivalent of noise of about 96 dB(A) measured with A-noise weighted scale) (Horonjeff and Mc Kelvey, 1994). In particular, this is the maximum noise level on: i) approach at a point of 1 nm (1.85 km) before the runway landing threshold, on the extended runway centreline when the aircraft is at a height of 379 ft (113 m) (nm-nautical mile); ii) on take-off at a point 3.5 nm (6.5 km) from the brake-release, regardless of the height; and iii) on take-off at a point to the side of the runway centreline at a distance of 0.25 nm (460 m) for two and three-engine aircraft (A330, B777, MD11) and 0.35 nm (650 m) for four-engine aircraft (B747, A340, A380) (Huenecke, 1997). In addition to regulation of the aircraft airworthiness with respect to noise, different models have been developed to evaluate the noise effects of these aircraft around airports. The main outputs from these models have consisted of definitions of the so-called 'noise contours' or 'noise footprints' whose shape indicates areas of constant noise. Figure 2.5 shows the generic noise contours for two different-sized aircraft including the relative location of the noise measurement points (Jenkinson, Simpkin and Rhodes, 1999).

The quieter (high-by-pass turbofan) jet engines have reduced the area of the 'footprints'. For example, a footprint area of the aircraft type B727-200, which entered commercial service in the 1960s was 14.25 km^2 with a constant noise level of 85 dB(A). This aircraft was equipped with the low by-pass engines. In contrast,

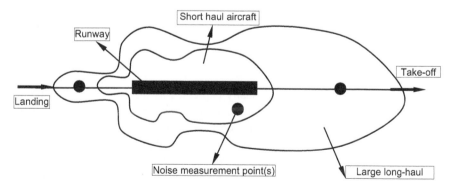

Figure 2.5   Simplified scheme of noise contours around a given airport

Airbus A320-200, in commercial service from the late 1980s, creates a footprint area of only 1.5 km^2 covered by the same constant noise level of 85 dB(A). Consequently, with relatively massive introduction of these and other similarly quieter aircraft, population expected to be exposed is much smaller.

2.2.2.3 Aircraft safety

Usually, aircraft safety is considered in both the national and international context. In general, safety encompasses the aircraft design, operations, and environmental aspects. In addition to the legislation and supporting national and international documents, safety has always been considered as highly dependent on all the actors involved in the air transport system operations. At the national scale, the legal responsibility for aircraft safety lies with the national authorities, which are responsible for issuing the aircraft airworthiness certificates according to given (national and international) standards. The institutions dealing with these standards are the national CAAs. In addition, the manufacturers should provide an aircraft design that meets these standards. The airlines should use the aircraft according to the prescribed procedures and within the prescribed limits respecting the standards of airworthiness. The airlines are obliged to report possible defects, which if systematic, can be eliminated initially at the manufacturers and/or during aircraft maintenance. These all enable the regulation authorities to set up the acceptable 'levels of safety,' which on the one hand provide confidence to the travelling public and are cost efficient on the other. For example, JAR (Joint Airworthiness Requirements) distinguishes four levels of safety in terms of the effects and frequency of the occurrence of accidents and incidents per unit of time (hour): i) minor; ii) major; iii) hazardous; and iv) catastrophic. In a given context, occurrence of a 'catastrophic' event resulting in multiple deaths and loss of the aircraft is expected to be almost improbable (less than 10^{-9} per hour). This is practically considered as an unlikely event in the operational life of a given aircraft (Jenkinson, Simpkin and Rhodes, 1999). The general philosophy of the safety management in this case is to completely prevent occurrence of the catastrophic events during the operating life of a given aircraft fleet and accept the occurrence of less hazardous failures with the probability being inversely proportional to their severity. Nevertheless, in practice, all types of hazardous events happen, which makes the safety issue an important component of the overall sustainability of the air transport system.

2.2.3 Operational Performances

2.2.3.1 Demand

General The airline demand is usually expressed as the number of transported passengers and/or volume of freight, as well as by the amounts of the revenue passenger (p-km) and/or revenue tonne-kilometres carried out during a given period of time (day, month, year). In both cases, one revenue p-km or t-km, respectively, implies carrying out one passenger or one tonne of cargo carried out over the distance of 1 km (instead of kilometres, statute miles are also frequently used: 1 mile = 1.609 km). In some cases, the revenue p-km and t-km are aggregated after converting about 10 revenue p-kms into one revenue t-km. In general, these volumes increase with an increase of any or both influencing factors simultaneously. In order to realise the large volume of p-km on the short haul distances, the airline should carry out a larger number of shorter flights, if the volume of passenger demand justifies it. The number of flights on the longer distances, again justified by the passenger demand, can be lower in order to realise the equivalent volume of output in terms of the revenue p-km.

Driving forces   The external and internal forces drive the airline demand. Historically, the external forces have been the economic growth at local-micro and global-macro level, and diminishing (in real terms) airfares. The economic growth has been materialised through growth of local (regional) and global Gross Domestic Product (GDP). Some analyses have shown that from about two-thirds to three-fourths of the growth of airline demand has been attributed to the growth of global GDP (Holloway, 2003). The growth and volatility of the airline demand has however been much higher than the corresponding figures for its economic development. Airfares in real terms have diminished significantly since the 1980s, primarily thanks to increasing aircraft and airline productivity through improvements in the aircraft technology and management of resources, increased competition, and expansion of the low-cost carriers with a radically different business model than the established so-called full cost (legacy or network) airlines. In many regions, they have been one of the main driving forces of airline demand. The internal demand driving forces have appeared to be flight frequency, sophisticated pricing, the revenue (yield) management, and different promotional and loyalty programmes. Current increases in fuel prices may become an important internal driving force, leading to increases in airfares, which in turn may slow down or even, in some cases, compromise the future growth of airline demand (Holloway, 2003).

In addition to the volume, the airline demand is characterised by the time fluctuation (seasonality and monthly, weekly, and daily peak and off-peak periods) as well as directionality. The latter is a common characteristic of freight demand.

Analysis and forecasting   In general, airline demand is usually expressed by the so-called demand function. This function inter-relates the volume of demand during a given period of time (day, week, month, year) and the main demand-driving forces. The demand function can be estimated for different levels of aggregation of demand. These can be different types of passengers and/or freight origins and destinations such as a pair of airports, cities, regions, countries, and even continents. They can also relate to the different segments of demand such as business or leisure. Consequently, the airline demand function in terms of the number of passengers or the volume of cargo carried out between a given origin (i) and destination (j) can be estimated as follows (Janić, 1997; TRB, 2002):

$$D_{ij} = a_0 GDP_i^{a1} GDP_j^{a2} T_{ij}^{a3} S_{ij}^{a4} AF_{ij}^{a5} O_{ij}^{a6} \dots \qquad (2.4)$$

where

GDP_{ij} is Gross Domestic Product of the air traffic origin (i) and destination (j), respectively (billion \$ US);

T_{ij} is the interaction between origin (i) and destination (j) in terms of the value of trade, services, and investments (million $ US);

AF_{ij} is the average airfare per passenger or freight shipment between origin (i) and destination (j) ($ US per passenger) (the airfare can be determined as the product of yield and distance);

S_{ij} Is the transport capacity offered between origin (i) and destination (j) (seats);

O_{ij} is the symbol embracing other variables that might influence the airline demand between origin (i) and destination (j) (it may also include dummy variables);

a_i is the coefficient of the demand function usually estimated by the regression technique (i = 1, 2, 3,..., 5,...).

Coefficients a_i (i = 1–5...) on the right side of Equation 2.4 represent the elasticity of airline demand subject to changes in the influencing factors – independent variables. The practical application of the demand function and its modalities has shown that the coefficients a_1–a_3 are usually positive, implying an increase in the airline demand with an increase in the related variables – GDP, level of interaction between the demand origin(s) and destination(s), and the offered transport capacities. Coefficient a_4 has always been negative, implying a decrease in airline demand with an increase in airfares. The coefficients of other variables might have different signs depending on the type of variable. For example, these variables can be dummy variables characterising the demand affected by some exceptional causes, the variables characterising attributes of performances of other competitive transport modes and so forth. In order to estimate the demand function 2.4, the historical time series data for both the demand side – dependent variables – and the influencing side – independent variables – has usually been needed. Then the regression technique based on the least-square estimates is applied to estimate particular coefficients, as well as the aggregate statistics enabling assessment of the quality and importance of the relationship, as well as particular independent (influencing) variables (TRB, 2002).

The demand function 2.4 can also be used to forecast future airline demand in a given origin and destination market. The pre-assumption is that the nature of influence of the demand driving forces (independent variables) will be the same in the future as it has been in the past. Under relatively stable conditions, this is almost acceptable. However, it appears rather difficult to accept this statement if the conditions of high uncertainty and volatile circumstances are expected. However, at least methodologically, this can be overcome by the scenarios-based prediction of the values of particular independent variables (TRB, 2002; Holloway, 2003).

The airline demand is usually forecasted for the future short- and long-term period. In the short-term, this can be a daily, weekly or monthly forecast. In the long-term, it is usually one or more years. The forecasting can be carried out for a single airline, airline alliances, or the airline industry of a particular country or region (continents). Regarding the geographical-market scale, this can be for a single airline route, part of or the whole airline or its alliance network, as well as for the entire airline industry of a given country or region. Figure 2.6 illustrates an example of the analysis and forecasting of the airline passenger demand for the US domestic and international

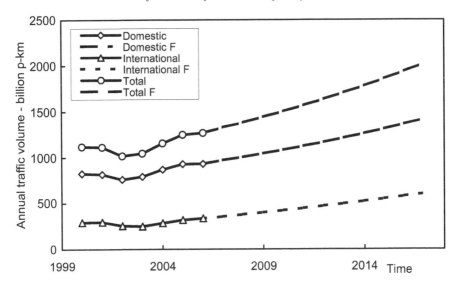

Figure 2.6   US scheduled passenger traffic: US scheduled carriers
Compiled from FAA (2005)

market in terms of the annual volumes of p-km for the period 2000–2017 (FAA, 2005). As can be seen, the forecasters expect relatively smooth and steady growth of passenger demand in both markets over this period.

The analysis and forecasting of passenger demand appears important for planning the airline resources – aircraft, energy, and staff, as well as the costs and resources. In addition, it also enables estimation of the external effects such as the average delays, noise, fuel consumption and related emissions of air pollution, i.e., the important factors influencing the air transport system's sustainability.

2.2.3.2 Capacity

General   The fleet, usually consisting of the aircraft of different types characterised by the seat- and cargo-volume capacity, is the main airline asset. The large full cost or the network (legacy) airlines usually operate large (global) networks consisting of routes of different lengths and passenger and freight volumes, which require deployment of different aircraft types in order to serve such diverse demand efficiently and effectively. In general, the smallest-regional, medium narrow-bodied and wide-bodied, and large narrow-bodied and wide-bodied aircraft serve the regional short, continental-medium, and intercontinental long-haul routes markets, respectively. Figure 2.7 shows an example related to the European airlines – members of AEA (Association of European Airlines), which can be used for illustrative purposes (AEA, 2005).

As can be seen, the diversity of aircraft types generally increases at a decreasing rate the number of aircraft in the fleet increases. This also implicitly indicates a higher diversity of routes regarding the length and volume of demand in the networks of larger airlines. The maximum number of different aircraft in the fleet is 13, the average number 6, while the average aircraft capacity has been 150 seats.

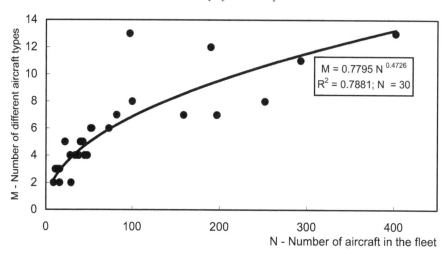

Figure 2.7   **Dependence on the number of different aircraft types on the airline fleet size: The AEA airlines**
Compiled from AEA (2005)

On the other hand the low-cost airlines usually operate in the short and medium-haul markets using the uniform aircraft fleet consisting mostly of a single aircraft type – in most cases the B737 or A319/320.

The airlines assign their fleets to serve the expected demand in their networks. The product of the number of seats and covered distances (i.e., length of route) represents the airline output (i.e., capacity) expressed by the aircraft seat- or tonne-kilometres (as-km and at-km, respectively) over a given period of time (day, week, month, year). The airline output in terms of the demand served is expressed by the revenue p-km. The available as-km and/or at-km are related to the p-km and t-km by the load factor, actually reflecting the percentage of the airline output (capacity) that is sold (Janić, 2001; Holloway, 2003). The load factor is always less than 100 per cent, thus enabling absorption of the short-term fluctuations (increases) in demand.

For the route between the demand origin (i) and destination (j), the load factor can be used for determining the flight frequency as follows (Janić, 2001; Bazargan, 2004):

$$f_{ij} = D_{ij}/n_{ij}\lambda_{ij} \qquad\qquad (2.5)$$

where
  D_{ij} is the passenger or freight demand between the origin (i) and destination (j) passengers or tonnes of freight;
  n_{ij} is the number of seats offered on the route (*ij*) (seats); and
  λ_{ij} is the average load factor on the route (*ij*).

Equation 2.5 indicates that the flight frequency and load factor have an inverse relationship. In addition, the flight frequency can be determined using other criteria. These can be: maximising the airline profits, minimising the airline costs, minimising the total generalised costs for both the airline and its users – passengers and freight shippers, and maximising the quality of service expressed by the passenger schedule delay, i.e., waiting time for the first available departure (Janić, 2001). The flight frequency and load factor can also be influenced by competition on given route (market).

The flight frequency can also be determined simultaneously for all routes of a given airline network. Usually this is carried out by using Operations Research (OR) optimisation techniques such as Linear and Integer Programming (LP and IP, respectively). The objective is minimising the total costs, and/or maximising the revenues and profits. In many cases, the airline fleet serving the network or its part is assumed to be rather homogenous, i.e., consisting of the same or very similar aircraft types in terms of carrying capacity, speed, fuel consumption, etc. Otherwise, a different fleet is assigned to different routes simultaneously with determination of the flight frequencies (I don't understand this sentence, in particular what you mean by 'with determination,' so I cannot correct it). The generic and simplest optimisation problem for determining the flight frequencies in a given airline network can be defined as the profit maximisation problem formalised by the objective (total profit) function and the specified constraints as follows (Janić, 2001):

$$\text{Maximise } \Pi = \sum_{i=1}^{M}(D_i p_i - C_i(n)f_i) \tag{2.6a}$$

subject to constraints:

$$f_i > D_i/n \tag{2.6b}$$

$$f_i \leq D_i/n\lambda_{imin} \tag{2.6c}$$

$$f_i > 0 \text{ for } \forall i \in M \tag{2.6d}$$

where
    M is the number of routes in the airline network;
    D_i is the demand on route (i) (passengers);
    p_i is the average airfare on route (i) ($ US per passenger);
    $C_i(n)$ is the average cost per flight on route (i) of aircraft of capacity (n);
    f_i is the flight frequency on route (i);
    λ_{imin} is the minimum ('break-even') load factor on route (i).

Constraints 2.6b ensure that the capacity offered by the airline satisfies demand on each route. Constraints 2.6c prevent scheduling of flights on routes without a guarantee of at least minimum (break-even) load factor. Finally, constraints 2.6d ensure that the flight frequencies are non-negative integers. This basic structure of the airline profit optimisation problem can be modified to include different aircraft types (i.e., seat capacity), constraints on the aircraft utilisation, and the constraints on the air route and airports capacity (Janić, 2001).

Route, network capacity, and fleet size   The flight frequencies, aircraft capacity, and length-distances of a particular route of a given airline network enables determination of the aggregate airline output, i.e., the capacity offered to serve the expected demand during a given period of time. The output expressed in as-km or at-km is as follows (Janić, 2001):

$$C = \sum_{i=1}^{M} f_i c_i d_i \tag{2.7}$$

where
$\quad d_i$ is the length of route (i) (km or miles).

Other symbols are analogous as to those in Equation 2.5. Equation 2.7 assumes that the airline network consists of different types of routes, each allocated with similar aircraft types in terms of capacity and other operational characteristics. Consequently, for a given cluster of similar routes the aircraft fleet size can be determined as follows (Janić, 2001):

$$N = (1/U) \sum_{i=1}^{M} f_i t_i \tag{2.8}$$

where
$\quad U$ is the average allowed utilisation of an aircraft of a given type over a given period of time (day, month, year) (hours);
$\quad t_i$ is the average block-time per flight on route (*i*) determined as the quotient of route length and the aircraft block speed (hours).

Other symbols are analogous to those in Equations 2.5 and 2.7. Obviously, the airlines operating the larger networks with longer route block times and more frequent flights need to deploy a greater number of aircraft. Since the flight frequency depends on demand, the size of an airline fleet will also directly depend on the volume of demand in the network, as well as on the capacity and load factor of each aircraft. In addition, if the average utilisation of an aircraft is higher, the number of aircraft required will be lower, and vice versa (Janić, 2001). The size of the airline fleet can be readjusted after determining the departure and arrival times of particular flights, and consequently the detailed itinerary of each aircraft in the network (Bazargan, 2004).

    Size, type, and utilisation of the airline fleet all directly influence the energy (fuel) consumption and associated emissions of air pollutants on the one hand, and the aircraft noise at airports of the airline network on the other. Therefore, minimisation of the size and maximisation of the utilisation of the deployed aircraft fleet and supporting resources to serve a given volume of demand over a given period of time is of crucial importance for the airline economic/business performances on the one hand and for decreasing the external impacts on people's health and environment on the other. In this sense, the airlines indirectly contribute to the sustainability of the air transport system through maintaining their economic performances and strength.

2.2.3.3 Quality of service

The quality of airline services can be considered as an outcome of a dynamic interaction between the demand and capacity. This quality may be expressed by different attributes related to the airline operations and schedule such as flight frequency, regularity, reliability and punctuality of particular flights or fleet type, and load factor (Janić, 2001).

The flight frequency on a given route implies the passenger schedule delay, i.e., the waiting time for the first available departure. The attitude behind this attribute of service quality is that passengers generally lose time while travelling but particularly while waiting for the service.

Regularity implies provision of flights on a given route on a regular basis, i.e., on certain days at the same time on each day (for example, these can be morning, midday and evening departures during weekdays). Reliability implies the percentage of realised flights as compared with the number of planned ones over a given period of time (day, week, year). In many cases, technical failures as well as the external disruptive events such as bad weather and terrorist threats can cause cancellation of flights, which are generally considered as unpopular and damageable for both users-passengers and freight shippers on the one side, and the airlines on the other, due to the increased costs.

Punctuality relates to the percentage of flights arrived according to the schedule and the length of the delay of delayed (unpunctual) flights. Arrivals of up to 15 minutes behind schedule are not considered as delays. Different causes similar to those influencing reliability as well as traffic congestion may affect the airline schedule and cause flight delays. However, in this case the intensity of the impact of particular causes is weaker, thus enabling flights to be carried out anyway (Janić, 2001). Delays represent loss of time for the affected actors – users-passengers, freight shippers, and airlines. They extend the block time of affected flights and thus require engagement of the greater number of aircraft and other supporting resources (staff). These might raise the costs and thus compromise the airline's overall profitability (Equation 2.8). Delays can also cause additional fuel consumption and associated emissions of air pollutants, which definitely qualify them always to be considered as an important negative factor of the sustainability of the air transport system.

The airline quality of service may also have some other attributes such as the quality of the in-cabin service, reliability of handling passenger baggage, loyalty programmes, and convenience of passenger handling at the departure and arrival airport gates. From the point of view of passengers and other actors involved, particular attributes of service quality may have different levels of importance, which enables the systematic ranking of airlines. The Airline Quality Ranking (AQR) programme in the US is one such example (Janić, 2001).

2.2.3.4 Safety

Airline safety is based on the aircraft safety and strict respect of the operational rules and procedures intended to guarantee safe operations. As applied to the aircraft fleet of a given airline, the safety of operations can be considered as an operational performance. Each airline has a safety record, which can be expressed by different absolute and relative measures. The former are the total number of aircraft incidents and accidents. The latter usually relate to the number of aircraft accidents and incidents per unit of the airline output-revenue p-km or t-km. These measures can also be established for airline alliances and the airline industry of a given region (country, association of countries and/or continent). Since the aircraft accidents and incidents result in loss of life as well as destroying and/or damaging properties, they play an important role as externality and are consequently an extremely important element of the sustainability of the air transport system. Recently, this importance has been institutionally verified by The Council of the EU (European Union), which has approved the creation of a 'common blacklist of the unsafe airlines'. These are those airlines which do not meet the common safety criteria and whose operations in EU airspace should be banned (CEU, 2005).

2.2.4 Economic Performances

The airline economic performances relate to the revenues, costs, and profits, which are dependent, among other factors, also on the type of airline economic/business model. In general, there are three such models: the full costs model of the so-called network (legacy) airlines, the low-cost model of the increasingly emerging schedule airlines, notably in the US, Europe and Asia, and the charter model of the charter airlines mostly operating in Europe.

2.2.4.1 Profits, revenues and costs

Profits   The most important attributes of an airline economic performance are profits Π as the difference between the total revenues R and total costs C at the end of a given period of time (one year). The generic relationship between these attributes is as follows:

$$\Pi = R - C >< 0. \tag{2.9}$$

If the profit is greater than zero at the end of a given period of time then the given airline is profitable, otherwise it has losses. Practice has shown that this has been one of the most important economic and financial characteristics of all contemporary airlines, which has influenced other performances – operational, social and environmental. Only if it is economically stable and strong, will the airlines be able and keen to offer attractive and safe services to its users and appropriately deal with other components of sustainability. Otherwise, they struggle to survive.

Revenues   The total airline revenues over a given period of time are the results of the output sold (p-km or t-km) at given prices. Under regular conditions, the total revenues increase as the sold output increases at a constant, increasing or decreasing rate. Two examples of the relationships for two types of airlines – the large full cost network airline (Lufthansa, Germany) and the large low-cost Southwest Airlines (US) – for the period 1996–2005 are presented as follows (Lufthansa, 2006; Southwest Airlines, 2006).

$$\text{Lufthansa: } R = 0.0264V^{1.402}, R^2 = 0.701; N = 10; 70 < V < 110 \tag{2.10a}$$

Southwest: $R = 0.1164V^{0.903}$, $R^2 = 0.968$; $N = 10$; $35 < V < 100$ (2.10b)

where

R is the total annual revenues (billion € at Lufthansa and billion USD at Southwest Airlines);

V is the total annual output (billion revenue p-km).

The annual output of Lufthansa is greater (70–110 billion revenue p-km) than that of Southwest (35–90 billion revenue p-km). Lufthansa's revenues have also been greater, reflecting the differences in the economic/business models of both airlines (10–20 billion € of Lufthansa and 3–7 billion USD of Southwest Airlines). In addition, the total annual revenues have increased with the increase in the output at an increasing rate for Lufthansa and at a decreasing rate for Southwest Airlines.

As expressed per unit of output sold, the price is called 'yield'. It comes up after dividing the total revenues by the total volumes of the output sold. Yield multiplied by the travel distance also gives the average airfare (or rate) expressed in monetary units per passenger or cargo shipment, respectively. At many airlines, airfares can be very diverse even across different flights in the same market (route) carried out at different times of the day, week, or month (morning, evening peak, non-peak, weekdays, weekends, etc.). In addition, they can be different for different categories of passengers, such as business and leisure, and the categories of airlines such as the network (legacy) and low-cost airlines. The philosophy behind the price diversification is to make them acceptable to different categories of passenger demand willing to travel at different times. Consequently, demand is always assumed to be sensitive (i.e., elastic) to prices. Independent of the demand category, the volume is always inversely proportional to the price, i.e., lower prices attract more passengers, and vice versa. The relationship between prices and demand is usually represented by the demand-price function, which can be set up for a given flight, a route, or a part of the entire airline network. Figure 2.8 shows a scheme of the down-sloped demand-price functions for three classes of passengers. For each class, if the price decreases the passenger demand will increase, and vice versa. The area(s) closed by the demand-price curves and given level of prices represents the passenger surplus. When the price is p_k the volume of demand will correspond to D_k ($k = 1,2,3$). In order to maximise revenues, the airline will allocate the capacity (seats) to D_1 passengers of class 1, $(D_2 - D_1)$ passengers of class 2, and $(D_3 - D_2)$ passengers of class 3. Such seat allocation brings the total revenues as follows:

$$R = \sum_{k=1}^{3} p_k(D_k - D_{k-1})$$ (2.11a)

Equation 2.11a implies that in order to maximise the revenue, the airline always assigns most of the available capacity to the most expensive passengers and leaves only a limited number of seats for the passengers who pay the least. Assigning capacity to different types of passenger demand aiming to maximise the revenue is called 'yield management' (Janić, 2001; Holloway, 2003). Once the aircraft type is

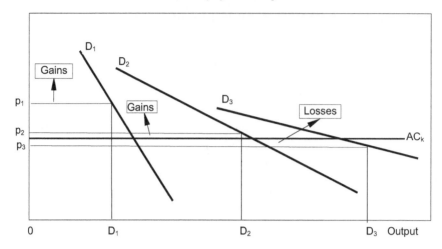

Figure 2.8   An example of the airline demand-price differentiation

determined, the average cost per seat is assumed to be fixed and not dependent on the volume of demand (horizontal line AC_k in Figure 2.8). The total profits on given route can be estimated as follows:

$$\Pi = \sum_{k=1}^{3} (p_k - AC_k)(D_k - D_{k-1})\qquad\qquad(2.11b)$$

As can be seen, the airline gains profits but also suffers losses from particular classes of passengers. However, it is important that the total profit is positive. Low-cost airlines diversify their prices much less if at all. However, their generally low prices based on the low cost of producing the output attract substantive demand. Returning to the cases of Lufthansa and Southwest Airlines, from Equation 2.10, the average and marginal yield per unit of the output sold (p-km) can be obtained. The average yield can be obtained by dividing the total annual revenues by the total annual volume of output. For Lufthansa the average yield is about twice that for Southwest Airlines and increases as the volume of output increases. For Southwest Airlines, the average yield decreases as the volume of output increases. The marginal yield can be obtained as the first derivative of the revenue function 2.10 with respect to the variable output. This yield is about three times greater for Lufthansa than for Southwest Airlines. In the former case, it increases and in the later case it decreases with as the volume of output increases. For Lufthansa, the marginal yield is greater than the average yield implying that producing each additional unit of output increases the average yield. At Southwest Airlines, the marginal yield is lower than the average yield implying that each additional unit of output decreases the average yield (Holloway, 2003; Lufthansa, 2006; Southwest Airlines 2006).

Costs   The costs must be incurred to get revenue from transporting passengers and cargo, i.e., the revenues cannot be generated without incurring costs. The main airline costs are those of resources such as capital (aircraft), energy (fuel) and labour

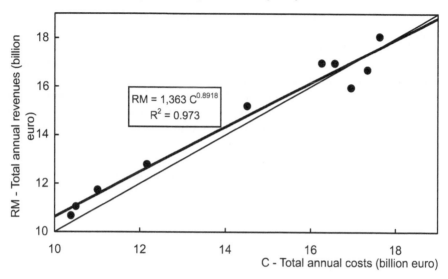

Figure 2.9   An example of dependability of the total airline revenues and costs

Compiled from Lufthansa (2006)

(staff). The total costs are proportional to the quantity and prices of the resources consumed, which increases with an increase in the volume of output. Figure 2.9 shows an example of the relationship between the total revenues and costs for Lufthansa during the period 1996–2005 (Lufthansa, 2006).

As can be seen, the total revenues have increased with the increasing in the total costs at a decreasing rate, indicating that more resources have been spent to achieve the satisfactory but lower revenues. An explanation can be found in losses recoded during particular years of the observed period due to disruptions such as the 11 September 2001 terrorist attack on the US, epidemic diseases (SARS in 2004), and increasing fuel prices.

Aircraft cost   The airline's total costs depend on the costs of the individual aircraft in their fleet and efficiency of their utilisation. In general, for each aircraft the cost could be classified into Direct Operating Costs (DOC) and Indirect Operating Costs (IOC). The former include all the costs associated with flying and direct maintenance of the aircraft. The latter include those costs, which are not directly attributable to a given aircraft, and its flying costs. Both costs can also relate to the individual aircraft, a fleet of the same aircraft types, and the entire airline fleet as well (Janić, 2001).

For an aircraft, the DOC usually includes the aircraft cost (depreciation, interest and insurance rate dependent on the overall aircraft price), maintenance and overhauling cost (airframe, engines and avionics), and flight operation cost (crew, fuel/oil, airport and navigation charges). The IOC is more diverse and may roughly include aircraft and traffic servicing, promotion and sale, passenger services, general and administrative overheads, ground property and equipment maintenance and depreciation expenses. In addition, there might be some other more or less detailed categorisation and specification of costs (Jenkinson, Simpkin and Rhodes, 1999; Janić, 2001). In general both the total DOC and IOC have been shown to increase with increasing aircraft size (i.e., seat capacity) and stage length. Due to their relatively high importance, the aircraft costs have been the matter of substantive empirical research, which has usually resulted in estimating the cost functions per flight or per unit of flight output (available s-km) on the basis of the non-stop flying distance and aircraft size. For such purposes, the regression technique is applied to the cross sectional data. One of the most recent examples looks like as follows (Swan and Adler, 2003):

$$C_1(S, D) = 2.44 S^{0.600} D^{0.750} \text{ for regional single-aisle aircraft (flights)}$$
$$C_2(S, D) = 0.64 S^{0.655} D^{0.912} \text{ for long-haul twin-aisle aircraft (flights)}$$
$$(2.12)$$

where
  $C(*)$ is the cost per flight;
  S is the aircraft seat capacity;
  D is the route length.

The cost per flight increases with increasing aircraft seat capacity and distance at a decreasing rate. This implies the existence of economies of seat density and economies of distance. The relationship(s) are generic but need to be adjusted for the different airline economic/business model (full, charter, low-cost), aircraft type, as well as changes of the cost of inputs.

Airline costs   The total airline costs have a different character than the cost of individual aircraft. They are usually expressed in monetary terms for a given period of time. Since being directly dependent on the airline output (available s-km), they can be expressed by the total cost functions (i.e., the cost-output relationship) estimated for a given period of time. Two examples are presented for Lufthansa (Germany) and Southwest Airlines (US) for the period 1996–2005 as follows (Lufthansa, 2006; Southwest Airlines 2005).

$$\text{Lufthansa: } C = 0.0098 V_0^{1.518}, R^2 = 0.680; N = 10; 90 < V < 150 \qquad (2.13a)$$

$$\text{Southwest: } C = 0.033 V_0^{1.077}, R^2 = 0.996; N = 10; 55 < V < 120 \qquad (2.13b)$$

where
  C is the total annual costs (billion € at Lufthansa and $ US at Southwest Airlines);
  V_0 is the total annual output (billion available s-km).

At both airlines the total costs increase with increasing output at an increasing rate – higher at Lufthansa than at Southwest Airlines. Dividing the total cost in Equation 2.13 by the annual volume of output produces the average cost per unit of output. In both cases, the average cost increases with increasing volume of output at a decreasing rate. This cost is lower for Southwest Airlines than for Lufthansa by about two and half times, which, intuitively, is to be expected. The marginal cost can be obtained as the first derivative of the cost function 2.13 with respect to the variable output. For both airlines this cost increases with increasing volume of output and is about three times higher for Lufthansa than for Southwest Airlines. The marginal cost is higher than the average cost for both airlines, reflecting conditions where the average costs increase with each additional unit of output. Such variations in the total, average and marginal costs as the airline output grows, as well as their relative relationships, indicate the lack of economies of scale at both airlines. This can be analytically proven by setting up the relationship between the average and marginal costs for the airline as follows:

$$l(V_0) = AC(V_0)/MC(V_0) \qquad (2.13c)$$

where

$AC(*)$, $MC(*)$ is the average and marginal cost, respectively (cents USD/s-km);

V_0 is the volume of airline annual output (billion available s-km).

In Equation 2.13c, $AC(V_0) = C/V_0$ and $MC(V_0) = \partial C/\partial V_0$. If $l(V_0) > 1$, there are economies of scale, if $l(V_0) = 1$ there is constant return to scale, and if $l(V_0) < 1$, there are diseconomies of scale. For Lufthansa, $l(V_0) = 0.658$ and for Southwest Airlines $l(V_0) = 0.923$, indicating diseconomies of scale at both airlines.

Comparison of the airline total revenues and costs (Equations 2.11 and 2.13) indicates that the average and marginal revenues have always been higher than the average and marginal costs, respectively, at both airlines, thus providing profitability for the airline during the observed period (1996–2005).

2.2.4.2 Economic/Business model

The airline profit, revenue and cost structure depends considerably on the economic/business model. There are three different models: full cost (legacy), low-cost, and charter airline model. In addition, many full cost (legacy) airlines form the airline alliances. The low-cost airlines operate individually. The charters usually operate in the scope of the tour-operator packages serving the leisure markets.

The full cost or the network (legacy) airlines   The full cost or the network (legacy) airlines usually operate a kind of hub-and-spoke network. This implies the star-shaped network on a spatial-horizontal layout. The hub airport is located in the centre of an area covered by the network, which is divided into one or more concentric circles defining the spatial rings in which the spoke airports are nearly uniformly distributed, at approximately equal distances from the hub (Jeng, 1987). Such diversity of distances as well as the volumes of passenger demand requires deployment of different aircraft types (capacity) in the given network in order to provide viable (profitable) airline operations. The incoming and outgoing flights connecting the hub to the spoke airports are scheduled as clusters at the hub airport in order to enable more efficient 'feeding' of each other with traffic and exchange of other resources such as aircraft and crews. This may increase the passenger volumes and density in particular routes and provide benefits from economies of scale and economies of density. Both occur when the average unit cost decreases with increasing volumes and density of demand. Such lower costs enable lower, more competitive prices, which can also be adjusted due to the inconveniences of changing flights at the hub airport.

Clusters of flights at the hub airport are called complexes, each consisting of an incoming and an outgoing bank of flights. The number of complexes and flights may vary. For example, Lufthansa (Germany) schedules four complexes of flights per day at its hub – Frankfurt airport (Germany) – each within the time windows of about three hours. In each bank there are about 45 incoming and outgoing flights. Some other European airlines such as KLM and Air France schedule four and six complexes of flights per day at their hubs – Amsterdam Schiphol (the Netherlands) and Paris Charles de Gaulle (France) airport, respectively, each within the time windows of about three hours. The average number of flights per complex for KLM is 92 (46 incoming and 46 outgoing) and for Air France 70 (35 incoming and 35 outgoing). The major US airline American Airlines schedules 10 complexes at its main hub Dallas/Fort-Worth. Delta Airlines operates 12 complexes at its Atlanta hub. The average number of flights per complex is 70 and 90, respectively.

In addition to operating their own hub-and-spoke networks, which could spread between countries and continents, some large full cost airlines tend to expand further by strengthening their position at the edges of their networks. The common practice during the 1990s for achieving such a goal was contracting an airline alliance. In particular, European airlines, confronted with the gradual liberalisation of the national and international EU markets, have contracted many alliances starting from the simple code share agreements to merging with their partners. The aim of alliances has been to create powerful airline groups that are able to offer comprehensive services, and dominate hubs and markets on the one hand, and avoid legal, political, and market constraints on the other. Particular reasons have been: expansion of the air route networks, feeding traffic between partners, cost efficiency, improved service quality, increased itinerary choices for passengers, and the advantages of being displayed at CRS (Computer Reservation System). Depending on the strength of the relationship, the alliance partners have shared the use of each other's resources, advertising and promotion, costs and revenues (Oum and Park, 1997).

At the beginning of the current decade, three large alliances have been consolidated, keeping about 46 per cent of the world's market share in terms of the number of passengers. These are: the Star alliance (leading airline: Lufthansa, Germany; hub airport: Frankfurt Main, Germany; 12 airlines; 19 per cent market share), Oneworld (leading airline: British Airways, UK; hub airport – London Heathrow; nine airlines; 13 per cent market share), and SkyTeam (leading airline: Air France, France; hub airport – Paris Charles de Gaulle; 14 airlines; 14 per cent market share) (EC, 2003).

Consolidation of airline alliances, including an additional concentration of demand and services, have created significant requirements for the expansion of the capacity (land take) as well as increased noise and air pollution at these airports. Consequently, the alliances should be considered as an important factor influencing the sustainable development of the air transport system.

The concentration of demand and services makes the hub-and-spoke networks operationally vulnerable to various disruptive events. In particular, bad weather, failures of the aviation system facilities and equipment, industrial action by aviation staff, and terrorist threats can all affect the hub airport for a relatively long time (several hours during the day) causing long delays and cancellations of many flights, which can be costly (Janić, 2005).

Low-cost airlines   Low-cost airlines operate the so-called 'point-to-point' networks. This implies scheduling direct flights between particular airports of the network. Most flights start and end at the airport representing the airline base. One airline may have a few such bases. For example, the largest low-cost Southwest Airlines (US) operates 10 such airports where the daily number of flight departures varies from about 80 to 195 (Southwest Airlines 2006).

Each flight is considered independently, without any intentional (scheduled) connections. The utilisation of each aircraft/flight individually can therefore be optimised, although this also relies on the short turnaround time. In addition, disruptions to the schedule can be more easily neutralised and kept relatively isolated to the individually affected aircraft and its flights alone. Most airports included in the network of low-cost airlines are smaller regional airports offering at least two advantages: no congestion and delays caused by traffic, and the short turnaround times of the 'independent' aircraft/flights. The short turnaround times are related to the individual aircraft only as it does not need to wait for a particular aircraft/flight, like in the hub-and-spoke networks of the full cost airlines. Higher utilisation of the airline fleet contributes to reducing the operational costs. An additional factor is the increased aircraft seat density, enabling better utilisation of the in-cabin passenger space. Since the low cost airlines operate rather homogenous fleets (mostly one aircraft type) along routes of very similar lengths, the aircraft utilisation can also be improved by increasing the number of flights per given period of time (day, year). In such a case, the following relation holds:

$$F_d = T/(t_0 + t_1 + v/D) \tag{2.14}$$

where
F_d is the number of scheduled daily flights per aircraft;
T is the period of time (available hours per day, year);
t_0 is the aircraft turnaround time at an airport of the network (hours);
t_1 is the duration of the taxiing in/out, and climbing/descending phase of flight (hours);
v is the aircraft/flight cruising speed including en route and terminal delays (knots – 1 knot = 1 nm/h; nm – nautical mile);
D is the average stage (route) length (nm-nautical mile).

The number of flights, which can be carried out in given time T increases as the flight block-time $(t_0 + t_1 + v/D)$ decreases. As has been mentioned, this can be achieved by shortening the first two components of the flight block time – the turnaround and taxiing time. For example, the full cost (legacy) airlines typically utilise an aircraft of comparable size to the aircraft of the low-cost airlines (B737) on the short- to medium haul routes for about eight hours per day. The low-cost airlines utilise the

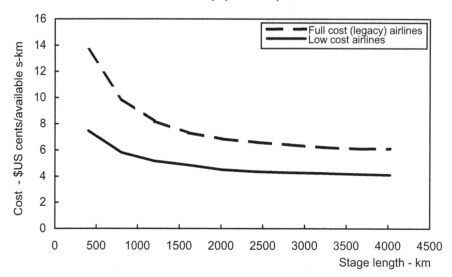

**Figure 2.10   The average cost vs route length of the US full and low cost
               airlines in the year 2003**

Compiled from GAO (2004)

same aircraft for about 10–11 hours per day, which is about 25–35 per cent higher. The main contributor to this distinction is the turnaround time, which is about one hour at the full and half an hour at the low cost airlines (Europe). In addition, the taxiing time of the full cost airlines may be longer due to the congestion at their larger hub airports. These and other factors related to the labour costs enable the low cost airlines to have generally significantly lower costs per unit of output (available s-km) than the full cost (network or legacy) airlines. Figure 2.10 shows an example for the US domestic market (GAO, 2004).

In this case the average unit cost of the low cost airlines amounts to about 55–66 per cent of the average unit cost of the full cost (legacy) airlines. For both categories of airlines this cost decreases more than proportionally as the route (stage) length increases, reflecting the existence of economies of distance. Such cost advantages have enabled the low cost airlines to set up lower more competitive airfares, growth and consequently an increasing share of the market from 23 per cent in 1998 to 33 per cent in 2003. During the same period the market share of the full cost (legacy) airlines decreased from 69 per cent to 65 per cent. This implies an increase in the number of passengers from 79.8 to 117.1 millions for the low cost airlines and a decrease in the number of passengers from 242.2 to 231.6 millions for the full cost (legacy) airlines (GAO, 2004).

In Europe, in 1998 the full cost (legacy) airlines carried about 203 and the low cost airlines about 7.7 million passengers. In the year 2003, the former airlines carried 227.5 and the latter 38.2 million passengers. The market share of low cost airlines has thus increased from 3.7 per cent to 14.5 per cent during the observed period. By the year 2010, the low cost airlines are expected to have gained a market share of about 35 per cent (MERCER, 2002; Heymann, 2004).

The growth of air traffic, with an increasing contribution by the low cost airlines, will generally result in increased fuel consumption and related air pollution. This may raise the question about the impacts of the low-cost economic/business model on the overall sustainability of the air transport system.

Charter airlines   Charter airlines have traditionally served the European leisure markets. Their economic/business model is characterised by i) higher seat density in the aircraft; ii) higher aircraft utilisation; iii) lower cost of seats thanks to relying on tour operators; and iv) regular flight frequencies during the high season for the longer stage lengths. For a long time, these characteristics enabled charter airlines to operate at about 30–40 per cent lower costs than the full cost airlines with aircraft of comparable size (Holloway, 2003). Nevertheless, this difference has disappeared on the long-haul intercontinental routes where the full cost (legacy) airlines have also achieved the similar cost performances mostly thanks to comparable or higher aircraft utilisation. Since the charter airlines operate larger but highly utilised aircraft their costs appear similar to the cost of low cost airlines.

As far as sustainability is concerned these airlines have usually been considered in the scope of sustainability of the tourist industry.

2.2.5 Social Performances

The airline social performances can be roughly classified into direct and indirect. Direct social performances can be represented by the direct employment of staff. The size of staff depends on the number of aircraft and their seat (and cargo) capacity. Generally, the larger the airline fleet the larger the staff team needed to operate it. In order to illustrate the possible existence of such dependability, three examples have been selected as follows: i) the US airline industry represented by the 19 biggest air airlines in 2002; Lufthansa airline as the large full cost network (legacy) airline during the period 1996–2005; and iii) Southwest Airlines (US) as the large low-cost airline during the period 1996–2005 (BTS, 2003; Lufthansa, 2006; Southwest Airlines, 2006). The results are given as follows:

$$\text{US airline industry (2002):} \quad \begin{aligned} N_e &= -6.695 + 0.126 N_a \\ R^2 &= 0.976; \ N = 19; \ 100 < N_a < 820 \end{aligned} \qquad (2.15a)$$

$$\text{Lufthansa (1996–2005):} \quad \begin{aligned} N_e &= -32.067 + 0.318 N_a \\ R^2 &= 0.824; \ N = 10; \ 280 < N_a < 440 \end{aligned} \qquad (2.15b)$$

$$\text{Southwest (1996–2005):} \quad \begin{aligned} N_e &= 0.309 N_a^{0.781} \\ R^2 &= 0.960; \ N = 10; \ 220 < N_a < 420 \end{aligned} \qquad (2.15c)$$

In the above equations, N_e is the number of employees (thousands) per year, N_a is the number of aircraft per year, and N is the number of observations. These equations enable estimation of the total, average and marginal number of employees per aircraft. Firstly, they confirm the increasing number of employees with increasing size of the airline fleet.

For the US airlines, the average number of employees per aircraft has varied from about 59 for the smallest to about 117 for the largest airline. This is lower than the marginal number of employees at an average airline, which is 126.

At Lufthansa, the number of employees has increased with the size of the fleet at a constant rate. The average number of employees per aircraft has also increased – from 203 to 245. The marginal number has been constant (318) and has always been greater than the average number of employees per aircraft. This indicates that adding each new aircraft has always contributed to increasing the average number of employees per aircraft.

At Southwest Airlines, the number of employees has increased at a decreasing rate with the increasing number of aircraft. Consequently, the average and marginal number of employees has decreased with the increasing fleet size, i.e., from 95 to 82, and from 74 to 64, respectively. Since the marginal number has always been lower than the average number of employees per aircraft, adding a new aircraft to the existing fleet has contributed to decreasing of the average number of employees per aircraft.

The indirect social performances can be the airline's contributions to the indirect employment in the non-aviation businesses such as investments, trade, and tourism at airports and associated regions of the airline network. This can be combined with the provision of choice to users as well as the exchange of education and culture opportunities between distant regions, which otherwise would not be possible. In some cases, contributions may imply creating or keeping jobs at the aerospace manufactures thanks to the regular orders for new aircraft.

2.2.6 Environmental Performances

The overall airline business strategy usually includes improvements in the environmental performances in line with growth. Long-term profitability is however a precondition for implementing the environmental policies. The most important environmental performances of many airlines are noise, local air quality around airports, climate change, waste, and resource use. They are usually expressed as the impacts aiming to be mitigated over time, with some short- and long-term targets. It should be mentioned that one of the main characteristics of these impacts is their specificity for different airlines and airports.

2.2.6.1 Noise

Noise is one of the most important environmental performances of airlines, and a matter of concern. In general, most airlines have become aware of the problems they create to the local population around the airports. In order to mitigate the impacts, they have adopted the so-called 'Balanced Approach' recommended by ICAO (International Civil Aviation Organisation). This approach stipulates investments in quieter aircraft, and undertaking local airport measures such as operational procedures and land use planning (BA, 2004).

Investment in quieter aircraft means increasing the proportion of quieter aircraft in the airline fleet. In many cases, airlines take part in research related to reducing aircraft noise at source.

The operational procedures for reducing noise at airports include the so-called 'track-keeping' and 'Continuous Descent Approach'. The former includes the strict and precise following of the three-dimensional approach and departure paths designed to mitigate noise around a given airport. The latter implies the aircraft's continuous descent from the cruising altitude to the altitude where the final approach and landing start. This has been made possible through introducing new technologies enabling the so-called aerial navigation RNAP-P. For example, at London Heathrow Airport (UK), over the past few years the dominant airline BA (British Airways) has carried out about 95 per cent of all arrival and departing flings 'on track' and about 91 per cent arrivals using CDA (BA, 2004).

The airlines might also be interested in the long-term planning of land take (and use) around airports, trying to resolve the noise problem through sufficiently separating the local population and the airport. In the scope of such efforts, they propose the introduction of the so-called 'noise compatibility airport zones'. On the one hand, these zones aim to protect given areas around the airport and prevent them becoming populated. On the other, they identify the already affected population and look for solutions, including possible re-settling.

2.2.6.2 Local air quality

Local air quality is an important element of the airline environmental performances. In addition to the emission of CO_2 from the fuel burned, the particular air pollutants of concern are nitrogen oxides (NO_x) and the resulting building up of nitrogen dioxide (NO_2). The amounts of these pollutants are particularly important at large airports where road traffic already produces a significant amount of these emissions. Airlines attempt to mitigate this impact by investing in the aircraft fleet, meeting the ICAO standards on these emissions, improving the utilisation of ground vehicles at airports, and stimulating development and use of the non-road based transport modes connecting the airports and their catchment areas (rail links).

For example, in the year 2004, about 79 per cent of BA aircraft fleet met the ICAO standards on NO_x emissions. At the same time this fleet generated about 35 per cent of NO_x emissions at the airline hub – London Heathrow Airport (BA, 2004).

2.2.6.3 Climate change

Most airlines are aware about their impacts on climate change at a global scale. Therefore, they permanently monitor and improve their fuel efficiency in order to mitigate these impacts. At the same time, mitigation of the environmental impacts through improvement of fuel efficiency contributes to improvements in the airline's economic/business performances. Fuel efficiency can be improved by fleet renewal, air traffic control and airspace improvements, and improvement in the aircraft utilisation (seat configuration, load factor, and other operational measures). Figure 2.11 shows an example of the long-term improvement of the fuel efficiency of BA (British Airways, UK).

The average fuel consumption per unit of output has decreased more than proportionally with increasing volume of output over the past 30 years (1974–2003), mostly thanks to the renewal of the fleet in combination with other measures. In addition, the airline expects further improvements in fuel efficiency of about 30 per cent by the year 2015 (BA, 2004).

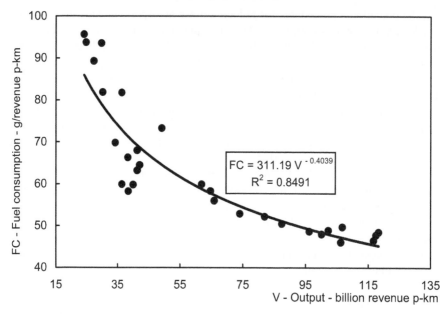

Figure 2.11 Dependence of the airline fuel efficiency on the volume of output
Compiled from BA (2004)

The additional emissions of the air pollutants (CO_2) are the airline properties using different energy sources for lighting and heating. Improvements in this segment also require setting up some targets.

Some airlines are also keen to participate in the different national and international policy initiatives aiming to constrain emissions of CO_2. One of the most recent initiatives has been consideration of some Association of European Airlines (AEA) members to join the EU Emission Trading Scheme (EUETS) (BA, 2004).

2.2.7 Institutional Performances

The airline institutional performances are broadly characterised by their internal and external attributes. The internal attributes mainly relate to the structure of airline ownership. Experience has shown that in most cases the airlines with higher private ownership have been relatively more efficient and effective on the one hand and very vulnerable to market disruptions causing loss of revenue and increasing of costs on the other. In addition, although mostly focused on the profitability, these airlines have also demonstrated rather positive short- and long-term concern for mitigating their own negative impacts on the environment. The external attributes relate to the level of regulation of airline markets, implying the freedom of market entry/exit, supply of capacity, setting up the airfares, and right of ownership of other airlines. Both types of institutional performances may indirectly influence a balance between the effects and impacts of the specific airlines, and consequently sustainability of the entire air transport system.

2.2.7.1 Internal attributes

The structure of ownership implies the number of stakeholders and the percentage of their stake in a given airline. Two types of stakeholders can own an airline: public and private. Public stakeholders are usually governments, their agencies, and public banks. Private owners are private banks, other airlines, other non-aviation business enterprises, and individual stakeholders. In both cases the percentage owned by a particular stakeholders can be quite different. In addition, the number of stakeholders across the airlines is quite diverse, without any analytical evidence that it might be conditioned by the fleet size and volume of output. Traditionally, governments have been owners of many airlines as the country's flag carriers. However, with deregulation and liberalisation of the air transport markets at the national and international scale, the governmental ownership and associated protection including subsidies have disappeared while competition by new private airlines has strengthened. New conditions have required more efficient and effective operations, which could be achieved by privatisation and building up the innovative economic/business models – hub-and-spoke networks, airline alliances, and low-cost models. Nevertheless, at many legacy airlines, governments have still retained some stake, considering them of great national importance. Relevant statistics for AEA (Association of European Airlines) in the year 2004 illustrates the diversity of airline ownership in Europe. The main attributes accounted for 28 airlines are i) the number of stakeholders, ii) percentage of public stakes; and iii) the number of airlines in which the airline has stakes (AEA, 2005).

The number of stakeholders　In the given sample, 12 airlines have two, nine four, five four, one seven, and one nine stakeholders. The average number of stakeholders per airline is two and the standard deviation 3.3.

Percentage of the public stakes　Regarding the percentage of public ownership, seven airlines in the sample have no public ownership; eight airlines have more than 90 per cent, and three airlines between 80 and 90 per cent. The average percentage of public ownership in an airline is about 59 per cent, and standard deviation is 40 per cent. The main stakeholders are from the country of airline registration or some other EU member state.

The number of airlines in which a given airline has stakes　The given sample indicates that 16 airlines have stakes in one or more other airlines. Eight airlines have stakes in two to four, two in four to eight, and two in 10–15 airlines. These airlines are mostly from the airline domicile country and rarely from abroad. The percentage of stakes also varies from the symbolic few to 100 per cent (complete ownership). The average number of airlines in which another airline has ownership is two and the standard deviation is 4.

2.2.7.2 External attributes

The external attributes of the airline institutional performances include market regulation and ownership of other airlines, this time between the airlines from different regions -continents. Three cases related to the European and US airline industry appear illustrative: deregulation/liberalisation of the internal markets; the 'open-sky' agreements; and a 'Single European Sky' concept. They have all aimed to mitigate or diminish the institutional barriers to the airline free market operations.

Deregulation of the domestic airline market in the US took place as a single act in 1978. In the EU (European Union) it was a gradual process of removing the institutional barriers between Member States on the one hand, and providing the airline industry with sufficient time to accommodate to the new conditions on the other (Janić, 1997). The market liberalisation process had taken place in the 1980s and at the beginning of the 1990s. The developments afterwards confirmed the success of these polices. In addition, partial liberalisation of the transatlantic market between two regions (i.e., Europe and the US) was carried out through contracting the so-called bilateral 'Open-Sky' agreements between the US and particular Member States (TBG, 2002; EC, 2003a). Although initially beneficial, these agreements have started to constrain the full market development between the two regions for the following reasons: the EU airlines can fly to the US only from their domestic countries; the airlines with the so-called 'Open-Sky' rights cannot contract the 'Open Sky' agreements with other non 'Open Sky' partners without losing US traffic rights; merging and ownership between the EU and US airlines is forbidden; US passengers are denied the benefits of other competitors. Therefore, the EU has started an initiative to remove the above-mentioned barriers through a two stage-process of negotiations: first, suspension of the bilateral 'Open Sky' agreement of its Member States (14 of 25); and second, negotiation of a single 'EU-US Open Sky' agreement. The objective is to create a single open market providing more efficient and effective air transport services within and between two regions including the free flow of investments in airlines. Some expected additional benefits are the non-restricted (increased) output, free flow of investments and labour between airlines, increasing of user welfare, and stimulation of the air transport suppliers. Negotiations are currently under way.

In addition to the benefits for the airline industry, the above-mentioned developments may stimulate even faster growth of air transport demand, which in the case of the lack of the appropriate mitigating measures may compromise the efforts to reducing the system's negative impacts on people's health and the environment, and consequently the required sustainability targets. This again shows the importance of having an integral and balanced approach in creating policies for different purposes in the contemporary airline industry.

2.3 Airports

2.3.1 Background

Airports are considered as part of the air transport system infrastructure. They can be of different sizes depending on the volume of accommodated traffic in terms of air passengers, aircraft movements-operations and airfreight (cargo) for given period of time (hour, day, month, year). Generally, airports consist of airside and landside areas. Their most important physical attribute is the size, reflecting the area of land taken for installing the fixed infrastructure – runways, taxiways, apron/gate complex, terminal and cargo buildings and ground access systems, respectively. In addition, the fixed, semi-mobile and mobile facilities, equipment and devices provide service to the users – aircraft, passengers, and cargo shipments. Both the infrastructure and service facilities and equipment are characterised by the service-processing rate, i.e., capacity, which depends on their constructive characteristics and the users' service rules and procedures. The total installed capacity depends on the volume and time pattern of demand for the service over a given period of time. Thus, demand and capacity appear to be the most important operational performances of given airport. In addition, this might also be quality of services as an outcome from permanent interaction between demand and capacity.

Handling the air traffic demand generally requires spending three types of resources: capital already described as the airport fixed infrastructure, facilities and equipment, labour represented by the staff preparing, executing and managing the service activities, and energy used for powering the facilities and equipment. The quantity and costs of these resources for serving a given volume of demand over a given period of time reflects one side of the airport's economic performances. The other side is represented by charges to users – aircraft, passengers and freight shipments – aiming to cover the costs. The difference between revenues and costs in both absolute (total) and relative terms (per unit of output-served user) reflects the airport's overall efficiency and effectiveness.

Social performances of airports are usually the direct and indirect employment of staff, which in turn strongly depends on the airport size, i.e., the volume of demand and corresponding airport capacity. The environmental performances include the land taken for building the airside and landside infrastructure, noise annoying the neighbouring population, and air pollution. In general, these performances again are strongly dependent on the volume of traffic but also on the efficiency of utilisation of the available capacity. The institutional performances reflect the airport's institutional structure in terms of ownership – public or private. In general, parts or all of the airport's fixed infrastructure and services might be either publicly owned, semi- or fully-privatised. In many cases, this performance might implicitly influence the airport's short- to long-term development, and other operational, economic, environmental, and social performances.

2.3.2 Physical and Technical/Technological Performances

2.3.2.1 Spatial layout
The main physical characteristic of an airport is its size, which can be measured by the area of land it occupies. The area of occupied land depends on the airport layout and the size of particular components in the airport's airside and landside areas. The size of particular components is governed by standards mainly related to the shape and the number of runways, taxiways, aprons and related facilities. Standards of different scales exist. Their selection depends on the size of relevant ('critical' or the biggest) aircraft and the expected volume of traffic. In addition, the number and orientation of the runways, the most land demanding components given the 'critical' aircraft, also depends on the required usability factor of the given airport with respect to the prevailing weather (wind and ceiling) conditions. Generally, this factor

should be not less than 95 per cent during the year. As well, the runways should be positioned in such a way that the approach and departure areas around the airport are free of obstacles, which could compromise smooth instrumental approach, landing, missed approach and departure (ICAO, 2004).

Respecting the above-mentioned recommendations, the airport's spatial layout should therefore minimise the area of taken land, which otherwise would be left as an intact 'green' area, or for some other commercial purposes (agriculture, other commercial buildings, etc.). This however is not valid for deserted land. Minimisation of land taken can be achieved using two general principles: i) sophisticated forecasting of airport demand including careful selection of the standards for particular components, i.e., the airport should not be built for aircraft which are unlikely to come there; ii) arrangements of particular components aiming to minimise the aircraft taxiing distance and time before take-off and after landing on the one hand, and enabling the efficient and effective movement of aircraft flows within the airport area on the other; this would also minimise the fuel consumption and associated emissions of air pollutants.

The above-mentioned principles can be materialised through six typical (theoretical) airport layouts (configurations) as follows (Horonjeff and McKelvey, 1994):

  a. A single runway;
  b. Two parallel runways – both used for landings and take-offs;
  c. Two parallel runways – one used for landings and another for take-offs;
  d. Two converging runways – each used for both landings and take-offs;
  e. Two parallel plus one crossing runway – each used for landings and take-offs; and
  f. Two pairs of parallel runways – the two outer used for landings and the two inner for take-offs.

Figure 2.12 shows the simplified layout of these configurations including the equations for a rough estimation of the land taken. Particular symbols are as follows:

  A is the area of land taken (ha or km²);
  d is the width of runway strip (m);
  h is the width of airport landside (terminal) area (m);
  l is the length of airport landside (terminal) area (m);
  d_0 is the distance between centrelines of two parallel runways;
  d_{01}, d_{02} is the distance between centrelines of the first and second pair of inner and outer parallel runway(s), respectively; and
  α is the angle between a pair of converging/diverging runways.

The minimal (standard) values of particular layout parameters for different airport categories (A–D) are specified as recommendations (ICAO, 2004). For example, the values of these parameters for airports handling the largest aircraft (Category D and E) are as follows: $d = 300$ m, $h = 500$ m, $l = 500$ m; $l = 4500$m; $d_0 = 2000$m; $d_{01} = d_{02} = 1,050$. Consequently, the area of land taken can be computed as: $A = 260$ ha for configuration a, 1,035 ha for configuration b and c, 878 ha for configuration d, 1,179 ha

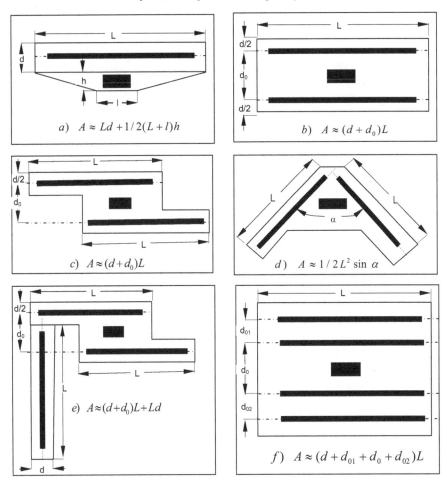

Figure 2.12   Idealised schemes of different airport configurations (layouts)
Compiled from Horonjef and McKelvey (1994)

for configuration e, and 1,980 ha for configuration f. The differences are obvious due to the increasing number of runways. As mentioned before, the size of land taken should be justified by the volume of traffic in terms of the annual, monthly and hourly number of aircraft, passengers and freight that can be accommodated safely, efficiently and effectively. This will in turn enable an appropriate utilisation of land taken. In such context, three remarks should be made. First, at most airports the size of the actual land taken is greater than the above-mentioned theoretical (ideal) scheme. The illustrative examples are the recently built off shore airports in Japan – Kansai and Centrair – which are located on the artificial lands of 560 and 470 ha, respectively. Second, the value of land used for the airport instead of for some other purpose needs to be estimated regarding the sustainable development of the airport in the short- and long-term. The last, some airports sometimes intentionally build new runways in order to mitigate (i.e., redistribute) the overall noise burden in which case the effects of reducing noise and the impacts of taking new land need to be evaluated. The new (sixth) runway at Amsterdam Schiphol Airport (the Netherlands) can be considered as the most recent example.

Each airport is physically connected with its catchment area by different surface transport access systems such as road-based buses, taxis, and cars, and the rail-based trains. These systems use fixed infrastructures also connected to the wider local, national and international road and rail networks. The size of an airport's catchment area differs at different airports and is usually measured by the distance and/or accessibility time for a given percentage of passengers and freight volumes. At most European airports these distances are from about 50 to 100 km or one to two hours of accessibility (travelling) distance and time, respectively (EC, 2002). Size of the airport catchment area is relevant for the assessment of the spatial impacts of the given airport on the air pollution, congestion and land take from the surface (ground) access systems (traffic) necessary for the complete airport operation.

2.3.2.2 Facilities and equipment

In addition to the fixed infrastructure, an airport is equipped with the various fixed, semi- and fully-mobile facilities and equipment needed to handle the aircraft (airside area), passengers and freight (landside area). In the airport airside area, there are generally two categories of facilities and equipment. The first category embraces the fixed components such as for example lighting systems near and on the runways taxiways and aprons enabling the aircraft smooth movement (landing, taking-off, and taxiing) under low visibility conditions (dark, dense fog, very low cloud, rain and snow storms). Navigational aids enabling approaching and departing the airport are not taken into account even though the lighting system might be part of them. The second category consists of vehicles serving the aircraft while entering and leaving the parking gates (stands) and those providing the aircraft servicing during the turnaround time. These are refuelling, catering, cargo, waste, passengers and their baggage delivering and collecting vehicles, and power supply vehicles. Almost all these vehicles consume fuel and consequently generate air pollution at the airport.

In the landside area, mobile and semi-mobile 'interfaces' enable the physical connection between the aircraft and terminals. Passengers may use airport buses (mobile units) in combination with the mobile stairs and/or air-bridges (semi-mobile facilities) to pass from the terminal to the aircraft, and vice versa. Buses consume fuel and generate air pollution.

The passenger (and freight) terminals are used to 'prepare' passengers and freight, respectively, for transfer from the airport ground access systems to the aircraft, and vice versa, i.e., for changing the transport mode. Different facilities and equipment are used within the terminals for processing passengers and/or freight. For example, for passengers these can be i) The passenger processing areas: for departing passengers, these are airline ticketing, check-in counters, security control desks and the outbound baggage space for sorting and processing baggage for departing flights. For the arriving passengers these are the immigration (and security) control desks, and the inbound baggage claim area consisting of both the passenger waiting space and baggage claim devices. In this area, the arriving passengers pick up their baggage; ii) Passenger lounges: these are the central hall for both departing and arriving passengers, departure and gate lounges, where passengers usually wait for the next phase of service; iii) Passenger circulation areas: this is a space for the circulation of passengers and visitors, which consists of the areas such as stairways, escalators, elevators and corridors. Passengers use them to pass from one part of terminal to other, or between two closed terminals; iv) Passenger service areas consist of post office, telephones, information, wash rooms, first aid, etc.; v) Concession areas include bars, restaurants, tax and duty-free shops, banks, insurance, car rental, etc.; vi) Observation decks and visitor lobbies include VIP facilities; and vii) baggage processing facilities and equipment (AACC/LATA, 1981; Horonjeff and McKelvey, 1994). The facilities and equipment processing passengers and baggage use mostly electric energy, which might be produced at plants using either renewable or non-renewable fuels. Consequently, the induced air pollution should be appropriately considered.

The interfaces on the external side of terminal building(s) (for example, those located towards the airport catchment area) consist of the static and mobile objects and equipment. For passengers, these are parking areas for private cars and taxicabs, and users' loading and unloading platforms; the rail and bus stations (terminals), escalators and moving walkways enabling direct and efficient movement of passengers from the parking areas, bus and rail terminal to the airport terminal, and vice versa; the intra-airport transport systems consisting of minibuses, standard buses, train sets, long moving walkways, etc., which usually operate at large airports enabling passengers efficient transfer between distant terminals. Also in this case, buses use conventional fuel while trains use electric energy obtained from non- or renewable sources (Janić, 2001).

The airport surface access systems comprise individual cars and taxicabs, and mass transport systems such as buses, and regional and national conventional and specialised high-speed trains. At many large airports passengers and others (airport employees and visitors) are usually offered services by all the above-mentioned systems. However, at smaller regional airports only services by road-based transport system may be offered. As far as the energy consumption ad related emissions of air pollutants are concerned, it appears important to establish an appropriate modal split, in which the use of mass (preferably rail-based) public transport mode is to be favoured. Table 2.1 gives an example for some European airports (RA/MVA, 2000).

Table 2.1   Existing and prospective market shares of the public transport modes at selected European airports

Airport	Share of public transport (1998) Passengers/employees (%)	Share of public transport (2020) Passengers/employees (%)
Frankfurt	37/22	50/40
Heathrow	33/20	40/40
Gatwick	32/11	40/40
Stansted	33/7	35/12
Paris Orly	23/16	30/40
Paris Charles de Gaulle	27/20	40/30
Amsterdam Schiphol	26/21	40/40

Source: RA/MVA (2000)

The main objective is to increase the market share of public transport to about 50 per cent for passengers and 40 per cent for the airport employees. Given the market share, the airport's annual number of origin-destination passengers and related number of employees, it is possible to estimate the number of users of public transport, and consequently the energy consumption and related adverse effects in terms of noise and air pollution.

2.3.3 Operational Performances

2.3.3.1 Demand
General   Demand has been shown to be one of the most important parameters for operating and planning an airport. It is interrelated with capacity and always in dynamic interaction in both the short- and long-term. This results in a given service quality for users – passengers, aircraft, and freight shipments – on the one hand and efficiency of airport operations on the other. Demand represents the volume of requests for services during a given period of time (hour, day, month, year). These requests in terms of the aircraft operations (air transport movements), passengers, and freight are satisfied by the airport's available capacity in both the airside and landside areas. In most cases, it appears to be very difficult to exclusively consider demand without considering capacity at the same time, and vice versa.

Figure 2.13 shows development of the air transport demand – aircraft movements, passengers, and freight – at four main European hub airports: London Heathrow (UK), Paris Charles de Gaulle (France), Frankfurt Main (Germany), and Amsterdam Schiphol (the Netherlands).

Such schemes could help in analysing the past trends and forecasting future development. For example, as can be seen, two sub-periods of development of airport demand can be distinguished during the observed period: that until the year 2001 and that after the year 2001 (the year of the 11 September terrorist attack on the US). During the former sub-period, the number of air transport movements (ATM) steadily increased at all four airports – approximately at the same rate at Frankfurt and Amsterdam Schiphol, slower and more steadily at London Heathrow, and at the highest rate at Paris Charles de Gaulle airport. Consequently, at the end of this sub-period, the number of ATM was the highest at Paris Charles de Gaulle airport, followed by London Heathrow, Frankfurt and Amsterdam Schiphol Airport. During the later sub-period, a significant drop in the number of ATM took place at the beginning at each airport, the greatest at Amsterdam Schiphol Airport, where also the recovery has been the slowest. Nevertheless, the growth trends have been re-established by the end of this sub-period.

The number of passengers at each airport followed a similar trend as the number of ATM during, i.e., the steady growth before, stagnation during and just after the year 2001, and recovery at the end of the period (Figure 2.13b). The highest number of passengers were accommodated at London Heathrow Airport, followed by Paris Charles de Gaulle and Frankfurt airport with nearly the same numbers, and then Amsterdam Schiphol. Given the number of passengers, ATM, and average load factor enables estimation of the average aircraft size.

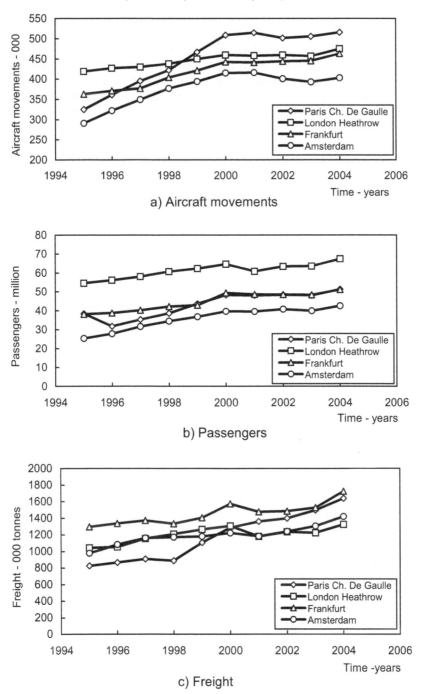

a) Aircraft movements

b) Passengers

c) Freight

Figure 2.13 Development of air transport demand at four main European hub airports: a) Aircraft movements; b) Passengers; c) Freight

Compiled from Schiphol Airport (2004)

As far as freight demand is concerned (Figure 2.13c), before the year 2001 it had steadily grown at all four airports – the fastest at Paris Charles de Gaulle and Frankfurt airport. In the year 2001 and just after it declined and stagnated at all except the Charles de Gaulle airport. Nevertheless, the recovery trend returned to all airports by the end of the observed period. Under given circumstances London Heathrow has fallen to fourth place while Frankfurt and Charles de Gaulle have become leading airports for freight demand by the end of the observed period (2004).

Airside area   Information on the airport demand in terms of the number of aircraft movements (ATM) from the past period could be used as an ingredient for forecasting the future demand as well as the long-term planning of the airport airside capacity. Figure 1.3 in Chapter 1 shows such an example for London Heathrow Airport (Janić, 2004). In that case, the future demand is forecasted by extrapolation of the past trends (growth rates) according to two scenarios, both assuming continuation of growth of the annual volumes of ATM at an annual average rate of 2.4 per cent and 1.4 per cent according to the 'optimistic' and 'pessimistic' scenario, respectively, over the period 2003–2020. In the former scenario, the annual number of ATM is expected to grow to about 567 in 2010, 638 in 2015 and 719 thousand by the year 2020. In the latter case, the annual number of ATM is expected to increase to about 519 in 2010, 556 in 2015, and 596 thousand by the year 2020. Such growth could be managed by implementing four possible solutions: 1) maintaining the current capacity; 2) extension of the airport operating time, 3) changing the runway operating mode from the 'segregated' to 'mixed' one; and 4) building a new (third) parallel runway and operated it in combination with solution 2 (Janić, 2004).

The airport airside demand in terms of ATM per hour (or 15 minute intervals) is usually used for estimating the aircraft delays given the airport's ultimate capacity, and consequently for declaring the airport's practical capacity (Janić, 2001). For such purposes, hourly fluctuations of both demand and capacity are taken into account enabling identification of the short periods in which the demand exceeds the capacity and causes aircraft queues and delays. Figure 2.14 shows examples of two types of hourly fluctuations during the day at two US airports – New York LaGuardia and Atlanta Hartsfield. As can be seen, the demand at New York LaGuardia Airport (Figure 2.14a) has been around or above the airport available capacity for most of the time during a given day (30 June 2001 as the beginning of the Independence Day holiday) (FAA, 2002). This was an indication that the congestion started early in the morning and was sustained for almost the whole day. At Atlanta Hartsfield airport, the fluctuation of demand during the same day was different from at LaGuardia Airport reflecting the hub-and-spoke operations of the main incumbent – Delta Airlines. In the morning and afternoon the realised demand was lower than the capacity, but in the late afternoon and evening (18.00–22.00 hours) it was considerably above capacity (Figure 2.14b). In this case, congestion was occurred in the late afternoon and evening hours.

Landside area   The airport demand in terms of the passenger volumes appears relevant for operation and planning of the airport landside area consisting of the passenger terminals, interfaces, and ground access systems.

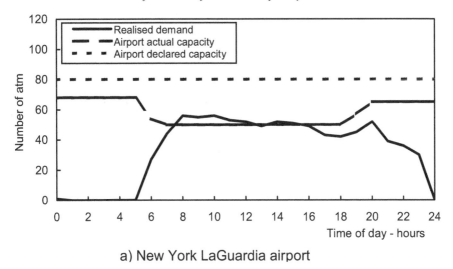

a) New York LaGuardia airport

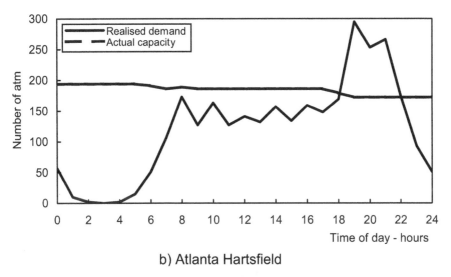

b) Atlanta Hartsfield

Figure 2.14   The relationships between demand and capacity at selected US airports: a) New York LaGuardia Airport; b) Atlanta Hartsfield Airport

Compiled from FAA (2002)

At the passenger terminals the demand, in addition to the operational, can also be used for planning purposes. Specifically, the annual volumes of this demand can be used for the long-term strategic planning of passenger terminal capacity. Figure 2.15 shows an example of possible scenarios of matching capacity of passenger terminals to growing demand at London Heathrow Airport (UK) (Janić, 2004).

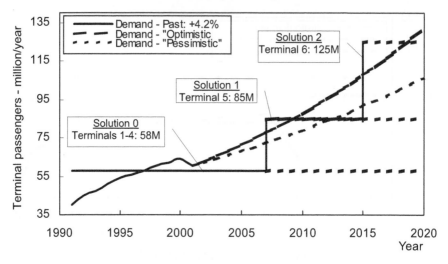

**Figure 2.15 Matching the airport passenger terminal capacity to demand at
           London Heathrow Airport**

Compiled from Janić (2004)

Forecasting the future passenger demand is carried out using the past trends (growth rates) in two future scenarios. The 'optimistic' scenario implies an annual growth rate similar to that of the past period (1991–2002) of about 4.2 per cent. This will bring the annual number of passengers to about 88 in 2010, 108 in 2015 and 132 million in the year 2020. The 'pessimistic' scenario implies an annual growth rate of 3.5 per cent, which will increase the annual number of passengers to 79 in 2010, 91 in 2015 and 106 million by the year 2020. Three solutions to expanding the passenger terminal capacity are proposed to strategically match such growing demand: solution 1 includes keeping the existing capacity of about 58 million passengers per year (this was already exceeded in the year 1998); solution 2 implies building a new terminal to accommodate the additional 25–27 million passengers (this project is currently under way as the Terminal 5 project); and solution 3 includes, in addition to Terminal 5, building a new terminal for the additional 40 million passengers after the new (third) runway is built. This latest solution would increase the airport passenger capacity to about 125 million per year. The building of a new passenger terminal together with a new runway would actually add a new fully developed single runway airport to the present location, which would require at least 300 ha of new land excluding the land for the airport surface access systems (Janić, 2004). Consequently, the population currently living on the settlement would need to be re-settled on the one hand while the population nearby the new infrastructure is likely to be exposed to the increased aircraft noise and air pollution on the other.

The predicted annual number of passengers can also be used as an operational-design parameter for sizing the passenger terminals and their components including determining the capacity of passenger processing facilities. These are ticketing/check-in, immigration and body check counters, baggage claim devices and surrounding areas, etc. For such purposes, the annual number of passengers needs to be converted into the hourly number, which is relevant for terminal sizing. This

Table 2.2   Dependence of the peak-hour passenger demand on the annual passenger volume

Annual passenger volume (million)	Per cent of the annual passenger volume	TPHP (Typical Peak-Hour Passenger) volume
< 0.1	0.200	< 200
0.1-0.5	0.130	130–650
0.5-1.0	0.080	400–800
1–10	0.050	500–5,000
10–20	0.045	4,500–9,000
20–30	0.040	8,000–12,000
> 30	0.035	> 10,500

Source: Ashford and Wright (1992)

can be carried out by different methods. One commonly used by airport planners around the world is the concept of the 30th highest hour of the year or Standard Busy Rate (SBR), which is frequently applied by European planners. In addition, the US FAA has recommended the Typical Peak-Hour Passenger (TPHP) concept based on the annual volumes of passengers and peak-hour passenger volumes being 0.03–0.20 per cent of the annual volumes (Ashford and Wright, 1992). Table 2.2 contains examples of TPHP based on the above-mentioned FAA recommendations.

The variations of TPHP in each class of the annual volumes of passengers are supposed to be significant, which offers a relatively low precision for sizing passenger terminals. Consequently, this approach has been useful only for the initial (preliminary) assessments of terminal size. For example, in the case of the above-mentioned Terminal 5 at London Heathrow Airport, the expected hourly peak number of passengers based on the additional annual number of about 25 million passengers per year would be about 10 thousand passengers per peak-hour, which, given the required space of about 14 m² per passenger makes the size of the terminal required to be about 112–162 thousand m², i.e., this is the area of the required land within and/or outside the present airport layout (Ashford and Wright, 1992; Horonjeff and McKelvey, 1994). In some cases, characteristics of the airline schedule, aircraft fleet mix and load factor need to be taken into account to modify (improve) the quality of this parameter. In such a context, the categorisation of passengers is important since different categories place different demands on the various components of a passenger terminal at different times. For example, passengers are broadly classified into originating, terminating and transit/transfer, then into domestic and international, and business and leisure passengers. At the hub airports, there is usually a high proportion of transfer/transit passengers. Some airports specialise in serving the charter traffic. These problems are usually resolved by using the scenario approach in determining the peak-hour demand relevant for the passenger terminal design (Ashford and Wright, 1992; Horonjeff and McKelvey, 1994).

2.3.3.2 Capacity

Airside area   Capacity is the parameter measuring the airport's capability to accommodate the expected demand. The capacity could be used as the operational and planning parameter for both airport airside and landside areas, and for different periods of time such as year, month, day and hour. The capacity is usually defined as the maximum number of units of demand, which can be accommodated during a given period of time under given constraints. In the airport airside area the aircraft operations – movements – represent demand. In the airport landside area passengers and freight shipments represent demand.

The airport capacity (if there are not other constraining factors) mostly depends on the operational factors such as 'safety constraints,' 'constant demand' for service and the 'average delay' per unit of accommodated demand. This capacity, as being used for operational purposes, is usually determined for one hour including an average delay per operation. For planning purposes it is determined for the period of one year. However, in many cases, different economic and environmental constraints may affect the airport operational capacity. In such cases, the concept of the airport economic and/or environmental capacity can be introduced. The airport economic capacity is mainly dictated by the short- and long-term economic constraints. In the short-term, these might be charges of airport services during peak and off-peak periods aiming to regulate access to the airport, covering the increased cost of services, and reflecting the type of users and their willingness to pay for services. They should also be compatible with ICAO recommendations and bilateral airspace agreements. In the long-term, availability of the investments for airport expansion usually determines the economic conditions and thus the prospective long-term capacity. The airport environmental capacity takes into account the environmental constraints in terms of the noise and air pollution intended to protect local people and the environment from adverse effects. In the short-term this capacity is expressed similarly as the operational capacity regarding given types of constraints. In the long-term, constraints on the land take may compromise the airport's spatial expansion, its capacity and consequently its growth (Caves and Gosling, 1999; Janić, 2001; DETR, 2002).

The capacity of the airside area is determined for the runway system, taxiways and apron/gate complex. The capacities of these components should be balanced in order to avoid 'bottlenecks' and consequent adverse effects such as delays, extra fuel consumption and related extra emissions of air pollutants. In addition, diminishing or even eliminating delays mitigates loss of time for both users and airlines-aircraft.

In practice, the runway system capacity appears to be the crucial element of the overall airside capacity. This is both the ultimate capacity and practical capacity reflecting the maximum throughput related to the average acceptable delay per operation – landing or take-off. The generic expression for the calculation of the airport runway system capacity for either type of operations is as follows (Janić, 2001):

$$\lambda_{a/d} = 1/\bar{t}_{a/d} \tag{2.16}$$

where

$\lambda_{a/d}$ is the runway arrival/departure capacity (operations per unit of time);

$\bar{t}_{a/d}$ is the average inter-arrival time between successive operations – arrivals/ departures – at the 'capacity calculating location' – usually the runway threshold (hours).

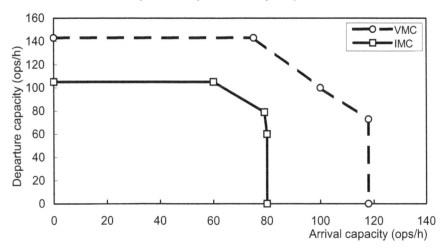

Figure 2.16 The capacity coverage curve at Chicago O' Hare Airport (US)
Compiled from FAA (2002)

In order to represent the possible operational patterns of different runway configurations, meteorological conditions, mix of the aircraft fleet in terms of the wake-vortex categories and the approach/departure speeds, and the mix of the arrival and departure demand, the ultimate capacity can be expressed by the so-called 'capacity coverage curve' (Janić, 2001). Figure 2.16 shows a typical example of such a curve.

Two facts appear important: i) dependency of the arrival and departure capacity, which in turn both depend on the mix of arrival and departure demand; and ii) relatively high sensitivity of the capacity to the meteorological conditions. The former is manifested through the shape of the capacity coverage curves, which generally consists of three segments: i) the horizontal segment represents cases when the arrivals can be realised (inserted) between the departures without interrupting them; ii) the steep segment represents cases where the arrivals and departures are directly dependent on each other and can be traded-off; and iii) the last vertical segment indicates cases where departures can be inserted between arrivals without interrupting them. The latter implies that an airport operates under different meteorological conditions: Instrumental Meteorological Conditions (IMC), and Visual Meteorological Conditions (VMC). They are based on the Instrumental Flight Rules (IFR) and Visual Flight Rules (VFR), respectively. In general the capacity under VMC conditions and VFR is always greater than the capacity under IMC conditions and IFR for any mixture of the arrival and departure demand, primarily due to the lower ATC/ATM separation rules.

Under such circumstances, changing the weather conditions from VMC to IMC and consequently diminishing the capacity usually causes significant aircraft delays accompanied by other adverse effects in terms of the extra fuel consumption and air pollution, if airborne. This frequently happens at the above-mentioned as well as other US airports operating under both conditions and rules.

Given the runway arrival capacity and the average aircraft apron gate/complex occupancy time enables the number of aircraft parking stands needed at the apron/gate complex to be estimated as follows (Janić, 2001):

$$N = \lambda_a \tau \tag{2.17}$$

where
λ_a is the arrival runway capacity obtained from the 'capacity coverage curve' (operations per hour);
τ is the average aircraft gate occupancy time (hours).

Equation 2.17 assumes that all incoming aircraft occupy the gates for approximately the same average time, implying that they are of the same category. This formula can, however, be easily adjusted to take into account differences in this time, particularly when considering different aircraft types (large, medium, small) and different operations such as the full cost airlines using the 'nominal' and the low-cost carriers using the 'fast' turnaround time. Given the number of parking stands, the size of each, and the type of parking enabling the aircraft safe and smooth manoeuvring, the area of the given airport apron/gate complex can be calculated, reflecting the land take. If the aircraft/parking manoeuvring is so-called 'push-out' the size of the single stand varies from about 1,100 to 5,000 m²; if it is 'taxi-out' the size of single stand can vary from about 200 to 7,200 m² (Horonjeff and McKelvey, 1994). Such differences imply that in addition to the number, the aircraft size, mix of different types, and flexibility of use of particular parking stands influence the size of the apron/gate complex, i.e., the land taken.

Landside area   The capacity of the airport landside area embraces the capacity of passenger and freight terminals. For passenger terminals, the capacity of particular components regarding their basic function can be determined as follows (Janić, 2001):

- Processors, i.e., servers for passengers and their baggage;
- Reservoirs, i.e., waiting areas for passengers (and their accompanies) and their baggage; and
- Links, i.e., the space equipped with facilities and devices connecting processors and reservoirs.

Processors serve passengers along their way from the airport surface access systems to the aircraft, and vice versa. Reservoirs provide space for passenger queuing and waiting for particular phases of the service process. Links designated as the long halls with the passageways, walkways and escalators connect particular processors and reservoirs. In general, the concept of 'static' and 'dynamic' ultimate capacity of these components exists. The 'static' ultimate capacity implies the maximum number of passengers (occupants) in an area of a given size. Each occupant is provided with the minimum space while being there. The 'dynamic' capacity implies the maximum processing/service rate of a given service facility. Each passenger can be given the maximum waiting time for service (Janić, 2001).

2.3.3.3 Quality of service

Airside area   In the airport airside area, the quality of service is usually measured by the average aircraft delay during landing and/or take-off. If related to the flow-throughput, this acceptable average delay represents an element of the airport's practical capacity, which at least in Europe is usually agreed among the particular actors involved such as airport, airlines, and air traffic control (Janić, 2001). In general, delay is defined as the difference between the actual and scheduled time of passing through a given 'reference location'. The threshold for either arrival or departure delay is 15 minutes behind the schedule time (AEA, 2001; BTS, 2001; EEC, 2002). Almost all delays longer than the acceptable limit of 15 minutes are considered as time losses for users-passengers and freight shippers, and airlines-aircraft. They occur when the demand for landing and/or take-offs exceeds the airport's available ultimate capacity. If carried out with the aircraft engines switched-on, these delays cause additional fuel consumption and the emission of air pollutants. In order to diminish the mentioned adverse impacts, the concept of ATFM (Air Traffic Flow Management) has been developed including the ground-holding strategy applied to many airports in the US and Europe. This strategy consists of carrying out foreseeable delays at the origin airport while the engines are switched off rather than in the air before landing at an airport expected to be congested. The key success factor has been shown to be the prediction, well in advance, of the capacity at the destination airport affected by changeable weather, i.e., before the operative decision to temporarily stay at the origin airport or depart towards the destination is made (Andreatta and Romanin-Jacur, 1987; Terrab and Odoni, 1993).

Airport delays are generally expressed as the average delay per any operation – ATM (flight) – and as the average per delayed atm (flight). In the former case the total delays are divided by the total number of atm; in the latter case the total delays are divided only by the delayed atm (FAA, 2002; EEC/ECAC, 2002). Figure 2.17 shows both types of delays at 32 US and 17 European congested airports in relation to the average annual demand/capacity ratio (i.e., the utilisation rate of the airport capacity).

In general, the average delay per flight – departure or arrival – is longer at the US than at European airports. At the US airports, the departure delays have generally been longer than the arrival delays. The former have varied between 10 and 20, and the latter between 5 and 15 minutes. At European airports, there is no significant distinction between the average arrival and departure delays. Almost all delays have been shorter than 15 minutes. In both regions, a slight increase in delays for both arrivals and departures as the demand/capacity ratio increases is noticeable. According to the 15-minute threshold, the sampled delays should not be considered as delays at all. This picture changes when considering the delays per delayed aircraft (flight). The average delay per delayed flight – arrival and departure – is again longer at the US than European airports, i.e., 40–60 minutes compared with 15–25 minutes, respectively. In both regions, the arrival and departure delays have been similar and very little influenced by the airport demand/capacity ratio. This implies that some

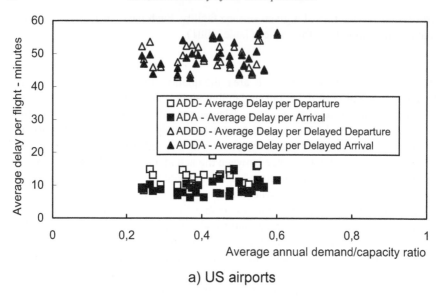

a) US airports

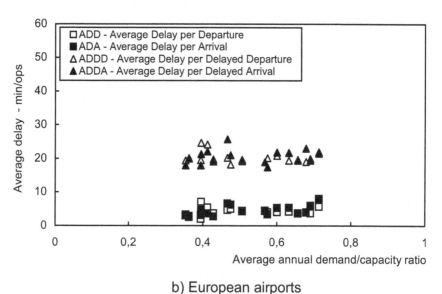

b) European airports

**Figure 2.17   The average delay vs the average annual utilisation of the airport
capacity: a) US airports; b) European airports**

Compiled from EEC/ECAC (2002); FAA (2002)

other factors have been the cause. At the US, bad weather at airports caused about
70–75 per cent and short-term congestion, mainly due to the airline hub-and-spoke
operations, about 20–30 per cent of these delays (BTS, 2001; FAA, 2002). In Europe,
severe weather has caused only 1–4 per cent and congestion about 30–40 per cent of
these delays (AEA, 2001; EEC, 2002).

Knowing the composition of the aircraft fleet affected by the above-mentioned delays, it should be possible to estimate the time and money losses of the passengers and airlines, as well as other external effects such as the additional fuel consumed and related extra emission of air pollutants. Consequently, options for reducing delays at airports need to be considered. On the one hand, this might include using the innovative technologies and operational procedures enabling more efficient utilisation of the existing airport capacity. On the other, it might be expansion of the airport airside area, which would take more land. Under such circumstances, a balance should be established between the two types of impacts, reflecting the need for also dealing with sustainability on the negative side.

Landside area   The quality of services provided to the passenger by the airport surface (ground) access systems is usually characterised by several attributes such as length and variability of access time, travel comfort, complexity of baggage handling, and travel cost. Expressed in terms of the generalised cost of users-passengers, airport employees and other visitors, these attributes influence the choice of the airport surface access system (mode). The importance of particular attributes may vary depending on the volume and structure of demand, and the trip purpose (Ashford et al., 1984). In any case, each mode should be sufficiently sized in order to enable a smooth, reliable and punctual service, which in turn raises the problem of land use for setting up the infrastructure (highway and railway lines), noise, fuel consumption and the associated emission of air pollutants.

The quality of service at the passenger terminals has appeared to be one of the powerful competitive tools of many airports. In general, the concept for measuring this quality under different circumstances has been based on the specified space standards for passengers while being in particular processing/service phases within the terminal. Some of these standards, widely applied at airports worldwide, are given in Table 2.3 (AACC/IATA, 1981; IATA, 1989).

Specifically, the space standard at level *A* provides 'excellent' quality of service; level *B* provides a 'high' quality of service. Level *C* guarantees a 'good' quality of service. Level *D* offers 'adequate' quality of service. Level *E* offers an unacceptable level of service. Finally, *F* denotes zero-level of service, which occurs when the service has completely broken-down.

In order to determine the size of particular areas in the passenger terminal guaranteeing a certain quality of service, the corresponding number of passengers

Table 2.3   The space standards for the passenger quality of service at the airport passenger terminal(s)

Area	Quality Level of Service[*]					
	A	**B**	**C**	**D**	**E**	**F**
1. Check-in, Baggage Claim Area	1.6	1.4	1.2	1.0	0.8	–
2. Holding w/o bags, Holdroom, Pre-inspection	1.4	1.2	1.0	0.8	0.6	–
3. Wait/Circulate	2.7	2.3	1.9	1.5	1.0	–

* *Square meters per occupant*
Compiled from: AACC/IATA (1981); Ashford (1988)

(occupants) expected to be simultaneously in that areas should be multiplied by the relevant (selected) space standards in Table 2.3. Then, the obtained space has to be enlarged for the space required for installing the appropriate service facilities and equipment (Janić, 2001). Obviously, the service quality appears to be the main determinant of the layout and size of passenger terminal and consequently strongly influences the area of land taken. In this context, a clue might be established between the passenger quality of service and the possible environmental impacts in terms of land take.

2.3.4 Economic Performances

An airport's economic performances are important because they directly influence the development of the airport and indirectly the external impacts. They embrace the airport revenues, costs and profitability, all dependent on the volume of traffic and spending resources for providing its adequate accommodation. In general, airports, as economic/business entities, tend to attract as much traffic as possible, i.e., to (permanently) grow, optimally use the available capacity, and thus achieve the economic objective – as high a profitability as possible.

Airport revenues are usually obtained from charging for services delivered to the aircraft, passenger and freight, as well as from other commercial activities in the airport. Typically, the airports charge the aircraft through landing fees for using the airside area (runways, taxiways and aprons), passengers and freight by fees for using the corresponding terminals, the aircraft by long parking fees (several hours or days), and by fees for other services (ACI Europe 2002; EC, 2002). These are the so-called aeronautical charges, which together with the non-aeronautical ones (i.e., from the airport's other commercial activities) provide revenues for covering the costs of capital, labour, and the energy spent for the provision of these services. In addition to attracting more traffic, the management of the airport's costs appears to be an important activity for achieving overall profitability. The structure of the airport costs generally corresponds to the type of input (resources) as follow: i) Capital cost includes the construction and overhaul costs of the airport's airside and landside area; this cost is registered as the annual depreciation and interests paid on bonds on investments; ii) Labour costs embraces the cost of staff deployed to handle passengers and aircraft; and iii) Operational cost embraces the current energy and maintenance cost of the airport facilities and equipment, and administrative costs. For example, the shares of particular components in the total costs at large US airports were 34 per cent operational, 44 per cent capital and 22 per cent labour cost (EC, 2002). At large European airports these shares were 34 per cent operational, 24 per cent capital and 42 per cent labour (Doganis, 1992). Such differences in the cost structure could be explained by differences in the structure of employment social costs and the higher number of employees at European airports, as well as by different type of financing of the US airports (higher interest rates on investments). Allocation of the total costs could also be carried out according to the so-called 'causality scheme,' where the total costs can be allocated in proportion to particular causal factors. For example, these can be the costs of the airside infrastructure and facilities, passenger processing areas, administration office space, commercial and

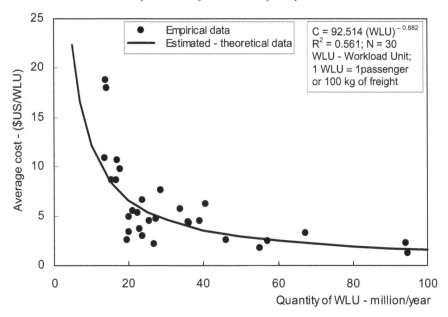

Figure 2.18 The average cost per workload unit vs the number of workload units at a given airport

Compiled from Janić (2001)

retail facilities, the aircraft refuelling facilities, land for providing refuelling services, land for providing ground services and freight handling storage facilities, sites for a cargo terminal and aircraft maintenance facilities, landside vehicle access, and waste disposal facilities (EECH, 2001).

In relative terms, the airport costs are expressed per unit of the airport output. Since the output is heterogeneous – aircraft, passengers and freight – a common unit called the workload unit (WLU) equivalent to one passenger or 100 kg of freight has been defined (Doganis, 1992). This unit has been shown to be convenient for the cost comparison between airports as well as for monitoring the cost efficiency in relation to the volume of output at a given airport. Figure 2.18 shows an example of the relationship between the average cost per workload unit and the annual volume of output (workload units) across different European airports. As can be seen, the unit cost per WLU decreases more than proportionally with the increasing volume of output indicating that economies of scale are present. This might be one explanation as to why the already large airports tend to become still larger.

A measure of the airport's economic stability and financial health has been also defined as the revenue-cost ratio. For a given airport or airports of a similar size (category), this ratio is usually related to the annual airport traffic, i.e., the volume of accommodated WLUs. At larger European airports this ratio varies from about 0.40 to 0.50 to 1.40 to 1.80 and increases at a decreasing rate with the increasing annual number of WLUs. At the smaller European regional airports, the profitability margin, when the ratio revenues-cost is equal to one, is achieved with an annual traffic volume of about 200 thousand WLU. At large European airports this marginal volume is usually counted in millions of WLU per year (EC, 2002).

Since profitability as an economic and financial category of the airport functioning seems to be inevitably related to the traffic volume, a balance between traffic growth and increasing of the associated external effects in terms of noise, fuel consumption and air pollution needs to be carefully considered in the scope of the airport's sustainable short- and long-term development.

The economic performances might also be dependent on the airport ownership structure whose main characteristics are described in the scope of the airport's institutional performances.

2.3.5 Social Performances

The social performances of an airport usually refer to employment in the broadest context.

In general, larger airports employ more staff and thus provide a greater contribution to the local and national GDP, and consequently overall welfare. In general, the number of employees is positively correlated to the airport size, i.e., the volume of traffic. This can be shown by the causal relationship between the total number of employees at and around an airport and the annual volume of airport traffic. Data related to the direct and total employment have been regressed as follows (ACI Europe, 1998):

  i. Direct employment

$$E_d(Q) = 1.4702Q - 4.209; R^2 = 0.901; N = 22 \qquad\qquad (2.18a)$$

  ii. Total employment:

$$E_t = 0.577Q^{1.493}; R^2 = 0.93; N = 22 \qquad\qquad (2.18b)$$

where
  Q is the annual number of passengers at a given airport (million).

Equation 2.18 indicates that the number of directly employed staff at a European airport is about fourteen hundred employees per million passengers, which is about 40 per cent higher than the commonly used ratio of one thousand. In addition, the number of total employees increases more than proportionally with the increasing volume of traffic.

2.3.6 Environmental Performances

The external driving forces of air transport demand, requirements for safe, efficient and effective operations and services, relatively convenient economic portfolios, and the overall social convenience certainly stimulate airports to grow. For such purposes, it is necessary to provide sufficient land for the airport's physical expansion.

Generally, as aircraft size and traffic volumes have increased, the runways have become longer, terminals expanded, and other supportive facilities and equipment aimed at preventing congestion, delays and consequent deterioration of the expected quality of service have been enlarged. In principle, sufficient land for the long-term development of a given airport should always be available. In such context, it is worthwhile being reminded of the main factors influencing an airport size as follows (Horonjeff and McKelvey, 1994):

•   The performance characteristics and size of aircraft fleet using the airport;
•   The anticipated volume of traffic;
•   Prevailing meteorological conditions; and
•   The elevation of the airport site.

The performances and size of the aircraft fleet directly influence the runway length and the corresponding area of land taken (ICAO, 2004). The anticipated volume of traffic and the requirements for its efficient and effective accommodation influence the configuration of airport airside area – runways, taxiways, and the apron/gate complex (see Figure 2.12). The direction and strength of the wind influence the number and orientation of the runways. The temperature influences the runway length, i.e., the higher temperature the longer runway, and vice versa. Higher airport elevation also requires longer runways. The airport's site should also be free from obstacles. This is particularly required in the airport turning zone and along the runway's extended centrelines. Actually, in this relatively large area – a circle with a radius of about 15 km – zoning for height restriction should be implemented (ICAO, 2004). National and international aeronautical legislation regulates the sizing of this area in order to prevent the deterioration of the efficiency of aircraft manoeuvring, compromising safety, unnecessary burning of fuel and associated air pollution.

The airport's site may be additionally characterised by the closeness of the airport to other airports, which is often the case in large urban agglomerations. Five London airports or three main New York City airports are illustrative examples. In such cases, the sufficient distance and airspace should be provided to enable safe, efficient and effective aircraft manoeuvring in and out of the airports without interference with each other.

Building up the airside and landside infrastructure and other properties within and in the areas adjacent to the airport imply long-term planning of the land use (take), known as the airport master plan. This plan must be coordinated with the objectives, policies, and programmes for the area surrounding the airport. However, in most cases, despite such planning efforts, there is a collision of interests between the airport and its neighbouring population, usually about aircraft noise. In such cases, the noise contours are drawn to estimate the compatibility of the current and future airport master plan (and operations) and land use (the noise contours are the areas of equal-intensity noise). When the area covered by noise contours is established, they can be used for implementing different noise protective measures.

These measures include the design of special noise mitigation procedures for the aircraft approach, landing and take-off, which may raise the complexity, time, fuel consumption and consequently air pollution. These impacts therefore need to be taken into account in any procedures undertaken. In addition, different insulation in the affected houses and, rarely, resettling people could be considered. If the area is undeveloped, the noise contours can be used for the careful planning of land use, i.e., controlling land use within the noise-affected areas. Usually, some industries, commercial and recreational activities can be located there. In summary, the principal objective for the planning of land use around an airport is minimisation of disturbances from noise by allocating land to particular (appropriate) activities according to their sensitivity to noise.

Air, water and soil pollution, industrial waste and disturbance of Nature – flora, fauna, and sightseeing – appear as other environmental issues related to the operation and development of airports. Air pollution is a matter of concern due to the cumulative effects of particular air pollutants, which stay around the airport or are blown by the wind to other locations, remaining in the atmosphere for a long time. The accidental and incidental leakages of aviation fuel, oil, de-icing, and other liquid chemicals can contaminate water and soil. In addition, considerable change to wildlife, habitats, and recreational areas may take place after the expansion of existing or building of new airports. Some of the impacts such as noise, fuel consumption and associated air pollution, and contamination of water and soil can be controlled by preventive protective measures and appropriately evaluated (quantified) in monetary terms as externalities. However, quantification of damage to natural habitats and wildlife seems to be almost impossible to express as externalities, at last from this more technically oriented approach.

The air traffic accidents and incidents taking place at airports due to compromising local safety rules might also be considered as relatively rare but still possible events causing social and environmental damages, and consequently externalities.

2.3.7 Institutional Performances

2.3.7.1 Privatisation and commercialisation
The institutional performances of airports are concerned with the structure of their ownership, which in turn may influence the control of assets, investments, operations, quality of services, and policies related to the protection of the local community and environment from adverse impacts. Generally, airports all over the world as economic-business enterprises have been publicly and privately owned. Public ownership refers to the governmental control of airport operations, maintenance and investments in the airside and landside areas, regulation of prices and quality of services, and collecting and distributing revenue. The act of transfer of ownership of the government's ownership to private investors (companies) is called privatisation. In some cases privatisation may involve only the transfer of some control functions from the government to private enterprises. Consequently privatisation can be carried out by transfer of ownership, transfer of control and/or of both. The new entity is usually a private company. In addition to privatisation, commercialisation – defined as an introduction of commercial objectives into publicly owned enterprises – has been applied to many airports worldwide.

2.3.7.2 Objectives, structure of ownership and trends

Privatisation and the commercialisation of airports is mainly driven by the following economic reasons (De Neufville, 1999):

- The need for raising the relatively large investments for development of the airport airside and landside area infrastructure. In many cases, the national governments as owners of airports have experienced difficulties in allocating taxpayer's money to these purposes on account of other equally if not higher priority social services such as for example health and education; and
- The expectation that a commercial approach combined with market forces will provide a greater efficiency and effectiveness of operations and services without compromising public welfare, safety and environmental protection. However, although private management may be more efficient, it can also perform in the quite opposite direction by sacrificing a lot of competitive benefits, public welfare and the appropriate protection of the environment on account of working exclusively for profitability.

In order to prevent undesired developments, governments dealing with the privatisation of airports have always considered them as the assets and enterprises of public interest, which should not be totally delegated to private investors. Two reasons support such acting: i) airports are central subjects in building up the local communities' welfare; and ii) as natural monopolists they might be very inclined to exploit the public (De Neufville, 1999). In other words, there is usually very strong public interest in the quality and price of airport services, the safety and security of operations, and possible effects on the local population and environment. Consequently, governments usually take part in the control of these airports either as the shareholders or through the regulation of both operations and the prices of services.

One illustrative example is Amsterdam Schiphol Airport (the Netherlands). Currently, the national government (75.8 per cent shares), local authorities of the city of Amsterdam (21.8 per cent shares), and the city of Rotterdam (2.4 per cent of shares) are owners of the airport. Under such circumstances, the national government is responsible for developing the airport as an asset of the highest national interest (the main port); the city of Amsterdam takes over the responsibility for local employment but also for the protection of local people and the environment from noise and air pollution; and the city of Rotterdam is responsible for developing the synergy between the airport and port of Rotterdam (also a main port in the Netherlands of the highest national interest). Privatisation of the airport, in which the national government would sell 49 per cent of its shares to private investors, has been temporarily removed from the agenda due to concerns over sacrificing some of the public's welfare.

Another example is the privatisation of the UK airports, started in the year 1987 with privatisation of the British Airport Authority (BAA) (Humphreys, Francis and Fry, 2001). Seven airports – London Heathrow, Gatwick, Stansted, Prestwick, Aberdeen, Edinburgh, and Glasgow – were set up as limited companies and subsidiaries of BAA plc. Since privatisation, these airports have been required to self-finance their future development and take over responsibility for the shareholders. In order to protect national interests, the UK Government has retained a substantial

Table 2.4   The structure of ownership of selected European airports

State/Airport	Owner (%)		
	National government	**Regional/local government**	**Private sector**
Belgium – Brussels	64.0		36.0
Denmark – Copenhagen	33.8		66.2
Germany – Frankfurt	18.0	53.0	29.0
Greece – Athens	65.0		45.0
the Netherlands – Amsterdam	76.0		24.0
United Kingdom – London airports			100.0

Source: EC (2002)

('golden') number of shares, thus giving itself the right to intervene if necessary. In addition, in order to protect the airport users against unfair pricing and other monopolistic exploitation, the government has authorised the CAA to regulate the charges and other accounts. CAA has also played a role in regulating the safety and operational standards through issuing an operational licence to each airport. The control of airport development has been left to the planning system. The most recent public enquiry on the expansion of capacity of UK airports has shown that there is still a very strong Government presence in the decision-making processes on investments (DETR, 2002).

In many other countries the privatisation of airports has escalated as a trend, which actually started in the USA. There the airports have never been fully publicly owned by the national government but mainly by local municipalities with the strong presence of the private sector. The trend continued in the UK, Australia, and Canada in the 1980s, and matured after privatisation in countries such as Argentina, Austria, Germany, Hungary, Italy, Mexico, the Netherlands, and South Africa (De Neufville, 1999; Humphreys et al., 2001). Table 2.4 gives some information on the structure of ownership of the selected European airports. The criterion for selection was mixed public and private ownership (EC, 2002)

Nevertheless, still as the main stakeholders, the national governments (the UK and other countries) and local municipalities (US) have remained in obligation to provide the majority of investments for the airport long-term development. Consequently, the public ownership with the very high level of commercialisation of operations and other businesses has shown as the most attractive form of the ownership for many airports. Such 'mixed' model seems to be effective in protecting the national and local (municipality) economic and social interests without compromising users' welfare (pricing and quality of services), safety, and people's health and environment.

2.3.7.3 Advantages and disadvantages
Privatisation and commercialisation of airports has both advantages and disadvantages (Holder, 1998; De Neufville, 1999; Humphreys et al., 2001).

The principal advantages have included increasing the profitability and productivity of operations, reducing the costs and prices of services, relieving pressure on governmental funds for investments, providing revenue to governments by selling the assets and then collecting taxes, and seemingly improvement in the quality of services.

The main disadvantages are the conflicts between the economic and environmental objectives. For example, at some privatised UK airports, the regulated generally lower charges have stimulated traffic growth, which, under conditions of constrained capacity, has increased external burdens such as congestion, delays, noise and air pollution and thus compromised the national and local environmental policy (DETR, 2002). This has become a rather serious problem due to the lack of national plans for guiding the economic and environmental interests of the national airport system.

To summarise, the airport's institutional performances, which materialised through the structure of ownership, might significantly influence the airport's physical, operational, economic and environmental performances, and therefore should be considered in the scope of dealing with sustainability.

2.4 Air Traffic Control/Management

2.4.1 Background

Air Traffic Control/Management (ATC/ATM) is considered part of the air transport system infrastructure. It includes the controlled airspace over countries, continents and oceans, radio-navigational facilities and equipment located on the ground and in space (satellites) and their complements in the aircraft, the operating staff (air traffic controllers), and operating rules and procedures for safe, efficient and effective air traffic. Understanding and assessing the contribution of ATC/ATM to the sustainability of air transport system implies elaboration of its performances. For the physical, technical/technological performances this embraces the analysis of airspace organisation, aircraft separation rules and the characteristics of existing and innovative radio-navigational facilities and equipment. The relevant operational performances embrace parameters such as capacity, demand, and the principles of their off- and on-line matching. This refers to the air traffic flow management strategies and tactics, safety, the aircraft/flight delays and fuel efficiency as the important elements of the air transport system's sustainability. The economic performances include the ATC/ATM revenues, costs and profitability. The social performances mainly refer to employment. The environmental performances embrace mitigation of noise around airports by innovative technologies and operational procedures, and reduction of the aircraft/flight fuel consumption in the en route airspace by allocating the fuel-optimal routes. Finally, the institutional performances include the structure of ownership, i.e., privatisation of the *ATC/ATM* and prospective implications for the contributions to the air transport system's sustainability.

2.4.2 Physical and Technical/Technological Performances

2.4.2.1 Organisation of airspace and aircraft separation
Air Traffic Control/Management (ATC/ATM) is established over a given airspace in order to provide safe, efficient and effective guidance of the air traffic (aircraft/

flights). Safety implies respecting the air traffic separation rules and serving users – aircraft/flights – without conflicts. The efficiency and effectiveness is concerned with providing the aircraft/flights smooth movement along the fuel-cost optimal trajectories connecting the origin and destination airports without deviations caused by the ATC/ATM.

Meeting the above-mentioned requirements (objectives) requires division (organisation) of the controlled airspace into smaller parts regarding the traffic intensity (density) and complexity. Such division is carried out at two levels: i) division of the airspace into the airport zones, terminal areas, low and high altitude en route areas; and ii) division of each of these areas into the smaller parts called the 'ATC/ATM sectors,' each under the jurisdiction and responsibility of one or a team of air traffic controllers.

The airport zones around airports provide management of the arriving and departing traffic flows. *Terminal airspace* is established above the airport zones in order to manage more intensive and complex arriving and departing traffic flows. This airspace covers an area with a radius of about 40–50 miles (one mile is equal to 1.852 km) around an airport. Usually, this area spreads vertically from ground level to FL (Flight Level) 100 (each flight level represents a constant altitude in 10^3 ft (1 ft ~ 0.305 m)) (ICAO, 1996). The aircraft fly through this area along the prescribed arrival and departure trajectories defined by the radio-navigational facilities and/or the ATC radar vectoring of aircraft. The *low-altitude* airspace is established around and above the terminal airspace spreading between 3,000 and 6,000 meters MSL (MSL – Middle Sea Level). Usually, the aircraft use this airspace to climb to and descend from their cruising altitudes, again along the prescribed trajectories as extensions of those in the terminal airspace. Generally, the climbing trajectories allow the aircraft to reach the desired cruising altitudes in the high-altitude airspace. The descending trajectories allow the aircraft to transfer from their cruising altitude to an altitude where they enter the terminal airspace. The *high-altitude* airspace continues above the low altitude airspace, i.e., above 6,000 meters MLS (Horonjeff and McKelvey, 1994). It covers the very large controlled airspace of countries and continents. Contemporary jet aircraft perform cruising phase in this airspace at specified flight levels along predefined airways. At cruising altitudes the minimum vertical separation between traffic flows (aircraft) on the closest flight levels is 10^3 ft (300 m) in Europe and US. In other parts of the world, the separation minima above FL 290 (FL290 = 29,000 ft) are 2×10^3 ft (EEC, 1996; ICAO, 2005). Figure 2.19 shows an example of the re-distribution of traffic flows between particular flight levels in the European airspace after introduction of the Reduced Vertical Separation Minima (RVSM) above FL290.

As expected, the traffic flows have re-distributed more evenly at the altitudes above FL290. Consequently, more aircraft/flights have been allocated newly available altitudes, which in turn has caused re-distribution of the fuel consumption for particular flight levels. Some estimates have shown that the consequent reductions of fuel consumption have amounted to about 1.5–2.3 per cent in absolute (overall) terms and 3.5–5.0 per cent in relative (only in given airspace) terms (EEC, 2002, 2005). In addition, the availability of new flight levels has increased the airspace capacity and the flexibility of its use, which in turn has contributed to the overall

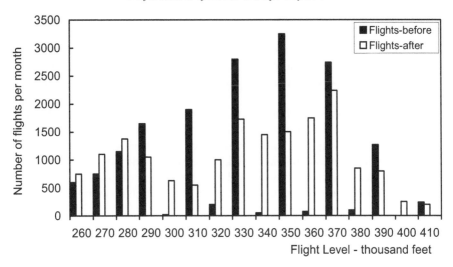

Figure 2.19 Distribution of the monthly traffic flows before and after reducing RVSM in European airspace
Compiled from EEC (2002; 2005)

reduction of en route delays. The overall effects have certainly contributed to a reduction in fuel consumption and the emission of air pollutants. According to some estimates, at the European scale the annual reduction of the air pollutants such as nitrogen oxides NO_x and sulphur oxides SO_x could amount to about 3,600 and 260 tons/year, respectively. In addition, reductions of carbon dioxide CO_2 and water vapour H_2O could be about 975 and 380 thousand tons, respectively. All the above-mentioned reductions could be considered as important contributions of the *ATC/ATM* to the sustainability of European air transport system (EEC, 2002).

The aircraft constituting traffic flows at the same flight level are horizontally separated by the ATC/ATM time and/or space separation rules (ICAO, 1996). These aircraft are characterised by performances such as the optimal cost-effective (the most economical) cruising speed and altitude, climb/descent rate, and the navigational equipment (avionics), which may influence the separation minima and consequently the air traffic control process.

A flight may be performed either according to the IFR (Instrumental Flight Rule) or the VFR (Visual Flight Rule). The IFR aircraft/flights are primarily responsible for maintaining the assigned flight paths while the ATC/ATM maintains the separation rules. The VFR aircraft/flights perform primary navigation and maintain safe separation rules between each other and between themselves and IFR aircraft/flights. In airspace shared by both IFR and VFR flights, the division of responsibility between the pilots and air traffic controllers to maintain safe separation is according to the IFR. The IFR flights are always carried out under IMC. The VFR flights can take place only under the so-called VMC. For example, in Europe, only IFR is exclusively applied for both IMC and VMC while in the US both are applied depending on the current weather conditions.

2.4.2.2 Facilities and equipment

The ATC/ATM uses two types of facilities and equipment for monitoring and controlling air traffic: i) communications and navigation facilities and equipment located both on the ground, in space and the aircraft, and ii) surveillance equipment with particular components both on the ground, in space and the aircraft.

• Communications facilities and equipment consist of communication channels for the transmission of information between pilots and air traffic controllers (VHF/UHF air/ground voice and non-voice communication links); communication links established between particular ATC/ATM control units; communication links providing the exchange of information between ATC/ATM and other actors involved in the air traffic control. One of the promising communications systems is VHF data link (VDL Mode 2 in the medium- and VDL Mode 3 and 4 in the long term). This data-link is used to automatically (without voice) communicate information on a whole range of flight parameters from the aircraft to the ground (air traffic controller). This substantially improves the situation awareness for the air traffic controllers and reduces the workload and consequently increases the system capacity, efficiency and effectiveness. In addition, Controller Pilot Data Link Communications (CPDLC) using the data link for ATC/ATM communications, Data Link Flight Information Service (D-FIS) enabling pilots to receive flight information through the air/ground data communications, and the satellite data-link enabling air/ground communication in cases when an aircraft is out of range of the ground communication systems have been developed (Little, 2000).

• The navigational facilities and equipment include the ground aids and the space satellites. Both allow the aircraft primary navigation. Regarding their location and primary function, they are classified into several groups as follows: i) external overall en route aids (VOR, DME, VOR/DME), ii) external overland terminal airspace aids (ILS - Instrumental Landing System; MLS – Microwave Landing System, and SLS – Satellite Landing System using information from Global Positioning System – GPS or Global Navigation Satellite System – GLONASS), iii) the airport external navigational equipment (approach lighting systems, slope indicators, surface detection equipment); iv) external over-water en route aids (OMEGA, Loran-C); v) the internal over-water en route aids (Doppler Navigation System; Inertial Navigation System (INS)); vi) internal overland en route aids (RNAV systems, which use the information about distance and azimuth by VORTAC stations as the inputs); and vii) internal overland terminal aids (RNAV systems used in terminal airspace) (Horonjeff and McKelvey, 1994; Little, 2000). The contemporary navigational procedures are primarily based on global navigation satellite-based (RNAV) systems providing the aircraft with direct routeing between origins and destinations rather than using the system of airways, which does not always follow the shortest-great circle distance. Such routeing allows more flexible utilisation of the available airspace and consequently increases the possibility of flying along the fuel-cost preferred routes.

- The surveillance facilities and equipment embrace different radar systems. Two radar types are available: primary and secondary (beacon) radar (Secondary Surveillance Radar (SSR) and Mode S). Radar enables the air traffic controllers to monitor the separation between aircraft on the radar screen (Horonjeff and McKelvey, 1994; Little, 2000). In the advanced ATC/ATM the surveillance capability is significantly improved by the introduction of the Automatic Dependent Surveillance Broadcast (ADS-B) system as a complement to SSR. The ADS-B allows the aircraft to periodically broadcast information about their position and other flight parameters relevant for the traffic controllers. Specifically at airports, A-SMGCS (Surface Movement Ground Communication System) is developed to enable surveillance, guidance and routeing of aircraft while on the ground under all local weather conditions.

- New communication, navigation, information and surveillance technologies allow the introduction of innovative procedures and consequently enhancement of the ATC/ATM system capacity on the one hand and improvement of the utilisation of existing capacity on the other. Such improved capacity should enable accommodation of the growing demand to be safer, more efficient and effective, implying the lack of air traffic accidents, less congestion and delays, lower fuel consumption and emissions of air pollutants, and less noise around the airports. Broadly speaking, innovative technologies can be classified into the following categories: i) those used for landing and take-off; ii) those supporting the aircraft manoeuvring around the airport; iii) those utilising the ATC/ATM infrastructure; and iv) those based on the concept of Free Flight and Autonomous Operations (Little, 2000).

  o The first group embraces the Arrival and Departure Management System (AMS and DMS), respectively, which, as the ground-based traffic management automation tools, optimise the incoming and outgoing aircraft flows at an airport. They are expected to relieve congestion by improving utilisation of the runway system capacity. The complement to the above-mentioned tools is the runway management system. These tools should reduce primarily the air traffic controller workload and consequently increase the system capacity. In addition, there is the Satellite Landing System (SLS), based on the information transmitted from either GPS or GLONASS, as an alternative to the ILS landing at congested airports. Finally, the ADS-B system enables simultaneous operations at closely parallel runways, which might increase their capacity, reduce congestion, and diminish the need for land take for new runways (see Chapter 5).

  o The second group enables the air traffic controller better surveillance of aircraft manoeuvring at airports (A-SMGCS). They are expected to improve efficiency of the aircraft ground movement through reducing the aircraft taxiing times and supporting timely delivery of departing aircraft to the runway. The latter appears to be particularly important under low visibility conditions.

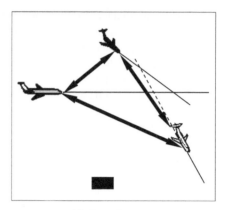

a) Conventional   b) Free Flight

**Figure 2.20   Conventional ATC/ATM and 'Free Flight' concept: a)
             Conventional; b) Free Flight**
Compiled from Hoekstra (2001)

○ The third group is mostly used for controlling the en route traffic such as VHF data-link (Mode 2, 3, 4), ADS-B, GPS, GLONASS, and CDTI (Cockpit Display Traffic Information). Individually or in combination, they allow direct/free routeing, which in turn should reduce flying distance, time, and consequently the fuel consumption and related air pollution without compromising the required level of safety.

○ Finally, the last group of new technologies includes the concept of 'Free Flight and Autonomous Operations'. The 'Free Flight' is defined as a 'safe and efficient flight operating capability under Instrumental Flight Rules (IFR) in which the operators have the freedom to select their paths and speed in real-time; the air traffic restrictions are only imposed to ensure the aircraft separation and safety, and prevent exceeding the airport capacity and unauthorised flights through the specially used airspace; any activity, which removes the restrictions represents a step forward towards "free flight"' (FAA, 1998). In such cases, some or all of the responsibility for the selection of routes and aircraft separation are to be shifted from today's ground-based air traffic controllers to the 'Free Flight System'. Figure 2.20 illustrates a scheme of such de-centralisation and shift of the aircraft separation function from the ground (ATC/ATM) to the air (aircraft crew).

This will enable freedom for aircraft flying outside the terminal airspace in selecting their preferred flight paths, self-separation, and minimal interaction with the air traffic controllers on the ground. This will significantly diminish the air traffic controllers' workload and consequently increase the en-route airspace capacity, reduce delays, improve the efficiency of flights and thus reduce the external effects such as fuel consumption and related air pollution. Some estimates have indicated that savings of the fuel burn in the US airspace could be about 5 million tons by

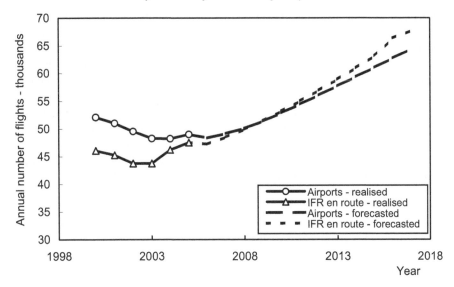

Figure 2.21   Development of the long-term ATC/ATM demand (flights) in the US

Compiled from FAA (2005)

the year 2015, which would be a decrease of about 6 per cent as compared to the consumption if the concept were not implemented. The savings in emissions of air pollutants would be between 9 per cent and 18 per cent, depending on the type of air pollutants (FAA, 1998; Hoekstra, 2001).

2.4.3 Operational Performances

2.4.3.1 Demand

As at an airport, the demand at a given ATC/ATM unit represents an important parameter of the operational performances. The demand is usually expressed as the number of flights aiming to pass through a given airspace over a given period of time. These flights have different spatial and time patterns. The space can be any of the ATC/ATM sectors. The unit of time can be an hour, day or year. For operational purposes and the short-term prediction of the air traffic controller's workload in a given airspace, the hourly number of flights is relevant. For the long-term strategic handling and demand, the annual number of flights is relevant. This number is also important for planning the long-term development of the *ATC/ATM* capacity. For such purposes, the long-term forecasting of the *ATC/ATM* demand in terms of the annual number of flights in given areas (regions) is carried out. Figure 2.21 shows an example for the US ATC/ATM for the period 2006–2017 (FAA, 2005).

In the above-mentioned case, flights handled at airports include the commercial air carrier, air taxi/commuter, general aviation and military flights. The en route flights include the IFR flights of the same categories. The annual number of realised flights over the past six year period (2000–2006) has been greater at the airports than in the en route airspace with a decreasing difference towards the end of the period.

In general, in both segments, the annual number of flights decreased until the year 2003 and then started increasing again. The main cause was the 11 September/2001 terrorist attack on the US The forecast for the next decade (2006–2017) indicates a steady growth of the annual number of flights at airports, which will result in the gradual increase in the number of en route flights. One of the explanations could be that more flights over the US will be carried out.

This flight demand is used for planning the ATC/ATM capacity at a strategic and tactical level. For example, at the strategic level, about 500 airport ATC/ATM units (towers) will be used to handle the airport demand in Figure 2.21 over the period 2006–2017 (FAA, 2005). At the tactical level during everyday operations, the larger number of 'sectors' of greater flight handling capacity will handle demand.

2.4.3.2 Capacity

The capacity is another important parameter of performance of a given ATC/ATM sector (unit). This capacity is usually expressed as the maximum number of flights/ aircraft served during a given period of time under given conditions. These are specified regarding: i) the size and configuration of airways in a given sector; ii) the facilities and equipment available to both ATC/ATM and aircraft, and iii) the intensity, volume, structure, time and space distribution, and continuity of demand. The ATC/ATM capacity embraces the capacity of particular components such as airspace, air traffic controllers, and communication links (Janić, 2001). In order to calculate the capacity of a given ATC/ATM sector based on the estimation of air traffic controller workload, the number, duration, and order of execution of particular control tasks-activities performed for the air traffic – aircraft/flights – passing through a given sector according to a given pattern during a given period of time should be considered.

The ATC/ATM controller workload is based on performing physical and mental tasks within a specified time framework. They aim to control the air traffic in the sector of his/her jurisdiction. In general, two types of mental tasks dominate: monitoring the current air traffic on a radar screen, and receiving, processing (decision-making) and delivering the oral and/or digital instructions to the aircraft. Monitoring the radar screen enables the memorising of a picture of the current traffic situation according to the internal 'mental' model. Having been memorised, this picture is stored in the controller's short-term memory and used later on as a basis for creating and executing particular control tasks. It is combined with information collected from the aircraft via the air/ground communication link, other screens, and other air traffic controllers from the neighbouring sectors. Research in psychology has shown that human operators (in this case the air traffic controller) are not able to simultaneously carry out two different types of tasks, i.e., monitoring of radar screen and executing the control tasks. He/she rather carries out the work on the principle of time-sharing, i.e., by executing partially or completely different types of tasks in a sequential order during very short time 'spots'. Such operations, and the necessary and available time for executing particular tasks, generate the air traffic controller's workload.

This workload can be quantified by self-assessment and through direct observation. Since both approaches have disadvantages such as the subjectivity and possibility of distrusting the working conditions, different analytical and simulation models for estimating the air traffic controller workload have been developed (Janić, 2001; Majumdar and Polack, 2001).

The most well known analytical models developed outside the academic community have been the Sector Design Analytical Tool (SDAT) developed in the US and the Capacity Indicator Model (CIM) used by EUROCONTROL (Europe). Both models estimate the workload in a given *ATC/ATM* sector based on the associated tasks. In addition they apply a probabilistic approach for predicting the expected number of conflicts including the resolution strategies.

The computer-based simulation models are more complex and detailed, and are thus able to capture the great complexity and dynamic behaviour of the ATC/ATM system. The most well known models are DORA TASK in the UK, and the Reorganised ATC Mathematical Simulation (RAMS) used by EUROCONTROL. Specifically, RAMS uses the detailed simulated data on the air traffic intensity and pattern (the aircraft/flight cruise, descent, ascent) and the individual and total flight time in given sectors as the independent variable, and the associated (calculated) air traffic controller workload as the dependent variable. Then, the causal relationship using the regression technique is established indicating the importance and influence of particular variables on the air traffic controller workload. In such a case, the 'threshold' on the controller workload can be set up and then the volume and structure of traffic determined in a way that the corresponding workload does not exceed a given 'threshold'. Using this evidence one can notice that the air traffic controller workload appears rather statistically derived and not considered sufficiently at a physical and/or physiological level. Under such circumstances, the analytical models could make more noticeable and generic relationships between the air traffic controller workload and related driving forces. For example, let the air traffic controller workload be expressed by the so-called 'workload coefficient' ρ (Janić, 1997a). Assuming that the air traffic controller behaves as a single server system operating with the constrained service rate (capacity) under conditions of constant demand for service, the workload coefficient can be expressed as follows (Janić, 2001):

$$\rho_w = S_t/S_a \tag{2.19a}$$

where
$\quad S_t \quad$ is the average service time of a control task of a given type (minutes),
$\quad S_a \quad$ is the average inter-arrival time of control tasks of all types, i.e., the available time for execution of a control task of a given type (minutes).

From Equation 2.19a, ρ_w may be considered as the server load factor similarly as the server 'utilisation rate' in the queuing system theory. Generally, the workload will increase if either the time for executing a control task, the intensity of the control tasks increase, or both simultaneously increase, and vice versa. The time of executing particular control tasks is the stochastic variable with a known probability distribution. The time between arrivals of particular control tasks depends on the intensity and complexity of air traffic in a given sector as well as the performances of the facilities and equipment supporting the execution of particular tasks. These can be i) internal and external coordination tasks, ii) flight data management tasks; iii) radio or telephone communication tasks; and iv) conflict planning and resolution tasks. The task of monitoring the radar screen is considered separately due to its different nature of execution and type of workload. The time monitoring the radar screen during one observation can therefore be quantified as follows:

$$t_1^* = \lambda_c t_f p_c H \tau_{01} + (1/2)([(\lambda_c t_f p_c)^2 - (\lambda_c t_f p_c)]\tau_{02}) \qquad (2.19b)$$

where

λ_c is the intensity of the aircraft flow entering a given sector (aircraft per hour);

t_f is the average flying time through the sector (minutes);

H is the total entropy the radar controller is confronted with while observing the radar screen aiming at identifying the status and intentions of specific aircraft;

τ_{01} is the time needed for processing a unit of information gathered visually from the radar screen during one observation (seconds per bit);

τ_{02} is the time to establish the appropriate relationship between any pair of blips on the radar screen depending on its quality (primary radar, beacon, and alphanumeric display may be available) (seconds); and

p_c is the proportion of blips (aircraft) on the radar screen during a single observation.

Since the traffic situation in the sector continuously changes, the air traffic controller should resume observation of the radar display after some time in order to update the traffic picture. Observations of the radar display will be more frequent if the traffic situation is expected to become more complex and change considerably and more rapidly. This particularly happens in the terminal airspace and congested low-altitude en route sectors. Under such conditions, the time between any two successive observations of the traffic situation on the radar display can be estimated as follows (Janić, 2001):

$$t^{2*} = [3\varepsilon_0 m(t_{0z} + t_{wz})]/[\Phi^{-1}(0.5 - p_0/2)] \qquad (2.19c)$$

where

m is the number of control tasks inducing significant changes of the actual traffic situation in comparison with that previously memorised;

t_{oz},   is the average time of creating and executing a control task (seconds);

t_{wz} is the time in which the control task can stay memorised in the air traffic controller's short-term memory (after time t_{wz} a given task will be forgotten, so updating should start) (minutes);

ε_0 is the maximum tolerable error during comparison of the successive pictures of a traffic situation obtained from the radar screen (this error is usually considered as the random variable with Gauss (normal) probability distribution);

p_0 is the probability that error ε exceeds a given threshold ε_0; and

Φ^{-1} is the inverse Laplace's function.

Consequently, the air traffic controlled workload generated during execution of both types of tasks stabilises after some time, implying that the corresponding workload coefficients satisfy the condition:

$$\rho_w = \rho_m + \rho_z = 1 \qquad\qquad (2.19d)$$

where

ρ_m is the workload coefficient caused by monitoring the radar screen ($\rho_m = t_1^* (t_1^* + t_2^*)^{-1}$);

ρ_z is the workload coefficient imposed by planning and executing other control tasks ($\rho_z = \lambda L t_{oz}$, where L is the number of control tasks performed for one aircraft while in the sector; other symbols are analogous to those in the previous expressions).

After simple mathematical manipulations with Equation 2.19, the polynomial of third degree can be obtained in which the variable λ_c is considered as unknown, i.e.:

$$a_0 + \sum_{n=1}^{3} a_n \lambda_c^{\ n} = 0 \qquad\qquad (2.20)$$

where

$a_0 = -[6\varepsilon_0 m(t_{0z} + t_{wz})]/(L t_{0z} t_f^2 p_c^2 \tau_{02}^2 \Phi^{-1}(0.5 - p_0/2))$
$a_1 = [6\varepsilon_0 m(t_{0z} + t_{wz})]/(t_f^2 p_c^2 \tau_{02}^2 \Phi^{-1}(0.5 - p_0/2))$
$a_1 = [2(H t_{01} - 0.5 - p_c \tau_{02})]/(p_c^2 \tau \tau_{02}^2)$
$a_3 = 1$

The real root of Equation 2.20 represents the capacity of a given sector based on the air traffic controller workload under a given traffic scenario (Janić, 1997a). Some results from the application of the model to the hypothetical ATC/ATM sector and traffic scenario are shown in Figure 2.22.

As can be seen, the capacity decreases with the increasing aircraft flying time through the sector (i.e., the size of the sector) more than proportionally. However, for a given sector, the capacity increases with the decreasing number and duration of control tasks as well as the duration of a single observation of the radar screen. This coincides with results obtained from the most complex simulation models. In addition, the results implicitly indicate that the partial or complete automation of particular ATC/ATM functions inevitably influences the structure and content of the air traffic controller workload, and consequently its (and sector) capacity. More automated and sophisticated components will put the radar controller more in the position of monitoring than executing control tasks. Such changes in the structure of the workload will certainly increase the capacity of a given sector. However, some researchers and the ATC/ATM experts are concerned that the automation itself

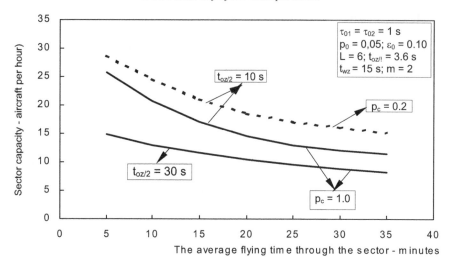

Figure 2.22   The ATC/ATM sector capacity vs aircraft flying time, duration of control task and strategy of monitoring the radar screen
Compiled from Janić (1997a)

may, under certain conditions, add to the workload. In addition, after introducing the 'free flight' concept, the currently 'centralised' workload will be de-centralised by passing the main tasks from the air traffic controllers to the aircraft/pilots. Under such circumstances, the consequences for the ground ATC/ATM controllers in terms of workload need to be further investigated.

2.4.3.3 Matching capacity to demand

Matching capacity to demand in the ATC/ATM including provision of the required safety and quality of services is achieved through the so-called Air Traffic Management (ATM), which implies continuously performing the ATC/ATM control functions at two levels: i) the operational level by providing safe separation between individual aircraft according to the prescribed separation rules and procedures; and ii) the tactical (and strategic) level by providing safe, efficient and effective handling of the air traffic flows in a given airspace under given conditions.

ATM functions   The most important sub-functions of the ATM are airspace management, air traffic services, and air traffic flow management.

   Airspace management deals with the provision of airspace to accommodate the requirements for optimal flight profiles. It includes the whole infrastructure, allocation and organisation of airspace, classification of services, required airborne capabilities, available routes and cruising levels, CNS/ATM facilities and services, and capacity determined in relation to the applied separation minima. An example of application of the airspace management relates to the North Atlantic (oceanic) airspace where traffic flows are kept on tracks (airways). A longitudinal separation of 10 minutes, lateral separation between tracks of 60 nm (nautical mile) and vertical separation of 2000 ft are applied between aircraft. Implementation of the vertical separation minima of 10^3 ft above FL290 provided six additional cruising levels resulting in a doubling of the current airspace capacity. The other opportunity consists of reducing the lateral and longitudinal separation from 60 nm to 30 nm, which will be possible by using RNAV-derived distance information. The implications of such increased capacity are the reduction of delays, flying at fuel-cost optimal altitudes and consequently a reduction of fuel consumption and associated air pollution.

The air traffic service as the main component of the Air Traffic Management function enables continuity in the provision of safe, efficient and effective air traffic.

The ATFM aims to optimise the air traffic flows through reducing costs of delays, and preventing system overloads. The ATFM uses technologies/decision support tools, which have already been described. The ATFM is usually applied at the strategic and tactical level. At the strategic level, the ATFM provides planning of the preferred routes and flight levels through traffic-orientation schemes. The strategic ATFM operates off-line implying the realisation of flights from several hours to several days and months in advance. At the tactical level, the ATFM implements the acceptance rates for the specific en route points and/or for destination airports accompanied by the departure slot allocation all in addition to the traffic-oriented schemes. The tactical ATFM operates on-line. For example, in the US ATC system, the aircraft position is updated through the nation-wide radar network every 3 min (Andreatta and Romanin-Jacur, 1987).

The main indicators of the quality of services in the ATC/ATM are safety, aircraft delays, and flight efficiency in terms of the additional fuel consumption due to flying along the fuel non-optimal trajectories.

Safety Safety can be measured by the risk of occurrence of an aircraft accident in the airspace of jurisdiction of a given ATC/ATM unit over a given period of time due to the unit's direct contribution. The accident occurs when either the prescribed distance or time-based separation rules between aircraft or between the aircraft and fixed objects on the ground deteriorate. Consequently, maintaining the prescribed separation rules reduces risk and guarantees safety of the air traffic operations. For example, relevant statistics for the European airspace indicate that the accidental events with direct contribution of the ATC/ATM have been relatively rare – seven accidents over the past 30 years (1972–2002). Some other evidence has shown that ATC/ATM has contributed to only 1.7 per cent of the whole world's air traffic accidents over the same period of time (EEC, 2003). However, the distinction between this and other causes of air traffic accidents should always be made.

Air traffic incidents imply events in which a high risk of evolution into an accident exists. The incidents are registered as AIRPROXs of category A and B. In the former case a serious risk of collision between aircraft exists. In the latter case, the safety of the aircraft involved is compromised. Both categories of events have been relatively infrequent. For example, in the European airspace the average rate of occurrence has been about 0.065 events per thousand flights during the period 1998–2001. In addition, the events with infringement of the separation minima have also been regarded as potentially risky. Again, in the European airspace, these events have decreased, from 0.12 per thousand flights in 1999 to about 0.095 per thousand flights

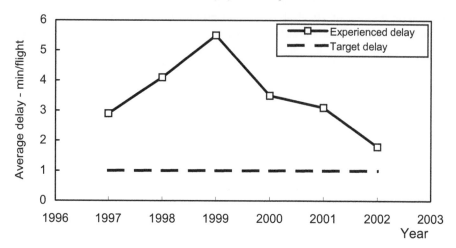

Figure 2.23 Evolution of the ATC/ATM delays in European en-route airspace
Compiled from EEC (2003)

in 2001. In both kinds of events – AIRPROXs and infringement of the separation minima, contribution of the *ATC/ATM* could not be found (EEC, 2003).

The ATC/ATM may set up qualitative and quantitative targets on its contribution to the air traffic accidents and incidents. The qualitative targets are to maintain or even reduce the current level of contribution to the accidents and incidents during a specified period of time. The quantitative targets set up the maximum rate of contribution. For example, EUROCONTROL has set up a target for contributing to a maximum of 0.623 accidents per year (i.e., three accidents in five years) in European airspace. This will certainly be revised, particularly when compared with the above-mentioned past period of 30 years when the contribution was about 1.165 accidents per five years (EEC, 2003). In addition, the national ATC/ATM have set up their own targets on their contribution to air traffic accidents and incidents. Despite being inherently complex to set up some quantitative targets, it has appeared useful for assessing whether the risk of air traffic accidents and incidents to which the *ATC/ATM* contributes is increasing or decreasing, and which strategic decisions and actions need to be undertaken to manage them.

Delays   Generally, delays to an aircraft/flight passing through a given ATC/ATM airspace occur as soon as the demand exceeds the capacity. Delays can happen at airports (as has already been discussed) and in the en route airspace. In addition to those imposed on particular flights because of congestion at the departure/arrival airports, the ATC/ATM can also impose delays, as soon as the demand exceeds the capacity of any ATC/ATM sector along the routes of these flights. In some cases, these delays can be realised at the origin airports together with the airport delays. However, delays by contribution of the *ATC/ATM* are often considered as an indicator of the quality of the system's service. In general, these delays are not particularly long. For example, Figure 2.23 shows an example of these delays expressed per individual flight in European airspace during the period 1997–2002 (EEC, 2003).

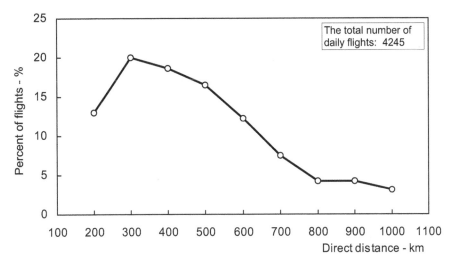

Figure 2.24 Distribution of flights by distance in European airspace
Compiled from EEC (2002a; 2003)

As can be seen, the average delay per flight increased until the year 1999 and decreased afterwards, thus approaching the target of 1 minute at the end of the observed period. As in the case of the airport-related delays, decrease of the en route (ATM/ATC) delays generally contributes to a decrease in the fuel consumption and related emissions of air pollutants. Indirectly, it contributes to a decrease in the airline's direct operating costs (fuel) and the costs of passengers' time. Under conditions of continuous traffic growth and increasing fuel prices, these seemingly minor improvements contribute to the large cumulative savings of the airline's operating costs, fuel and consequently air pollution.

Flight efficiency   Flight efficiency is usually measured by the difference in fuel consumption while flying along the assigned and required (fuel-optimal) 4-D (four dimensional) trajectories. For a given flight, the fuel consumption depends on the aircraft type, route length, weather conditions (wind) and the four-dimensional profile of the flight path. The route length, i.e., flying distance, can be extended due to many reasons such as the lack of a direct route, avoiding congestion on the shortest route, unexpected weather or unacceptable charges, tactical changes of flight paths, and following the standardised (sometimes non-optimal) instrumental arrival and departure routes. Delays due to congestion along such extended routes additionally increase the flying time and fuel consumption. In addition, flying at the fuel non-optimal altitudes again due to congestion in the en route airspace also increases fuel consumption. Thus, either individually or simultaneously, these effects contribute to the environmental (fuel consumption and air pollution) flight inefficiency.

In particular, short flights might be vulnerable to these impacts since they could spend almost the whole cruising phase at fuel non-optimal altitudes. Figure 2.24 shows an example of the distribution of flights in European airspace regarding their length, i.e., flying distance.

This case is used as the 'reference' case for estimating the contribution of the ATC/ATM to the difference between the actual and optimal flight trajectories. In Figure 2.24, the length of flights varies between 200 and 1,000 km; the most frequent flight lengths have been from 300 to 600 km. After comparing these with the optimal (the shortest) lengths, the average extension of about 8.9 per cent has been calculated. The increased fuel consumption has been about 9.6 per cent. The differences between both cases have been due to flying at fuel non-optimal altitudes. Realistically, improvements of the ATC/ATM guidance in European airspace could result in the shortening of actual flights and diminishing additional fuel consumption by about 5 per cent per year. For comparison, in the US airspace of comparative size, the extension of flights has amounted to 7 per cent, resulting in an additional fuel consumption of about 7.7 per cent (EEC, 2002a, 2003). Both figures roughly coincide with the corresponding figures for the world's *ATC/ATM* of about 6 per cent per year (IPCC, 1999).

In addition, there might also be a significant economic value to making the flight trajectories 'optimal' through the contribution of the ATC/ATM. For example, bearing in mind the average unit cost of flight is about 12–30 € per minute and the average cost of fuel is about 20 € per minute, the annual savings in European airspace could be of the order of a billion euros. With rising fuel prices these savings would be even higher, implying the growing importance of the ATC/ATM effectiveness (EEC, 2002a, 2003).

2.4.4 Economic Performances

The economic performances of ATC/ATM refer to the costs, revenues, and profits. Generally, the costs are the spending of capital (facilities and equipment), labour (staff), and energy to serve the given volume of air traffic during a given period of time (usually one year). Coverage of these costs is provided from revenues obtained by charging the airline flights for the ATC/ATM services (currently, these charges account for about 6 per cent of the direct operating costs of the European airlines). In addition, so-called aeronautical charges are made for the use of the airports. They include the runway, parking and air bridge use, terminal navigation and noise charges. Their share in direct operation costs of the European airlines is about 7 per cent (EEC, 1999).

In Europe, for example, the en route charges cover 90 per cent of the national-state and 10 per cent of the international – EUROCONTROL – costs of services. At the national level, the staff costs are 52 per cent, cost of depreciation of facilities and equipment 15 per cent, operating costs 22 per cent, and the interest costs 8 per cent of the total costs. Regarding the category, the cost of Air Traffic Services (ATS) is 64 per cent and the administrative cost 13 per cent. At the international EUROCONTROL level, the staff costs are about 67 per cent, operating costs 10 per cent, depreciation cost of facilities and equipment 18 per cent, and the interest costs 5 per cent. At both levels, the above-mentioned costs are usually expressed using the various indicators, which interrelate the costs of service and its production. Particular indicators are the cost per service unit, per flight, per kilometre, and the cost per sector.

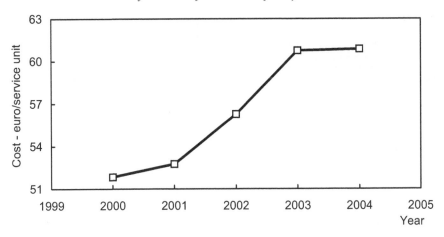

Figure 2.25 Cost of the ATC/ATM service unit in European en-route airspace
Compiled from EEC (2005a)

The cost per service unit is the total chargeable cost divided by the number of service units carried out. The number of service units is usually determined as follows (EEC, 1999):

$$N_{su} = (d/100)^*(MTOW/50)^{0.5} \tag{2.21}$$

where
  d is the flying distance (km);
  $MTOW$ is the aircraft Maximum Take-Off Weight (tonnes).

Considered from the user-airline perspective, Equation 2.21 can be used to calculate the average price per service unit for a range of aircraft types (categories) operating in European airspace. In such sense, this indicator appears to be more of a price than a cost indicator. Figure 2.25 shows the average cost per service unit in European airspace over time.

As can be seen, this cost has generally increased from about €50–60/unit during the past five years. The cost per flight is mainly related to flying distance (i.e., the size of the country). In principle it increases with the increasing flying distance, reflecting the higher air traffic controllers' workload and spending of their more costly resources. For example, in Europe, this cost has varied from about €50/flight in Slovenia to about €300/flight in countries like France, Italy, UK and Turkey (EEC, 2003).

The cost per kilometre is usually used for comparison of the costs of different ATC/ATM service providers in different countries. Throughout Europe this cost varies from about €0.35 to about €1.2/km, mainly depending on the air traffic complexity, density, and the conflict probability.

The cost per sector reflects the cost of all simultaneously operating ATC/ATM sectors. If the number of sectors is maximal, this cost reflects the cost of the total deployed ATC/ATM capacity. In Europe, the cost per sector amounts to about

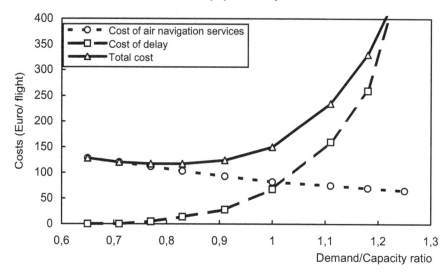

**Figure 2.26 The total ATC/ATM cost per controlled flight vs the demand/
capacity ratio**
Compiled from EEC (2005b)

€2,000–3,000 in Slovenia and Croatia to about €20,000 in Turkey. In particular
sectors, some figures show that the average total cost per IFR flight in the year
2001 was €704 and the cost per hour of IFR flight €560. For a comparison, these
costs in the US were €454 and €323, respectively, which is for about 60–70 per cent
lower than in Europe (EEC, 1999, 2002a). In addition, the cost of delay per flight as
an indicator of the traffic complexity and the ATC/ATM flight guidance efficiency
could be added to the above-mentioned unit cost. The total cost then reflects the
price indicator of a given ATC provider (EEC, 1999).

Consequently, the total costs per flight comprising the cost of navigational
service and the cost of delays can be used. The former cost depends on the cost of
deployed resources – facilities, equipment and labour. The latter cost depends on
the relationship between the ATC/ATM demand and capacity. In this context, delays
are measured as the difference between the actual and planned duration of the flight.
Figure 2.26 illustrates an example of changing the total costs per flight depending on
the demand/capacity ratio in Maastricht Upper Area Control Area (Europe).

This ATC/ATM unit controls air traffic in the upper airspace (above FL245 –
24,500 ft) of Belgium, the Netherlands, Luxembourg, and North-West Germany. In
the year 2004, a total of one million and 300,000 flights were served. The total delays
imposed on these flights were about 700 90 minutes. Consequently, the average delay
per flight was about 0.61 minutes, which is lower than the EUROCONTROL target
of 1 minute per flight. Such performance was achieved by continuous adaptation
of the ATC/ATM capacity to the growing demand over the period 1990–2004.
Nevertheless, despite increasing the capacity through deploying more operating staff
(air traffic controllers) combined with modernisation of the facilities and equipment,
the demand capacity ratio has continued to rise, causing a more than proportional
increase in the average delay cost and a rather constant decrease in the average cost

of navigation services. Under such circumstances, the total cost as the sum of two cost components has initially decreased, reached the minimum and then continued to increase in line with the increasing demand/capacity ratio. For a demand/capacity ratio of 0.85, the total cost was minimal, indicating an optimal operational level for a given ATC/ATM unit. However, in the given case, this optimal operational level was reached in the year 2004. Reducing aircraft delays and the related costs in a given en route airspace certainly diminishes the losses of time to passengers and airlines, associated extra fuel consumption and related emissions of air pollutants. This again might require balancing the full positive effects and negative impacts (i.e., the sustainability approach) while managing the further development of the given ATC/ATM unit.

2.4.5 Social Performances

The volume of services influences the number of staff employed at a given ATC/ATM service provider. As mentioned above, in Europe, the staff cost is more than 50 per cent of the total costs of a given state's *ATC/ATC* provider.

The staff mainly relates to the operational staff – air traffic controllers responsible for controlling air traffic in the airspace of their jurisdiction. They are organised into teams of one, two or three persons. Division of tasks and responsibilities within the team is precisely determined. Given the size of the controlled area, the number of sectors, and the number of team members per sector, the total number of staff simultaneously engaged in direct control of air traffic during a given period of time can be estimated. It is not easy to judge whether the ATC/ATM is labour intensive activity or not. Some figures on employment in the European ATC/ATM may give an idea. In the year 2001, the controlled area was about 10.8 million square kilometres divided into 58 Area Control Centres with about 600 sectors (i.e., 10 sectors per centre), 157 Airport Approach units, 341 Tower control units, and 70 Aeronautical Flight Information Service units. They employed in total approximately 44,600 staff, of which 12,760 were the air traffic control officers and operational staff. Of these 7,730 were working in Area Control Centres (i.e., about 133 per centre and about 13 per sector), and the remaining 5,030 at Airport Approach and Tower control units, i.e., about 10 per unit.

The first group (Area Control Centres) processed about 8.1 million IFR flights in the year 2001. The second group processed about 14.5 million IFR and 5 million VFR flights. For some comparison, in the US a controlled area of comparative size, about 13.8 million square kilometres, was divided into 21 Area Control Centres and 780 en route sectors (i.e., 37 sectors per centre). The unit employed a total of 34,532 staff of which 17,341 were the air traffic controllers. Of them, 7,724 were employed at the Area Control Centres (about 368 per Centre and about 10 per sector), and 9,617 at Airport Approach and Tower control units. In the year 2001, they processed about 17.7 million flights (EEC, 2003).

2.4.6 Environmental Performances

Air Traffic Control/Management may significantly influence the short- and long-term impacts of the air transport system on people's health and the environment and consequently its sustainability. Two examples are illustrative: i) mitigation of the aircraft noise around airports using innovative technologies and operational procedures; and ii) reduction of the airborne aircraft delays, consequent fuel consumption and associated air pollution around the airports and in the en route airspace.

2.4.6.1 Mitigation of aircraft noise at airports
The mitigation of noise around airports could be achieved by carrying out the so-called 'noise mitigation landing and take-off procedures'. Innovative technologies are used for safe operations and maintaining the throughput of traffic flows at the level of the runway system capacity. One example of such a procedure is the 3° Decelerating Procedure in which the aircraft, using RNAV and GPS, follow any straight or curved approach path from any point in the airspace to a given runway threshold (Clarke, 2000). In addition, for the final approach and landing, they use ILS in combination with GPS or DME (Distance Measurement Equipment). Figure 2.27 shows the vertical profile of the current 'baseline' and innovative 'advanced' procedure and related noise effects.

According to the baseline procedure the aircraft descend to an altitude of about 2,500 ft at a distance of about 19–20 nm from the landing threshold. At that time their speed is about 210 kts (knots). They continue a horizontal flight up to a distance of 6 nm from the threshold where they intercept the ILS glide slope of 3° at a reduced speed of about 135–140 kts. From that distance, they continue common ILS landing procedure.

According to the advanced procedure the aircraft intercepts the ILS glide slope of 3° at an altitude of about 6,000 ft at a distance of about 19 nm and speed of 250 kts. They continue descending with idle thrust up to an altitude of 2,500 ft and a distance of 6 nm from the threshold. At that moment their speed is about 135–140 kts (Figure 2.27a). From that point they follow a conventional ILS approach and landing procedure. The advanced procedure brings some noise reduction below the aircraft flight path. An example is shown in Figure 2.27b, which are the results of experiments by the US NASA (National Aviation and Space Administration) using B757 testing aircraft (Clarke, 2000). Despite significant noise reduction, the main obstacle to implementation of this procedure is the inability of the air traffic controllers and pilots to automatically maintain the precise aircraft sequencing and spacing under heavy traffic conditions. Currently, this is resolved by fixing the aircraft speeds and separation, which in the case of the 3° Deceleration Procedure is not possible due to the need for the aircraft to decelerate. Therefore, 'buffers' in separation need to be introduced to compensate the differences in aircraft deceleration rates, which in turn increases the average separation and consequently diminishes throughput, i.e., utilisation of the available runway capacity. This is an obvious case in which a compromise needs to be made between the operational (capacity) requirements, potential of technology (avionics), and potential for diminishing noise burden around airports.

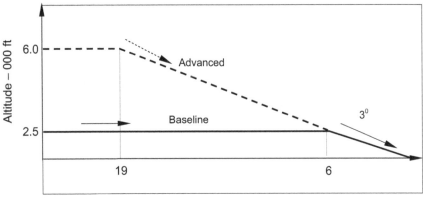

a) Baseline and advanced noise abatement

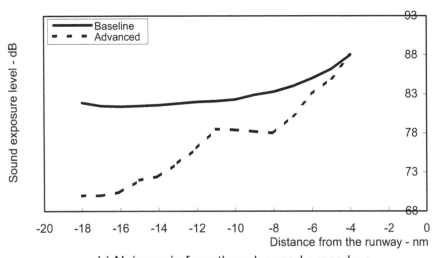

b) Noise gain from the advanced procedure

Figure 2.27 **An example of the noise abatement procedures: a) The vertical layout of the baseline and advanced procedure; b) Noise gains from the advanced procedure**

Compiled from Clarke (2000)

2.4.6.2 Reduction of the en-route fuel consumption and air pollution
Reduction of the flight/aircraft fuel consumption has partially been discussed in the scope of increasing the ATC/ATM service quality for the short- and medium-haul flights in Europe. The present discussion looks at the potential contribution of the ATC/ATM to the environmental efficiency of long-haul flights. Also in this case, the ATC/ATM can contribute to reducing the aircraft fuel consumption and

associated air pollutions by providing as many aircraft/flights as possible the fuel optimal trajectories (paths) between origin and destination airports regarding the given (weather and traffic) conditions. Two elements appear important: length of the aircraft/flight trajectory, which again should be as short as possible (i.e., the great circle one), and the cruising altitude, which should be the fuel-optimal for as long as possible. The technologies such as GPS, ADS-B, and RNAV offer such flexibility whilst maintaining the required level of safety. In particular, they help in economising flights along the long distance routes with frequently volatile weather (wind). Flying between Europe and US over the North Atlantic is an illustrative example. This airspace has no conventional radio-navigation facilities and fixed airways, which the aircraft/flights can always follow. Instead, they change routes from case to case aiming to take advantage of the wind, which in the airspace above FL 300 where most flights take place can reach speeds of 100 kts (Graves, 1998). The effect of wind on the flight fuel consumption can be estimated as follows:

$$F = fT = f\frac{d}{v \pm w} \qquad\qquad (2.22)$$

where

f　　is the aircraft/flight fuel consumption per unit of time (tons/h);
T　　is the duration of the flight (or its cruising phase) (h);
d　　is the length of the flight (i.e., flying distance) (nm or km);
v　　is the aircraft/flight airspeed (kts or km/h); and
w　　is the wind speed, which takes a positive sign for a tail wind and a negative sign for a head wind.

Equation 2.22 is applied to data from the North Atlantic airspace. The average route length is $d = 3,000$ nm, the aircraft speed $v = 450$ kts, the wind speed $w = 50$ kts, and the aircraft fuel consumption $f = 6.2$ tons/h (two engine aircraft of the capacity of *300 seats*). As a result, the total fuel consumption is $F \sim 41$ *tons* without wind, $F \sim 46.5$ *tons* with head wind, and $F \sim 37$ tons with tail wind. Therefore, this is a rough indication that selection of an appropriate route with respect to the wind might save up to about 10 tons of fuel and 31.8 tons of CO_2 per flight. Multiplying this amount by the daily, monthly and/or yearly number of flights may give an estimate on the scale of impact and eventual fuel and air pollution savings, provided by the contribution of the ATC/ATM. Therefore, the role of the *ATC/ATM* in these cases appears to be very important for the economic and environmental airline performances and consequently sustainability of the entire air transport system.

2.4.7 Institutional Performances

The institutional performance of a given *ATC/ATM* relates to the institutional environment it operates and the structure of ownership.

2.4.7.1 Institutional environment

Institutional environment implies the regulation of providing the air traffic services in a given region. If the region is a country, the services are provided by the single governmental or semi-private agency. However, the emergence of increasingly integrated political and economic areas such as, for example, the EU has raised the question of coordinating the air traffic navigation services in order to increase the safety, efficiency and effectiveness of flights throughout the EU area. Consequently, the European Parliament has approved a policy package entitled 'Single European Sky' in which the European Commission is responsible for institutional and EUROCONTROL in cooperation with particular national ATC/ATMs for technical/ technological, and operational implementation (EC, 2004). The specific aims for the ATC/ATM are as follows: more flexible airspace use, classification and design of airspace, harmonisation of charging policies; interoperability of facilities, equipment, and operational procedures in the upper airspace (above FL290 – 29,000 ft); and research and development. Certainly, implementation of the package will contribute to increasing the system capacity, reducing delays, improving safety, and reducing the aircraft/flight fuel consumption, which qualifies the package as an important operational/policy contribution to improving the sustainability of the European air transport system.

2.4.7.2 Structure of ownership

Historically, most countries worldwide have allocated an exclusive responsibility for provision of the air navigational services to their governments and their agencies. However, over the past two decades (from 1987 to 2005), about 38 governments have shifted this responsibility or part of it from their agencies (civil service departments) to outside organisations – the independent air navigation service providers operating as business and performance-based organisations (GAO, 2005). This process of transformation is called 'corporatisation' or 'commercialisation'. In general, this term can cover a range of different forms of independence of new organisations from the related governments, i.e., from those under direct governmental control to those, which have been completely privatised and are thus independent. Nevertheless, in all cases the government has retained responsibility for the provision of air traffic services, but authorised commercialised bodies to perform these services on their behalf. In such context, these bodies take over the responsibility for carrying out safe and efficient services, as well as having the freedom to undertake different business actions to achieve such objectives. In particular, being sufficiently flexible market entities, they are expected to be able to access the capital markets and provide investments for modernising and updating technology. This is expected to enable them to cope with the growth in air traffic demand more safely, more efficiently and effectively. After having been established at the national level, these bodies have established an international organisation called the Civil Air Navigation Services Organisation (CANSO) operating as their trade organisation.

The performances of new organisations providing the air traffic services could be analysed through the following attributes of performances (Majumdar and Ochieng, 2004; GAO, 2005):

- Funding;
- New technologies and project management;
- Price regime;
- Safety;
- International opportunities; and
- Customer responsiveness.

The above-mentioned attributes of performances have been analysed using the experience of five 'commercialised' cases from New Zealand (the Airways Corporation of New Zealand), Canada (NAV CANADA), UK (the Airline group-National Air Traffic Services Ltd.), Australia (Airservices Australia), and Germany (Deutsche Flughsicherung GmbH). The summary is as follows:

Funding The 'commercialised' operators have been given free access to the capital markets. This in turn has provided sufficient funding for the major capital projects aiming to maintain safe, efficient and effective services under conditions of growing air traffic demand.

New technology and project management These operators have been able to deliver new technologies on time and within a given budget. The project management has significantly improved, primarily thanks to the opportunity of appointing managers at different levels of organisation who have been capable of influencing the strategic decisions.

Safety Seemingly, the 'corporatised' operators have not compromised air traffic safety. They have been looking after safety through setting up their safety plans in accordance with the national safety regulations.

Pricing regime In general, these operators have negotiated charges for their services with users – commercial airlines. This has generally resulted in lowering charges, which has made users very satisfied. In addition, the preferences and interests of other stakeholders have been mediated through different committees established for just such purposes.

International opportunities The 'corporatised' operators have developed an increasingly important international dimension based on the development of technology and management of air traffic, often in close cooperation with the manufacturers of *ATC/ATM* facilities and equipment at home and abroad.

Customer responsiveness The 'corporatised' operators have established good relationships with their main customers – commercial airlines. On the one hand, this has reduced charges for the customers. On the other, improvements of technologies, procedures, and air routes have enabled more efficient and effective air traffic, which has contributed to the reduction of other costs components of the customer airlines.

Based on the above evidence, one might conclude that 'commercialisation' of the ATC/ATM has generally created benefits to almost all the actors involved in the process. Regarding its contributions to the sustainability of the air transport system, there have also been some benefits, mainly through improvements of technologies enabling safer, and more efficient and effective management and control of air traffic at the strategic, tactical, and operational level under conditions of growing air traffic demand (the number of flights). This has in turn contributed to reducing the fuel consumption and related emissions of air pollutants, and protected the required level of safety. The latest has appeared to be very important since the safety issue has been a matter of main concern before undertaking reorganisation in many cases.

References

AACC/IATA (1981), *Guidelines for Airport Capacity/Demand Management* (Geneva: AACC/IATA).

ACI Europe (1998), *Creating Employment and Prosperity in Europe, A Study of the Social and Economic Impacts of Airports* (Brussels: Airport Council International Europe).

ACI Europe (2002), *Airport Charges in Europe, Airport Council International Europe – Report* (Brussels: Airport Council International Europe).

AEA (2001–2005), *Yearbook 2000/4* (Brussels: Association of European Airlines).

Andreatta, G. and Romanin-Jacur, G. (1987), 'Aircraft Flow Management under Congestion', *Transportation Science*, **21**, 249–253.

Archer, L.J. (1993), *Aircraft Emissions and Environment*, EV 17 (Oxford: Oxford Institute for Energy Studies).

Ashford, N. (1988), 'Level of Service Design Concept for Airport Passenger Terminals: A European View', *Transportation Research Record*, **1199**, 19–32.

Ashford, N. and Wright, P. (1992), *Airport Engineering* (New York: John Wiley and Sons).

Ashford, N. et al. (1984), *Airport Operations* (New York: John Wiley and Sons).

Ausubel, J.H. and Marchetti, C. (1996), 'Elektron: Electrical Systems in Retrospect and Prospect', *Daedalus*, **125** (summer), 139–170.

BA (2004), *Social and Environmental Report 2003/2004* (London: British Airways).

Bazargan, M. (2004), *Airline Operations and Scheduling* (Aldershot: Ashgate Publishing Limited).

Boeing (1998), *Evolution of the World Fleet: Time Line* (Seattle: Boeing Aircraft Company). http://www.boeing.com.

BTS (2001), *Airline Service Quality Performance Data* (Washington, DC: US Department of Transportation, Bureau of Transportation Statistics, Office of Airline Information).

BTS (2003), *Airline Statistics* (Washington, DC: US Department of Transportation, Bureau of Transportation Statistics, Office of Airline Information).

Button, J.K. and Stough, R. (1998), *The Benefits of Being a Hub Airport City: Convenient Travel and High-Tec Job Growth*, Report, Aviation Policy Program (Fairfax VA: George Mason University).

Caves, R.E. and Gosling, G.D. (1999), *Strategic Airport Planning* (Amsterdam: Pergamon, Elsevier Science).

CEC (2001), *European Transport Policy for 2010: Time to Decide*, COM [(2001) 370] (Luxembourg: Commission of the European Communities, Office for Official Publications of the European Communities).

CEU (2005), *The Council Approves the Creation of a Common Blacklist of Unsafe Air Carriers*, Doc. 15360/05 (Press 343) (Brussels: Council of European Union).

Clarke, J.-P. (2000), 'A System Analysis of Noise Abatement Procedure Enabled by Advance Flight Technology', *AIAA Journal of Aircraft*, **37**, 266–273.

De Neufville, R. (1999), 'Airport Privatization: Issues for the United States', *Transportation Research Record*, **1662**, 24–31.

DETR (1999), *Oxford Economic Forecasting: The Contribution of the Aviation Industry to the UK Economy* (London: Department of the Environment, Transport and Regions).

DETR (2000), *The Future of Aviation, The Government's Consulting Document on Air Transport Policy* (London: Department of the Environment, Transport and Regions).

DETR (2002), *The Future Development of Air Transport in the United Kingdom: South-East*, A National Consultation Document (London: Department for Transport).

Doganis, R. (1992), *The Airport Business* (London and New York: Routledge).

Doganis, R. (2002), *Flying of Course -The Economics of International Airlines* (London, UK: Routledge).

EC (2002), *Study on Competition between Airport and the Application of State Aid* (Brussels: European Commission, Directorate-General Energy and Transport).

EC (2003), *Analysis of the European Air Transport Industry – 2002* (Brussels: European Commission-DG TREN).

EC (2003a), 'EU-US Enter "Open Sky" Negotiations', *EU News*. EurActiv.com.

EC (2004), 'Regulation (EC) No 549/2004, No 550/2004 and No 552/2004 of The European Parliament and of the Council', *Official Journal of the European Union*, **47**, 1–30.

EEC (1999), *Cost of the En-Route Navigation Services in Europe*, Eec Note No. 8/99, Project Gen 4-E2 (Bretigny-sur-Orge: EUROCONTROL).

EEC (2002), *The EUR RVSM Implementation Project: Environmental Benefit Analysis*, EEC/ENV/2002/008 (Brussels: EUROCONTROL Experimental Centre).

EEC (2002a), *Environmental Key Performance Indicators*, EEC/ENV/2002/002 (Bretigny-sur-Orge: EUROCONTROL Experimental Centre).

EEC (2003), *Performance Review Report: PRR 6* (Brussels, Belgium: Performance Review Commission, EUROCONTROL).

EEC (2004), *Defining Sustainability in the Aviation Sector* (Bretigny-sur-Orge: EUROCONTROL Experimental Centre).

EEC (2005), 'Aviation and Environment', *EUROCONTROL Summit, (17 March), Geneva, Switzerland*, 12.

EEC (2005a), *Report on the Operation of the En Route Charges System in 2004* (Brussels: EUROCONTROL).

EEC (2005b), 'Maastricht UAC – Europe's Only International ACC', *Regional ATC* (Brussels: EUROCONTROL), 64–75.

EEC/ECAC (2002), *ATFM Delays to Air Transport in Europe: Annual Report 2001* (Brussels: EUROCONTROL).

EECH (2001), *Airport Cost Allocation, Report for the CAA by Europe Economics* (London: Europe Economics Chancery House).

FAA (1998), *The Impact of National Airspace Systems (NAS) Modernisation on Aircraft Emissions*, FAA Report No. DOT/FAA/SD-400, 98/1 (Washington DC: Federal Aviation Administration).

FAA (2002), *Airport Capacity Benchmarking Report 2001* (Washington, DC: Federal Aviation Administration).

FAA (2005), *FAA Aerospace Forecasts: Fiscal Years 2006-2017* (Washington, DC: US Department of Transportation, Federal Aviation Administration, Office of Policy and Plans).

GAO (2004), *Commercial Aviation: Legacy Airlines Must Further Reduce Costs to Restore Profitability*, Report No. GAO 04-836 (Washington, DC: US Government Accountancy Office).

GAO (2005), *Air Traffic Control: Characteristics and Performance of Selected International Air navigation Service Providers and Lessons Learned from Their Commercialisation*, Report No. GAO 05-769 (Washington, DC: US Government Accountancy Office).

Graves, D. (1998), *UK Air Traffic Control: A Layman's Guide* (London: Airlife Publishing Ltd.).

Heymann, E. (2004), '*Perspective of Low Cost Carriers in Europe*', AER Hearing *'The Possible Impact of EU Guidelines on the Cooperation of Regional Airports and Low-cost Airlines. Is Regional Development at Stake?'* (Barcelona: Deutsche Bank Research).

Hoekstra, M. (2001), 'Designing for Safety: the Free Flight Air Traffic Management Concept'. PhD thesis, Delft University of Technology, Delft, the Netherlands.

Holder, S. (1998), *Privatization and Competition: The Evidence from Utility and Infrastructure Privatization in the UK*, 20th Plenary Session of OECD Advisory Group on Privatization (AGP), 17–18 September (Helsinki: Organisation for Economic Co-operation and Development), 27.

Holloway, S. (2003), *Straight and Level: Practical Airline Economics* (Aldershot: Ashgate Publishing Limited).

Horonjeff, R. and McKelvey, X.F., (1994), *Planning and Design of Airports,* 3rd edn. (New York, USA: McGraw Hill Book Company).

Huenecke, K. (1997), *Jet Engines: Fundamentals of Theory, Design and Operations* (Shrewsbury: Airlife Publishing Ltd.).

Humphreys, I., Francis, G. and Fry, J. (2001), 'Lessons from Airport Privatization: Commercialization, and Regulation in the United Kingdom', *Transportation Research Record*, **1744**, 9–16.

IATA (1989), *Airport Terminal Reference Manual*, 7th edn (Montreal: International Air Transport Association).

ICAO (1996), *Rules of the Air and Air Traffic Services: Procedures for Air Navigation Services, DOC 4444-RAC/501, Thirteen Edition* (Montreal: International Civil Aviation Organization).

ICAO (2004), 'Aerodromes, Annex 14', *International Standards and Recommendation Practices, 4th Edition* (Montreal: International Civil Aviation Organization).

INFRAS (2000), *Sustainable Aviation, Pre-Study, MM-PS* (Zurich: INFRAS).

IPCC (1999), *Aviation and the Global Atmosphere, Intergovernmental Panel on Climate Change* (Cambridge: Cambridge University Press).

Janić, M. (1997), 'Liberalisation of the European Aviation: Analysis and Modelling of the Airline Behaviour', *Journal of Air Transport Management*, 3(4), 167–180.

Janić, M. (1997a), 'A Model of Air Traffic Control Sector Capacity Based on Air Traffic Controller Workload', *Transportation Planning and Technology*, **20**, 311–335.

Janić, M. (1999), 'Aviation and Externalities: The Accomplishments and Problems', *Transportation Research-D*, **4**, 159–180.

Janić, M. (2001), *Analysis and Modeling of Air Transport System: Capacity, Quality of Services and Economics* (London: Taylor & Francis).

Janić, M. (2003), 'An Assessment of the Sustainability of Air Transport System: Quantification of Indicators', *2003 ATRS (Air Transport Research Society) Conference*, 10–12 July, Toulouse, France.

Janić, M. (2003a), 'Modelling Operational, Economic and Environmental Performance of An Air Transport Network', *Transportation Research D*, **8**, 415–432.

Janić, M. (2004), 'Expansion of Airport Capacity: Case of London Heathrow Airport', *Transportation Research Record*, **1888**, 7–14.

Janić, M. (2005), 'Modelling Consequences of Large Scale Disruptions of an Airline Network', *Journal of Transportation Engineering*, **131**, 249–260.

Jeng, C.Y. (1987), 'Routing Strategies for an Idealized Airline Network'. PhD dissertation, Institute of Transport Studies, University of California, Berkeley, USA.

Jenkinson, L.R., Simpkin, P. and Rhodes, D. (1999), *Civil Jet Aircraft Design* (London: Arnold).

Little, D.A. (2000), *Study into the Potential Impact of Changes in Technology on the Development of Air Transport in the UK, Final Report for Department of the Environment Transport and Regions (DETR)* (Cambridge: DETR).

Lufthansa Group (2006), *Annual Report 2005* (Frankfurt: Lufthansa).

Majumdar, A. and Ochieng, W. (2004), 'From "Our Air is Not for Sale" to "Airtrack": The Part Privatization of the UK's Airspace', *Transport Reviews*, **24**, 135–176.

Majumdar, A. and Polack, J. (2001), 'Estimating Capacity of Europe's Airspace Using a Simulation Model of Air Traffic Controller Workload', *Transportation Research Record*, **1744**, 30–43.

MERCER (2002), *Impact of Low Cost Airlines* (Munich: Mercer Management Consulting).

Oum, T.H. and Park, J.H. (1997), 'Airline Alliances: Current Status, Policy Issues, and Future Directions', *Journal of Air Transport Management*, **3**, 122–144.

RA/MVA (2000), *Landside Accessibility and Ground Transport, Common Options For Airport Regions – COFAR*, Report, Resource Analysis (Delft and London: MVA Limited).

Southwest Airlines (2006), *Annual Report 2005* (Dallas: Southwest Airlines).

Spangenberg, J.H. (2002), 'Institutional Sustainability Indicators: An Analysis in Agenda 21 and a Draft Set of Indicators for Monitoring Their Effectivity', *Sustainable Development*, **10**, 103–115.

Swan, W.M. and Adler, N. (2004), *Aircraft Trip Cost Parameters: A Function of Stage Length and Seat Capacity*, Air Transport Research Society World Conference (1–2 July) (Istanbul: Air Transport Research Society World Conference), 28.

TBG (2002), *The Economic Impact of an EU-US Open Aviation Area, Report for European Commission (Directorate General-Energy and Transport)* (London: The Brattle Group).

Terrab, M. and Odoni, A.R. (1993), 'Strategic Flow Management in Air Traffic Control', *Operations Research*, **41**, 138–152.

TRB (2002), *Aviation Demand Forecasting: A Survey of Methodologies*, Transportation Research E-Circular No.E-C040 (Washington, DC: Transportation Research Board, National Research Council).

Chapter 3

The Sustainability of the Air Transport System

3.1 Background

Many environmental problems originate from energy conversion and energy end use. This implies that energy sources become polluting when they are converted into the mechanical energy and put into the end use, one of which is the air transport system. The simplified scheme of the process of converting energy within the air transport system is shown in Figure 3.1.

According to the scheme the main driving force for energy conversion is air transport demand, mainly driven by the system's external socio-economic driving forces. The energy obtained from the jet fuel as a derivative of crude oil (fossil fuel) together with the capital (aircraft and air transport infrastructure) and labour (employed staff) represents the input for the economic process of serving demand, that is carrying out flights between given origin and destination airports. These flights provide services for the end users – passengers and freight shippers, that is demand. Pollution occurs at all stages of this process, influencing the greenhouse effect and consequently climate change at a global level, and creating local air pollution, noise, congestion, and waste around airports. The additional local impact is the land take (use) for building and/or expanding the air transport infrastructure – airports and other facilities (buildings). This impact appears particularly severe when the local population needs to be resettled and the landscape (flora and fauna) destroyed.

Global effects are caused mainly by the global nature of the air transport services (particularly due to the increase in the number of long-distance flights) and dynamism of transporting and depositing particular air pollutants in the atmosphere. Very often they are deposited far away from the polluting sources.

This chapter deals with the physics, quantification, evaluation and trading-off between the effects-benefits and impacts-costs of the air transport system on the environment and society. For such purposes, we need to understand and be able to estimate i) the type and quantity of particular emissions/burdens, ii) their spatial and time concentration, iii) prospective damages they might cause to people's health and the environment (landscape, flora and fauna), and iv) the cost of preventing, mitigating and/or repairing these damages.

Identification and quantification of particular emissions/burdens is possible using the information about the air transport system technologies and operations. Estimating the spatial and time concentration of particular emissions/burdens is carried out using the volume, time and spatial concentration of the system operations and activities in combination with the related fuel consumption and air pollution rates. The damages

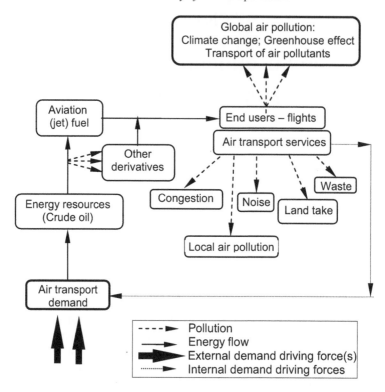

Figure 3.1   Pollution from energy conversion in the air transport system

caused by the emissions are estimated mostly by applying different models usually calibrated using the data from local measurements. The costs of damages have been estimated using different methods from economic theory, which have put the values (costs) on damages according to different criteria. The effects on GDP caused by reduced productivity or illness due to exposure to the excessive noise and air pollution and the concept of the 'willingness to pay' for avoiding escalation of impacts have often been applied. Consequently, the concept of the external costs or externalities has been developed. At present the market does not recognise the external costs of the above-mentioned damages, that is they are not internalised. However, sooner or later internalisation of these costs will have to be made for at least two reasons: i) continuous increasing of the emissions/burdens in absolute terms and the actual and prospective damages inflicted to the environment and society; and ii) prevention of the market distortion. That implies that those who actually cause damages and create the needs for protecting others should bear the costs of their activities. These costs could be expressed as charges or taxes based on the estimated costs of damages, protective, and/or mitigating measures. Despite being theoretically sound, the concept of externalities has been very complex to implement due to many barriers within the air transport system. Some of the strong ones have seemingly been the complexity of understanding the physics of impacts, and rather vague estimates of particular damages and their costs in the short- and long-term. An additional complexity has been the selection of measures for quantification of particular emissions, impacts, and related costs either in absolute or relative terms. Some policies have suggested charging particular impacts by charges (fuel and emissions taxes) based on the marginal costs of damages and/or protective and mitigation measures. These charges would inevitably contribute to an increase in the total costs of the air transport operators and consequently airfares. As a result, recognising the importance and evident positive influence of the air transport system on the overall socio-economic development on the one hand, and the obvious damages it causes to the environment and society on the other, the concept of trading-off between the costs of impacts and positive effects-benefits has been introduced as the policy of sustainability. In general, implementation of the sustainability policy requires monitoring the system's development by using convenient tools, which objectively reflect the system's performances relevant for particular actors involved such as: users-air passengers and freight shippers, airlines, airports, ATC/ATM, aerospace manufacturers, investors (banks), insurance companies, local communities around airports, and local and central governments (policy makers).

In this chapter, principles for quantifying particular emissions/burdens, their concentration and the cost of damages derived from particular methods are described. The methods themselves are not analysed since they are, due to their size, out of the scope of this book.

3.2 Physics of Impacts

3.2.1 Air Pollution

3.2.1.1 Aircraft fuel consumption

The burning of aviation fuel as the derivative of crude oil, in addition to other man-made emissions, contributes to the increase in concentration of the so-called greenhouse gases in the atmosphere and consequently to the climate change called 'global warming'. In addition, it depletes the reserves of fossil fuels (crude oil) as non-renewable energy resources. Therefore, many reasons have driven the contemporary air transport system worldwide, and particularly the aircraft manufacturers, to reduce fuel consumption. At the beginning, the main reason was the economic efficiency of flights, which in turn has enabled lower prices and consequently cheaper air travel. The reduction of travel prices has been an important driving force in the significant and constant growth of air transport demand in terms of the volumes of passengers, passenger and freight kilometres, and aircraft departures. More recently, with the increasing awareness of the depletion of the reserves of crude oil and the consequently potentially limited availability of jet fuel as its derivative at given prices as well as on the harmful impacts of the products of burning this fuel on people's health, natural habitats, and the earth's atmosphere, aerospace manufacturers have made a lot efforts to improve jet engine fuel efficiency, and consequently reduce the emission of air pollutants. The design of such engines has consequently simultaneously embraced solving a range of complex problems. The most complex have included balancing the engines' propulsion and thermal efficiency. Better propulsive efficiency has

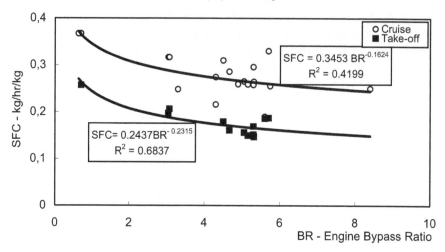

Figure 3.2   Specific Fuel Consumption vs jet- engine bypass ratio
Compiled from Jenkinson et al. (1999)

provided a greater propulsive power from the combustion process while the improved thermal efficiency has generated a higher overall engine pressure ratio and turbine temperature using the same amount of fuel (energy). Other problems have related to proper balancing between the engine weight, drag, noise, and emissions.

In order to obtain a higher propulsive efficiency it has been necessary to reduce the waste energy in the engine exhaust stream, which has decreased the jet velocity. Since the engine thrust is the product of exhaustive mass flow and its velocity, if this velocity was reduced the mass of flow would be increased to retain the desired level of thrust. This implies an increase in the bypass ratio defined as the rate between the amount of air flowing round the engine core and the amount of air passing through the engine itself. The engines with the higher bypass ratio usually have lower Specific Fuel Consumption (SFC), defined as the ratio of the fuel burned per hour per tonne of the net thrust (Janić, 1999). The SFC of the most contemporary jet aircraft engines amounts to about 0.25–0.30 kg of fuel/kg of thrust/hour. This is likely to be the case until the year 2015 and that it will then further diminish to about or less than 0.184 kg of fuel per hour per kg of thrust. The SFC relates to the jet engine bypass ratio (BR). The nature of this relationship is illustrated using data for 20 engine types produced by the different airspace manufacturers such as CFM Company (joint corporation of Snecma (France) and General Electric Company (USA)), Rolls-Royce (UK), Pratt & Whitney and General Electric (USA) and IAE (International Aero Engines AG made up of the engine manufacturers Pratt & Whitney, Rolls-Royce, MTU (Europe) and Aero Engine Corporation (Japan)) The regression relationships in which the bypass ratio (BR) is considered as the independent and the SFC as the dependent variable is given in Figure 3.2.

As can be seen, SFC has, independently of the phase of the flight (take-off or cruising), decreased nearly more than proportionally with the increasing engine bypass ratio, which might be useful information for estimating the trend of development of jet engines for commercial aircraft (CE Delft, 2002).

Improvements to aircraft aerodynamic performances have also played an important role in the improvement of aircraft fuel efficiency. The case of the development of the most recent Boeing B777-300ER (ER – Extended Range) appears illustrative. On this aircraft, the more fuel-efficient 'raked wing tip' design is replacing the winglets option used previously on other B777 versions as well as on the B747-400 and B737 NG (Next Generation) aircraft. The new winglets are expected to improve the short-field climb performances and fuel efficiency by about 1–2 per cent for longer flights. If the aircraft is supposed to be utilised for about 5,000 hours per year, the average fuel consumption is about 7.5 tons per hour, and the fleet size is 20 aircraft, each with an average depreciation period of about 20 years, the total fuel consumption will be about 15 million tonnes of fuel, and 47.7 million tonnes of CO_2, and about 17.7 million tonnes of water vapour (H_2O) would be produced. Fuel savings of about 1 per cent is therefore a quantity of 150 thousands tonnes, and consequently 477 thousand less tonnes of CO_2 and about 177 thousand less tonnes of H_2O (water vapour) would be produced. On the other hand, if the average price of fuel was about USD500 per ton, the monetary saving would amount to about USD25mn (CE Delft, 2002; West, 2005).

Parallel to improving the engine/aircraft fuel efficiency, two groups of air traffic and transport operational measures have been implemented in the airline industry. The first group aims at cutting the overall fuel consumption by improvements to the air traffic control system, that is by carrying out more direct flights and consequently reducing the total travel distance per passenger, optimising the aircraft climb/descent profiles in terms of the fuel consumption and emissions of particular air pollutants such as for example NO_x, reducing cruising speed(s), harmonising fuel prices, improving the load factor, and reducing the long taxiing and towering of aircraft at airports. Another group of measures includes optimising the cruising altitudes and removing restrictions on the flight routings, reducing the number of flights during daylight, widening (or narrowing) the flight corridors and repositioning the flight corridors in order to avoid extreme weather conditions leading to excessive fuel consumption. Some of these measures have been described in Chapter 2. Figure 3.3 illustrates an example of the combined efforts of the US airline industry.

As can be seen, during the period from 1960 to 2005, the fuel consumption in kg per tonne-kilometre decreased more than two times with increasing output expressed by the annual volume of tonne-kilometres – from about 0.92 to about 0.43 kg/t-km. Some forecasts for the next decade (2006–2017) expect the average fuel consumption to remain at the present level of about 0.42–043 kg/t-km despite further growth in the volume of output (FAA, 2005).

In addition, the general trend in terms of aircraft fuel consumption and energy efficiency during the period 1990–2015 has been illustrated using regression analysis where the average fuel consumption per available tonne-kilometre expressed in grams of fuel consumed per tonne kilometre is considered as the dependent variable FC whose values have been estimated by assuming an annual improvement in the fuel efficiency of about 2.5 per cent during the observed period. Two independent variables have been time T (in years for the period 1990–2015) and the annual volume of output TK in million tonne-kilometres. This variable is estimated by taking into account the prospective renewal of the aircraft fleet and relatively stable growth of the passenger and freight transport demand at an average annual rate of 5 per cent

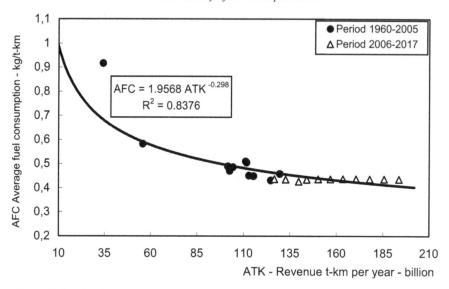

Figure 3.3   **Fuel efficiency vs the volume of output in the US airline industry (1960–2017)**

Compiled from BTS (2001); FAA (2005)

and 6.5 per cent, respectively. One tonne-kilometre is adopted to be equivalent to 10 passenger-kilometres. The average load factor is assumed to be 0.55. The resultant equation is as follows (Janić, 1999):

$$FC = 30690.22 \ TK^{-\ 0.355} \ T^{-\ 0.017}$$

$$\text{(143.67)} \qquad \text{(-7.234)} \quad \text{(-62.390)} \tag{3.1}$$

$$R^2 = 0.993; \ F = 10023.563; \ DW = 1.515; \ N = 25$$

The values of t-statistics in parenthesis below the coefficients and F-statistic indicate that the independent variables and the equation are significant at the level of 1 per cent and 5 per cent, respectively. R_2 statistics confirm the strong explanatory power of the selected independent variables. The DW statistic indicates the lack of auto-correlation of the first order between variables. Based on this relationship, it is noticeable that the average fuel consumption has decreased by about 18 per cent during the period 1990–2000, and will continue to decrease further by about 65 per cent over the period 1990–2015 (Janić, 1999). This estimate coincides with estimates of the IPCC (Intergovernmental Panel of Climate Change) of 70 per cent reduction in the fuel consumption per unit of output during the period 1960–2000. Nevertheless, there have been some doubts about further annual reductions of fuel consumption of about 2.5–3 per cent until the year 2020 (Peeters, Middel and Hoolhorst, 2005).

3.2.1.2 Air pollutants
The air pollutants are substances that in much higher concentrations than in the natural environment (that is the hypothetical environment without human influence) cause

damage to plants, animals and people. They may exist in the air, water and soil and are permanently moving from one place to another. In the case of the air transport system, such movement has both local and global character. In addition, some pollutants, for example those from burning aviation fuel, can interact with each other or with the natural substances causing so-called secondary pollutants. In general, the major pollutants from the jet-A aviation fuel used by the commercial aircraft worldwide are the 'greenhouse gases' carbon dioxide (CO_2) and water vapour (H_2O), nitric oxide (NO) and nitrogen dioxide (NO_2), which together are called NO_x, sulphur oxides SO_x and smoke. The emission rates of pollutants such as carbon dioxide (CO_2), water vapour (H_2O) and sulphur dioxide SO_2 are relatively constant – 3.18 kg/kg of fuel, 1.23 kg/kg of fuel, and up to 0.84 g/kg of fuel, respectively (EEC, 2001). According to the currently available evidence, the emission rate of nitrogen oxides (NO_x) changes, that is increases with increasing jet engine pressure ratio, which in turn increases the jet engine thermal efficiency. The engine pressure ratio is defined as the ratio of the total pressure at the compressor discharge and at the compressor entry. The other jet engine performances such as thrust, fuel consumption and efficiency also depend on this ratio. For contemporary turbofan engines, this ratio amounts to about 10–50, which originates from the typical design of the combustion chambers of these engines (Hunecke, 1997). Experiments to investigate the relationship between the engine emission index of NO_x, compressor outlet temperature, and pressure ratio have resulted in a regression equation as follows (RAS, 2001):

$$EI_{NO_x} = 0.17282\, e^{0.00676593T_s} \tag{3.2}$$

where

EI_{NOx} is the engine emission index of NO_x expressed in g/kg of fuel; and

T_s is the compressor outlet temperature (°K) ranging between 280 and 1,080 °K.

Equation 3.2 indicates that the jet engine emission index of NO_x increases as the compressor outlet temperature increases, but at a decreasing rate. Assuming that a flight takes place in tropopause at the speed of M0.85 (M – Mach Number), the compressor inlet temperature will be about 250 °K. Given the compressor efficiency of 0.9, this can be written:

$$T_s = 250r^{(2/(7*0.9))} = 250r^{(1/3.15)} \tag{3.3}$$

where

r is the engine overall pressure ratio ranging from 10 to 50.

Combining Equations 3.2 and 3.3, the emission index of NO_x becomes:

$$EI_{NO_x} = 0.17282\, e^{(1.69158*r)(1/3.15)} \tag{3.4}$$

Equation 3.4 confirms that the emission index of NO_x increases with increasing engine pressure ratio at an increasing rate, that is more than proportionally.

Therefore, a trend towards increasing the engine pressure ratio of larger aircraft might compromise and even diminish other effects obtained by reduction of the fuel consumptions and related emissions of CO_2 and H_2O.

The chemistry and direction of impacts of particular pollutants from burning aviation fuel have been relatively well explained (ICAO, 1993a; Archer, 1993; IPCC, 1999). Therefore, only the most important pollutants and their perceived impacts, which might be relevant for the assessment of sustainability of the air transport system are mentioned as follows: nitrogen oxides (NO_x), sulphur oxides (SO_x), carbon oxides (CO and CO_2), water vapour (H_2O), and Non-Methane Hydrocarbons – (NM)HCs.

Nitrogen oxides (NO_x)   The symbol NO_x implies the nitrogen oxides NO and NO_2 ($NO_x = NO + NO_2$). They are produced by any combustion in which air in the form of N_2 and O_2 is brought to a high temperature by burning fuel. NO_x is formed in the flame at a temperature of a few thousands °K and generally its formation increases as the burning temperature increases. Aviation fuel burns at such temperatures within contemporary jet engines. The second source of NO_x comes from its presence in aviation fuel of about 1 per cent. The remedy for NO_x is generally twofold: i) reducing the fuel-burning temperature, which generally reduces the jet engine fuel-burning efficiency; and ii) lowering the available oxygen for combustion. Commercial jet engines generate NO_x according to Equation 3.3. The amount is the greatest during take-off and climbing when the engine burning temperature is the highest, a little bit less during cruising, and lowest during the approach and landing phases of a flight.

Sulphur oxides (SO_x)   Crude oil and its derivative – jet aviation fuel – may have considerable amounts of sulphur. In a chemical reaction with the water vapour in the atmosphere it creates acid rain, damaging trees other dependent natural habitats. In order to diminish its presence in jet engine fuel and exhaustive gases after its burning, catalysts are added.

Carbon oxides – CO and CO_2   Carbon monoxide CO is always produced during the burning of fossil fuels of which aircraft jet fuel is one derivative. It reacts with the oxygen O_2 in the atmosphere and forms carbon dioxide CO_2, which appears to be one of the most important greenhouse gases as described before. The rate of emission of CO_2 from jet fuel is almost constant – 3.18 $kgCO_2$/kg of fuel, which makes it easier to estimate the emitted quantities at both a local and global scale, based on the quantities of fuel consumed. Emissions of CO_2 have a long lifetime in the atmosphere (about 100 years). There is no remedy for reducing the quantity of emissions of CO_2 by improving the fuel burning process in the jet engine simply because of the chemistry of fuel. The only option remains reducing the amount of fuel burned and development of new technologies using some other type of fuel. The former option has already been taking place (see Figure 3.3 and Equation 3.1). Developments have contributed to the reduction in the emissions of CO (and consequently CO_2) per unit of the system output (that is the amount of revenue tonne-miles per year) over a given period of time. Figure 3.4 shows an example for the US airline industry.

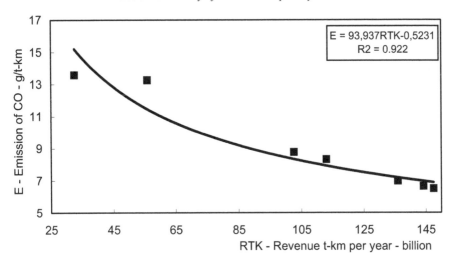

Figure 3.4   Emission of CO vs the volume of output in the US airline industry (1970–1999)

Compiled from BTS (2001)

As can be seen, the emissions of CO have decreased more than proportionally with the increasing amount of the system output over a given period of time, which is an obvious illustration of the achieved progress in improving the jet-engine emission efficiency.

The latter option – changing the fuel type – is a technological breakthrough for the future, probably after developing new aircraft and engines using liquid hydrogen as the main energy component. This fuel would certainly eliminate emissions of CO_2 but would also increase emissions of the water vapour (H_2O). Therefore, its net contribution to the sustainability of the air transport system still needs to be investigated.

Water vapour (H_2O) – Contrails   Water vapour (H_2O) emitted after burning the jet fuel influence climate change through the formation of contrails in the troposphere (10–12 km) where most commercial aircraft perform the cruising phase of their flights. The contrails are the icy clouds formed behind the aircraft flying at the high altitudes. They are often visible from the earth's surface when there are clear skies. They form as follows: behind an aircraft, the warm exhaust gas containing particles of soot, ash, and other pollutants expands and mixes with the colder and dryer air. If the amount of water in the air is at saturation level, water droplets are formed. Due to the low ambient temperature (at altitudes of about 10–12 km, the temperature is about −40 °C or lower), these droplets rapidly freeze and form ice crystals, which under conditions of sufficient water vapour build up very quickly into the persistent and visible contrails (clouds). The layer of the atmosphere where this process occurs is called the contrails producing layer. Some estimates have shown that contrails cover about 0.1 per cent of the earth's surface and contribute to the earth's overall coverage by high clouds (Williams et al., 2002).

In general, clouds near the earth's surface affect the atmosphere by reducing the amount of solar radiation returning to space and by increasing the amount of solar radiation reflected from the atmosphere. Consequently, the surface becomes warmer in order to keep the radiative forces in balance. According to some estimates contrails contribute to radiative forcing by 0.007–0.06 W/m^2 with the expectation that they will increase to about 0.04–0.4 W/m^2 with the projected air traffic growth until the year 2050 (The contribution of CO_2 to radiative forcing has been estimated to be about 0.02 W/m^2) (IPCC, 1999).

However, recently the contrails have been identified as one of the causes for diminishing the intensity of sunlight reaching the Earth's surface, thus causing the cooling of temperatures during the daytime and the warming of temperatures during the night. This phenomenon is called 'global dimming' and can be explained as follows: As polluted clouds, contrails consist of smaller and more droplets than the atmospheric clouds due to maintaining the constant amount of water vapour in the atmosphere. This structure changes the optical properties of contrails by making them more reflective (higher albedo) to the sun's rays. At the same time they also absorb the infrared radiation, which actually warms up the earth surface. The net effect is still unknown. What is however certain is that contrails certainly contribute to global warming directly or indirectly, that is with their presence the earth's surface can warm up due to the absorption of the infrared radiation; with their absence the sun's radiation reaches the earth's surface more intensively, also causing an increase in the temperature.

Unlike CO_2, contrails have a particular impact in the regions with intensive air traffic such as Europe and North America. After being recognised, two proposals have come up to eventually diminish this impact as follows: i) scheduling flights during the sunrise and sunset periods when the impacts of contrails in terms of blocking the earth-emitted radiation is smaller; and ii) restricting the flight cruising altitudes to those where the contrails have a smaller impact. The latter option seems ambiguous for several reasons: first, restricting the aircraft fuel-optimal altitudes might increase fuel consumption and consequently the emissions of CO_2, which in turn contributes to an increase of the total quantities of greenhouse gases in the atmosphere; second, restriction of the usable altitudes certainly diminishes the airspace capacity and consequently increases delays, which again might result in an increased fuel burn and emissions of air pollutants; finally, the workload of air traffic controllers might be increased, which would require more staff to maintain the required level of air traffic safety, efficiency and effectiveness.

Non-methane hydrocarbons – (NM)HCs   The hydrocarbons (HCs) contribute to the formation of smog and global warming. However, the amounts emitted by burning jet fuel have not been recognised as particularly worrying as compared with other types of air pollutants. Nevertheless, the contribution of HCs to global warming appears to be relatively important through i) the production of ozone O_3, ii) extending the lifetime of methane CH_4, and iii) their conversion into carbon dioxide CO_2 and water vapour H_2O, as the most important greenhouse gases (Archer, 1993).

3.2.1.3 Contribution to climate change

Physics   Generally, life on earth is related to some physical properties of the sun-earth system. The surface temperature of the sun is about 5,800 °K (°K − Kelvin degree), which results in an emission spectrum with a maximum wavelength of 500 nm* (nm* − nano meter; 1 nm* = 10^{-9} meter). This makes the solar temperature of the exact magnitude to induce photochemical reactions. Depending on the radius of the sun and the earth, their distance and the above-mentioned surface solar temperature, one can estimate that the earth receives energy of 1,379 W/m² (watts per square meter), although the solar constant is always taken a bit lower, that is $S \approx 1{,}370$ W/m². With the support of some gases in the earth's atmosphere, this energy appears sufficient to maintain an average temperature on the earth's surface of $T = 288$ °K (+15 °C). A part of the received energy is reflected from the earth's surface back into the space. This is called albedo *a* from the Latin term 'albus' meaning 'white'. Astronomers usually use albedo to express the brightness of the earth as seen from space. Consequently, the energy equation can be set up as follows: $(1 - a)\pi R^2 S = 4\pi R^2 \sigma T^4$, where R is the earth's radius (6,400 km) and σ is the Stefan-Boltzmann constant (σ = 5.672 × 10^{-8} $Wm^{-2}K^{-4}$). With an estimated value of albedo $a = 0.34$, one can obtain a temperature of the earth's atmosphere of $T = 250$ °K (usually it is taken to be $T = 255$ °K). This is lower than the earth's surface temperature (288K), which is mainly due to the presence of gases such as carbon dioxide (CO_2), ozone (O_3), nitrogen dioxide (NO_2), methane (CH_4), and water vapour (H_2O). Otherwise, this temperature would be lower by about 30 °K. In general, these gases absorb most of the heat radiation from the earth and reemit it back towards the earth surface, which is the process called the 'greenhouse effect'. The mentioned gases are therefore called 'greenhouse gases'. Currently, the concentration of greenhouse gases is continuously increasing due to both natural and human causes. For example, the concentration of CO_2 has increased by about 25 per cent over the past 200 years; the level of CH_4 has doubled during the last hundred years, while the concentration of NO_x has been increasing by about 0.25 per cent per year. The air transport system (flights) has made its contribution.

Increasing the concentration of greenhouse gases might strongly influence the climate by increasing the average temperature on the earth surface. During the past hundred and 30 years, the average global temperature has increased by about 0.6 °K. The speed and scope of the process is still not precisely known. For example, according to some past estimates, air transport has contributed to about 2 per cent of the total man-made emissions of CO_2 in 1992 – a small but significant contribution to the total impacts of 'greenhouse gases' (IPCC (1999). CO_2 has a very long residence time in the atmosphere where it mixes well with other gases. Some simple estimates can show that, for example, an instant doubling of the concentration of CO_2 relative to the present concentration would increase the average temperature on the earth's surface by about 1.4 °K. This phenomenon can be explained as follows: increasing the concentration of CO_2 will reduce the earth's long wavelength radiation at the top of the atmosphere by a certain amount and consequently reduce the inward flux there by the same amount. The energy balance at the top of the atmosphere requires a constant flux. Therefore, the earth's surface temperature should rise in order to compensate such an imbalance. This effect is called the radiative forcing. Some estimates have shown that the air transport system might contribute to increasing

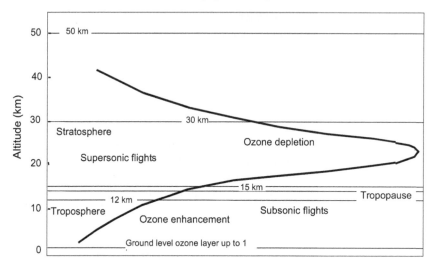

Figure 3.5   Distribution of ozone concentration and the impact of air transport system (mid-latitude)
Compiled from NASA, GARE (1996)

of the radiative forcing by about 0.02 W/m². An example of contributions of the radiative forcing to the increase of the earth's surface temperature is presented below. Such or any increase in the global temperature might cause additional effects – increasing or mitigation of the concentration of CO_2 as the reversible process. Some estimates suggest that the current concentration of CO_2 in the Earth's atmosphere is around 382 ppm and the tendency is at it will increase by an annual rate of about 1.2 ppm over the next 40 years (until the year 2050) (ppm – parts per million). Some other estimates indicate that when the total known reserves of crude oil of about 1650 billion (10^{12}) US barrels are exhausted by the end of the twenty-first century, the concentration of CO_2 will contribute to the increasing of the average global temperature by about 2.5 °K (Boeker and Grondelle, 1999).

Equally important gas in the earth's atmosphere is ozone O_3. Its presence protects the earth from the harmful solar UV radiation by absorbing all light with a wavelength less than 295 nm* (nanometer). The layer of O_3 in the earth's atmospheres is relatively thin – about 0.3–0.4 cm – under constant temperature and atmospheric pressure. The gas is present throughout the atmosphere but it is maximally concentrated in the stratosphere at the altitudes of about 20–26 km from the earth's surface. It is permanently formed through the reaction of the molecular oxygen O_2 and the atomic oxygen O influenced by the solar UV radiation. Most of the ozone is formed above the equator where the amount of UV solar radiation is maximal. From there, it moves towards the poles where it is 'accumulated' up to a thickness of about 0.4 cm during the winter period. Figure 3.5 illustrates an example of the changing concentration of the ozone layer with the distance from the earth surface.

As can be seen, the ozone concentration increases with increasing altitude up to about 20 km and decreases further above it. However, ozone is sensitive to the free radicals such as the atomic chlorine CI, nitric oxide NO, and hydroxyl radicals

OH. They are formed from the water vapour (H_2O) and chlorofluorocarbons (CFCs), products of burning aviation fuel, which escape from the troposphere (10–12 km from the earth surface) where most commercial flights take place to the stratosphere where the ozone layer is formed. At these altitudes free radicals including NO_x lead to depletion of the ozone layer. Those that do not escape remain extremely stable in the troposphere where they, together with NO_x, contribute to thickening of the ozone layer. The residence time of No_x in these regions increases with altitude. Therefore, NO_x affects the ozone layer regionally if injected into the troposphere and globally if injected into the stratosphere (IPCC, 1999). In any case, the increased concentration of the above-mentioned pollutants might generally cause depletion of the ozone layer with inevitable impacts. For example, depletion of this layer by about 10 per cent may cause an increase in the UV radiation by about 45 per cent, which certainly inflicts damages to almost all biological cells and in particular causes skin cancer in people who expose their skins.

The scale of contribution The contribution of the air transport system to climate change, that is global warming, could be roughly estimated by using a zero-dimensional greenhouse model (Boeker and Grondelle, 1999). The model is based on considering the total energy flux at the top of atmosphere. Under equilibrium conditions the radiation flux vanishes, that is inward and outward radiation are in balance. This balance can be disrupted by a reduction in the earth's long-wave radiation at the top of the atmosphere by an amount equal to that caused by the increase in greenhouse gases (for example, CO_2). Consequently, the outward radiation will decrease by the same amount. Since the energy balance at the top of atmosphere requires a constant flux, the temperature at the earth's surface will increase by amount Δ_{Ts} in order to compensate for the reduction of the earth's long-wave radiation Δ_I. This phenomenon, known as radiative forcing, puts the variables Δ_I and Δ_T into the following relationship:

$$\Delta I = \frac{\partial I}{\partial T_s} \Delta T_s \tag{3.5a}$$

The term $\partial I / \partial T_s$ can be approximated as follows:

$$\partial I / \partial T_s = 4 / T_s (1-a) S / 4 \tag{3.5b}$$

where

a   is the albedo of the earth as viewed from space ($a = 0.34$; otherwise, at the earth's surface $a = 0.11$);

T_s is the earth's surface temperature ($T_s = 288\ °K$);

S   is the solar constant ($S = 1.370 \times 10^3\ J/sm^2$ (Joules per second per square meter)).

Replacing the values of parameters by their common values gives an estimate of Equation 3.5b of $\partial I / \partial T_s = 3.1 W/m^2 K$ and its reciprocal value of $G = 0.32 m^2 K/W$. From the Equation 3.5a follows $\Delta T_s = G \Delta I$. Many models of climate change specify the values of radiative forcing at about $\Delta I = 4 - 4.6 W / m^2$ as contributions of the man-made emissions. Some reports suggest that the air transport system might participate in this total maximally by about 3.5 per cent, that is its radiative forcing would be: $\Delta I_a = 0.035 \Delta I = 0.14 - 0.16 W / m^2$. Applying this to equation $\Delta T_s = G \Delta I$ gives $\Delta T_{s/a} = G \Delta I_a = 0.3x(0.14 - 0.16) = (0.042 - 0.048)K$ (IPCC, 1999). Some other inputs on parameter Δ_I suggest that the air transport could contribute to the increase in the earth's surface temperature of $\Delta T_{s/a} = (0.052 - 0.096)K$ between the year 2010 and 2050 (Boeker and Grondelle, 1999; IPCC, 1999). Both results mean that the earth's surface temperature would not significantly increase due to the air transport system's operations under given circumstances. In other words, the air transport system seems not to significantly contribute to global warming. However, more investigations are needed to confirm or reject this hypothesis particularly due to the high sensitivity of the available models to inputs, which are also the outputs from some other very complex models of climate change.

3.2.2 Noise Pollution

3.2.2.1 Noise nuisance

Noise nuisance is a phenomenon to which some people are more sensitive than others. In addition, any noise could be considered as an unacceptable sound level. Many sensitive people try to mitigate the unpleasant and damaging impact of noise by insulating their homes against noise penetration, scattering noise away and/or by absorbing noise. Noise is usually expressed by the so-called pressure level L_p as follows:

$$L_p = 10 \log_{10} \frac{p_{rms}^2}{p_{ref}^2} = 20 \log_{10} \frac{p_{rms}}{p_{ref}} \qquad (3.6)$$

The pressure mean square value p_{rms} in Equation 3.6 may vary in practice from between 10^{-5} and 10^3 Pa (Pa – Pascal). The reference pressure is usually taken to be: $p_{ref} = 2 \times 10^{-5}$ Pa, which is the pressure just audible to the human ear at a frequency of 1,000 Hz (Hz – Hertz). The unit of sound pressure is decibel – dB. Therefore, the threshold of hearing at the frequency of 1 kHz corresponds to $p_{rms} = p_{ref} = 1$, thus giving $L_p = 0$. According to this definition, doubling p_{rms}^2 implies an increase in L_p of about 3 dB. However, these relationships have not always appeared to be particularly appreciated or clearly understandable. In addition, the human hearing system does not recognise only a large range of fluctuations in the sound pressure levels but also an enormously wide range of frequencies – on average from 20 Hz to 20 kHz. Nevertheless, for the purpose of internal standardisation the upper limit of 'effective' noise is considered to be just 1 kHz.

Another important aspect of dealing with noise is the distance from the source. In order to assess the influence of distance, the noise power is expressed in W (watts). Under such circumstances the noise power can vary between 10^{-9} and 10^4 W, a scale similar to the one for p_{rms}^2 and twice as wide as the one for p_{rms}. Taking that into account the sound power level can be expressed as:

$$L_w = 10\log_{10}\frac{W}{W_{ref}} \qquad (3.7)$$

The value of $W_{ref} = 10^{-12}$ W(watts). Consequently the sound pressure level $L_p(R)$ at distance R from a point source can be approximated as (Boeker and Grondelle, 1999):

$$L_p(R) = L_W - 10\log_{10}(4\pi R^2) + 0.14 \qquad (3.8)$$

where L_W is determined from Equation 3.7. In general, Equation 3.8 implies that the distance from the noise source is important and that the sound power decreases as the distance from the source increases. For example, by doubling the distance from R to $2R$, the sound power L_p decreases by about 6 dB.

It is known that human tolerance to noise nuisance is much greater during the day than during the night, which means that sounds ignored during the day become annoying during the night. Some investigations under laboratory conditions have indicated that noise levels below 50–60 dBA do not cause noise-induced sleep disturbance. In addition, the number of awakenings by noise is about 10 per cent of the number of people living in a given residential area. The maximum allowable noise level, dependent on the number of night-time exposures, has been empirically estimated as follows (Ruijgrok, 2000):

$$L_{A\max} = (-0.09 + 0.129N + 0.0018N^2)^{-1} + 53.16 \quad dB(A) \qquad (3.9)$$

where
    N is the number of awaking events during the night.

Since the exposure to excessive noise has been shown to be damaging to the human hearing system many standards based on the upper limit of 90 dBA in order to avoid permanent damage to the human hearing system have been introduced. In many cases, these standards have allowed a 3 or 5 dBA increase under conditions where the duration of the exposure to such maximum noise is halved.

3.2.2.2 The nature of noise pollution

The standards on noise limits, based on laboratory experiments, have been monitored by noise measurement at the affected locations. Methods for noise measurement and exposure have been developed for such purposes. These methods are mainly based on the human reaction to 'loudness'. One of the examples is the widely used A-weighted dBA scale. In the air transport system, two sources of noise are from the aircraft engines machinery and primary jet noise. The machinery noise originates from the rotating engine parts such as the fan, compressor and turbine. The noise from the fan and compressor is spread forward of the engine. The noise from the turbine is spread backwards from the engine. The mixing the high-speed gas exhausted from the engine with the surrounding air generates the primary jet noise. The fan exhaust also generates noise, particularly at high rates of thrust, for example during take-off.

At the same time, the presence of the fan exhausts muffles and reduces the primary jet noise. In summary, the major source of noise during take-off is primary jet noise; machinery noise appears the major source of noise during landing (Horonjeff and McKelvey, 1994).

Noise from aircraft engines has been shown to be of a particular nature in terms of its wide-ranging and variable spectral character, transience, and fluctuating intensity over time. Under such conditions, the A-weighted scale has been shown to be inconvenient for measurement. Instead, special scales have been developed based on levels of annoyance rather than on loudness. These scales include the specific spectral characteristics and persistence of noise. Specifically, for aircraft noise the perceived noise scale (PNdB) and effective perceived noise scale (EPNdB) are used. Nevertheless, due to the inherent complexity of the EPNdB scale the dBA scale is also applied for the noise certification of propeller-driven light aircraft, plotting the noise contours, and setting up local airport restrictions. The PNdB and EPNdB scales have been used for the noise certification of all jet-powered commercial aircraft (Smith, 2004).

Aircraft noise was regulated for the first time in 1959 by setting limits to the sound generated by aircraft operating at particular airports. The acceptable noise limit was 112 PNdB. The ICAO established the international certification standards for commercial jet aircraft in 1971. In the late 1970s new noise restrictive standards were included in Chapters 2 and 3 of the ICAO Annex 16, Vol. 1 (Environmental Protection) (Walder, 1993). These standards have been applied to all jet aircraft that have entered service since October 1977. More recently, they have been reconfirmed in Chapter 4. In addition, many descriptors of aircraft noise annoyance have been developed at the national scale. Currently, 18 such different descriptors are used throughout different countries.

One of the first such descriptors at the national scale was the Noise and Number Index (NNI) developed for London Heathrow and Gatwick airports (UK). This index tends to inter-relate annoyance to both the individual aircraft (peak) noise levels and the frequency of operations (that is the noise intrusion) in order to produce a single indicator of noise annoyance. The NNI uses PNdB as the basic noise unit but adjusts the average of PNdB levels over a given period of time by a correction factor $15log_{10}N$, where N is the number of aircraft operations, in order to account for the repeated intrusion. Some other indicators usually use the factor $10log_{10}N$. The NNI can be expressed as:

$$NNI = 10\log_{10}\left[\frac{1}{N}\sum_{j=1}^{N}10^{\frac{L_{MAX}(j)}{10}}\right]+15\log_{10}N-80 \qquad (3.10a)$$

where

$L_{max}(j)$ is the maximum perceived noise level EPNL of the event (J) – an aircraft landing or taking-off; and

N is the number of events – aircraft landings or taking-offs during a given period of time.

For many years, the equivalent noise exposure level in the US had been the noise exposure forecast (NEF) with the descriptor EPNdB. Later, it was replaced by the day-night sound level (LDN), which uses the simpler dBA unit. In comparison to EPNdB, dBA is about 10–15 dB. The LDN fully recognises the duration element of noise by using the integrated dBA of the sound exposure level (SEL). It adjusts the SEL by adding arbitrarily 10 dBA to noise generated during late evening and night hours (Horonjeff and McKelvey, 1994).

The third indicator of noise exposure level proposed is the equivalent continuous sound level L_{eq} (ECAC, 2004). It has been shown to be quite satisfactory when correlated with the sleep disturbance. This is the noise energy-averaging index using dBA units derived by summing the noise pattern over a given period of time and then normalising it to the unit time as follows:

$$L_{eq} = \overline{L_{AE}} + 10\log_{10} N - 10\log_{10} T \tag{3.10b}$$

where
$\overline{L_{AE}}$ is the average of the maximum noise levels L_{amax} (dBA) generated by N noise events – aircraft operations – during the period T;
N   is the number of aircraft operations during period T; and
T   is the period of time.

The fourth measure of noise exposure has been developed in the Netherlands for controlling the noise around Amsterdam Schiphol Airport. It uses the so-called Kosten units and takes into account the perception of noise by the affected population during different periods of the day as follows (Ruijgrok, 2000):

$$B = 20\log_{10}\left[\sum_{j=1}^{\overline{N}} w(j)10^{\frac{L_{a\max}(j)}{15}}\right] - 157 \tag{3.10c}$$

where
$L_{amax}(j)$ is the maximum weighted sound from the event (j); and
$w(j)$ is the time of day weighting factor due to the noise event – flight (j).

Finally, the EU has recently proposed a measure for the exposure of the local population around airports to the aircraft noise in terms of daily and night noise as follows (EU 2003):

$$L_{den} = 10\log_{10}\left(\frac{1}{86400}\sum_{i,j}(N_{d/ij} + 3.16N_{e/ij} + 10N_{n/ij})10^{SEL_y/10}\right) \tag{3.11a}$$

and

$$L_{night} = 10\log_{10}\left(\frac{1}{T_n}\sum_{i,j}N_{n/ij}10^{SEL_y/10}\right) \tag{3.11b}$$

where

    N_{dij} is the number of movements of the j-th aircraft group on the i-th flight path during the day period on an average day;

    N_{eij} is the number of movements of the j-th aircraft group on the i-th flight path during the evening period on an average day;

    N_{nij} is the number of movements of the j-th aircraft group on the i-th flight path during the night period on an average day;

    T_n is the duration of the night period (seconds); and

    SEL_{ij} is the sound exposure level from the j-th aircraft group on the i-th flight path.

This seems to be a more refined measure, which explicitly takes into account different aircraft groups in terms of noise performances as well as different incoming and outgoing aircraft/flight paths around an airport.

The most important characteristics of the above-mentioned noise exposure measures – indicators – are the number of operations in and out of the airport and the recorded noise level of each particular operation at the noise 'reference locations' (measurement points).

3.2.3 Congestion

People travel by air at some times more than at others. The airlines schedule flights to satisfy such needs. When the number of flights exceeds the airport capacity congestion occurs. The obvious consequence of congestion is delay to the aircraft/ flights and passengers. The time period in which congestion happens is called the peak period. This can be one or several hours during the day, a week, and/or year. During congestion different aircraft/flights interfere with each other. The measure of such interference is the demand/capacity or the capacity utilisation ratio defined as the quotient between the intensity of demand and the capacity. This ratio can be lower, equal or greater than one. Specifically, if the number of aircraft/flights at an airport is equal to the capacity, this ratio is equal to 1.0 or 100 per cent (Newell, 1982; Janić, 2005).

Substantive investigation has been carried out to determine the aircraft delays for different values of the demand/capacity ratio. In general, long delays occur when this ratio comes close to, equalises or even exceeds the value '1' over a longer period of time T, which is let's say a few hours during the day. For such conditions the delay to an aircraft/flight entering either the arrival or departure queue at given airport at time (t) ($t \in T$) can be estimated as follows:

$$w(t) = Q(t) / \mu \tag{3.12}$$

where

    Q is the length of the aircraft/flight queue the given aircraft joins at time (t); and

    μ is the airport capacity, that is the average service rate (the aircraft/flights per unit of time).

The queue $Q(t)$ in Equation 3.12 is the difference between the cumulative number of aircraft/flights demanding service and the cumulative number of aircraft/flights served by time (t). The delay $w(t)$ increases in proportion with an increase in the queue $Q(t)$ and decreases with increasing airport capacity μ (Levison et al., 1996; Janić, 2005).

Congestion causes the aircraft/flight delays and internal-private costs. However, each flight/aircraft also imposes delays and related costs on the other flights/aircraft taking part in congestion. These very often-neglected delays are called the marginal delays and are considered as an externality. These are discussed in detail in Chapter 4.

3.2.4 Air Traffic Accidents

3.2.4.1 The concept of risk and safety

In contemporary life everyone, either individually or as a member of various societal groups, has been permanently exposed to many risks threatening his or her life. The risk can be defined as the probability of hazardous events, specific activities or actions occurring in a more or less random manner, whose outcome may be the individual's (or group's) death or injury. Frequently, the risk is related to the statistically expected value of loss. In other words, it represents the statistical likelihood of random exposure of an individual to some hazardous event, in which case a measure of the probability of severity of the impacts is involved. In such a context, safety represents the acceptable level of risk.

Various classifications of risk throughout the literature have been made (Sage and White, 1980; Kuhlmann, 1981; Kanafani, 1984; Evans, 1996). For example, this can be voluntary risk, which individuals take on themselves and involuntary risk, which individuals do not elect to assume. For example, using the air transport system where an air traffic accident may take place represents exposure to a voluntary risk. On the other hand, living nearby an airport where the aircraft accident (crash) may happen represents a case of exposure to an involuntary risk. Any type of risk may involve an objectively or subjectively known or assumed probability of exposure, which possesses three attributes: space, population and time dependency.

Generally, the space attribute implies that the probability of exposure to risk can come from the occurrence of the very local to the global hazardous events. For many types of risks, a certain group of the population may bear the specific risk. The attribute dependency implies that the probability of exposure to risk may be continuous, periodic (non-cumulative) and cumulative. It is dependent on the duration of the hazard and its effects in terms of cumulative or non-cumulative exposure.

According to Sage and White (1980), four types of societal risk might be identified as follows:

- Real risk, which may be determined based on the circumstances after they have fully developed;
- Statistical risk, which may be determined by the available statistics about a particular type of accidents;

- Predicted risk, which may be predicted analytically from the dedicated models; and
- Perceived risk, which may intuitively be felt by individuals.

The air transport system is an example where all the above-mentioned types of risks are inherently present. They are related to the occurrence of aircraft accidents. For example, for companies providing insurance to the airlines, flying constitutes a known statistical risk of the occurrence of an accident. At the same time, for passengers purchasing insurance before undertaking a trip, flying represents a perceived risk. In this case, the perceived risk usually exceeds the statistical risk. To the ATC/ATM, anticipated changes in the air traffic patterns and implementation of new technologies involve a predicted risk.

3.2.4.2 The nature of air traffic accidents

The concept and influencing factors   Air traffic accidents have some specific features as compared with accidents that happen in other transport modes as follows:

- They may happen at any time and/or in any space implying that individuals over large areas can be exposed to both individual and global hazard;
- Air passengers and aircraft crews are primal target groups exposed to the risk of an accident. Individuals on the ground are also exposed to the same risk but to a much lesser extent due to their spatial concentration;
- Each accident, although a relatively rare event, usually results in severe consequences such as numerous deaths and/or severe injuries, and destroyed properties;
- As an aircraft movement – flight – is considered as an inherently risky event, its accident could be considered as highly unlikely (although still possible) event according to probability of occurrence; and
- The risk of an accident permanently exists.

Evidence has shown that most air traffic accidents are caused by the complex system of mutually dependent events happening in a sequential order. These events can be classified according to two criteria. Firstly, according to the current state-of-knowledge they can be 'known and avoidable' and 'unknown and unavoidable'. The term 'unknown and unavoidable' should be considered only conditionally, since just after an accident the real causes might not be known. As the accident investigation progresses, causes often become 'known and avoidable' except in some rare cases. Second, according to type, the causes of air traffic accidents can be classified into 'human error,' 'mechanical failure,' 'hazardous weather,' 'sabotage,' and 'military operations' (Janić, 2000).

   Evidently, 'human error' can be managed and reduced by proper training of air transport system staff. The other factors such as the hazardous weather, 'hidden' mechanical errors, sabotage, and military operations are usually uncertain, and thus much less avoidable and controllable. Since these factors are permanently present, apparently one can never be certain that an air traffic accident will not occur. Does it mean that the air transport system will never be absolutely safe? The answer is negative since the occurrence of air traffic accidents does not necessarily mean that the system is unsafe. In order to properly judge this matter, safety should be considered regarding the basic causes of air traffic accidents, that is if they occur due to reasons that are already known and avoidable, the system should be considered as unsafe. Otherwise, it should be considered as safe (Kanafani, 1984).

Measurement The number of air traffic accidents and the number of fatalities (deaths and/or injuries) on the one side, and the volume of the system output carried out during a given period of time on the other are usually used for the assessment and quantification of risk and safety in the air transport system. Usually, the figures are compared with some targets. For example, ICAO has specified the target of the probability of an air collision at about 5×10^{-9} per hour (ICAO, 1988). The historical records on the number of air traffic accidents, deaths, and injuries are useful for estimating the safety trend, particularly needed for legal cases, that is whether the system is safer (less risky) now than it used to be in the past. Besides the national aviation authorities and airlines, ICAO has collected data on the safety records for commercial, domestic and international aviation. In these records, the number of accidents and their causes have been recorded and described. In addition, the corresponding volumes of output have been collected in terms of the annual and monthly volumes of aircraft kilometres, flying hours, number of departures and passenger-kilometres. In addition, the information has been synthesised per aircraft type, per airline, and per region.

In order to estimate the risk and safety trends using the above databases, a few measures can be used. The first one is the number of passenger fatalities per 100 million passenger-kilometres (that is the fatality rate). The other one is the number of air traffic accidents per 100 million aircraft kilometres (that is the accident rate). In addition, the number of accidents per given number of flights/departures of particular airlines or aircraft types is used as a measure. These measures are usually considered at the level of airlines and aircraft types through causal relationships with the volume of the corresponding output and time. For example, the number of deaths per passenger-kilometre (that is the fatality rate) is assumed to be the dependent variable F_R dependent on the total number of people who died per aircraft accident (crash) and the annual volume of passenger kilometres as the independent variables N_d and PKM_A. The regression equation has been obtained using data for the period 1981–1999 as follows (Janić, 1999):

$$F_R = 3.801*10^{-10} + 4.196*10^{-11} N_D - 2.095*10^{-16} PKM_A$$
$$ 2.983 (10.674) (3.446) (3.13)$$
$$R^2 = 0.901; F = 69.296; DW = 1,617; N = 19$$

Statistically, the relationship is significant at both the 1 per cent and 5 per cent level (F-value). The coefficients of the independent variables have the expected sign and are also significant (*t*-statistics in parenthesis below them) with high explanatory power

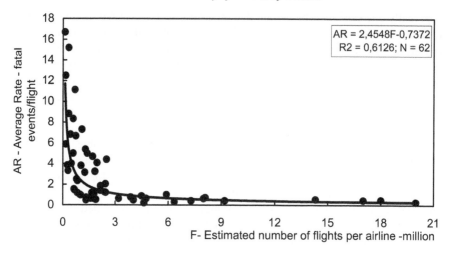

Figure 3.6   Dependence of the average accident rate on the number of airline flights (1970–1999)

Compiled from Janić (2000)

(R^2), and without auto-correlation of the first order (DW-statistic). Therefore, it could be used for explanatory purposes. The signs of coefficients of the independent variables indicate that the fatality rate has increased with the increasing number of people onboard the affected aircraft and decreased with the increasing annual volume of output. On the one hand, this indicates an increased risk for a greater number of passengers being onboard larger aircraft. On the other hand, these aircraft by their number, frequency, speed and size have increased the system output by flying increasingly reliably.

Another example relates to the risk and safety of the individual airlines, which has been estimated by regressing the number of air traffic accidents per million airline flights (the dependent variable AR counted as the accident rate) and the total (cumulative) annual airline flights (independent variable F). The relevant accident data relate to the airlines operating in different regions all over the world covering the period between the establishing of an airline until the present time Figure 3.6 shows the general trend.

As can be seen, the accident rate per airline decreases with the increasing number of flights more than proportionally. Since larger airlines have performed a larger number of flights they could be, according to this figure, considered as safer. In such a context, the US and European airlines have recorded significantly lower accident rates than airlines from other continents.

The safety of particular aircraft types can be estimated by regressing the number of accidents per aircraft type (that is the accident rate related to given aircraft type AR), the number of flights performed FL (million), and the average age of the aircraft type AGE (years). Data relate to the period between the time aircraft type entered service and the year 1992. For that period the participating aircraft types were: Fokker F28, Fokker F70/F100; Airbus A300, A310, A320; Lockheed L1011; British Aerospace BAe146; Boeing B727, B737-1/200, B737-3/4/500, B747, B757, B767; McDonnell Douglas DC-9, MD80. The regression equation is as follows (Janić, 1999):

$$AR = -1.206 + 1.713FL + 0.900AGE$$
$$(0.692) \quad (6.355) \qquad (3.887)$$
$$R^2 = 0.929; \ F = 84.64; \ DW = 1.823; \ N = 16$$

(3.14)

Equation 3.14 is statistically feasible. The independent variables have high explanatory power (R^2); the equation and particular coefficients are significant at both the 1 per cent and 5 per cent level as the corresponding F and t statistics indicate; and there is no auto-correlation of the first order between the independent variables (DW-statistics). The signs of particular coefficients of independent variables indicate that during the observed period of time the number of accidents of particular aircraft types increased with the increasing number of flights performed and their ageing, that is accidents happened more frequently for aircraft that were older and used more. The latter fact suggests that the cause of some air traffic accidents was generally the so-called 'geriatric problem,' which escalated faster for these aircraft.

3.2.5 Land Take (Use)

The air transport system takes land for settling its infrastructure – airports and ATC/ATM buildings, facilities and equipment. The greatest area of land is taken for installing airports – airside and landside area. As shown in Chapter 2, the size of the area of land used can be substantive – hundreds and thousands of hectares. In addition, one of the important characteristics in this context is the intensity of use of the land taken, which is characterised by the volume of activities performed on the given land over a given period of time. This is usually the umber of air transport movements and the number of passengers (and freight) accommodated at an airport during a given period of time (year). In many cases, these heterogeneous outputs are unified into the so-called Workload Unit (WLU) as an equivalent for one passenger and his/her baggage and/or 100 kg of freight (Doganis, 1992).

Given the fixed size of the area of land an airport occupies, the intensity of land use increases with the growth of traffic. When the existing infrastructure reaches saturation new infrastructure (runway, terminal building) is added, in which case the intensity of land use temporarily falls before recovering again with the continuing rise in traffic. Figure 3.7 shows the variations of the intensity of land use at Amsterdam Schiphol Airport (the Netherlands) during the period before and after the building of the new (sixth) runway in the year 2002.

As can be seen, the intensity of land use rose due to increasing air traffic before the year 2001 (the year of crisis caused by the September 11 terrorist attack on the USA). Over the next three years it stagnated due to a combination of factors such as the stagnation of traffic growth and the opening of the new runway (2002), which actually increased the area of land used by the airport. Later on, the intensity of land recovered again thanks to recovering traffic and the continuation of its growth.

The intensity of land use is closely related to the size and value of the land. Since the higher intensity of land use increases both the airport revenues and costs, it is not always quite clear if the additional land taken for building and/or expanding an airport should be considered as a pure externality. The intensity of land use at a given airport can also be expressed in terms of the volume of p-km or t-km realised per unit of land taken (ha).

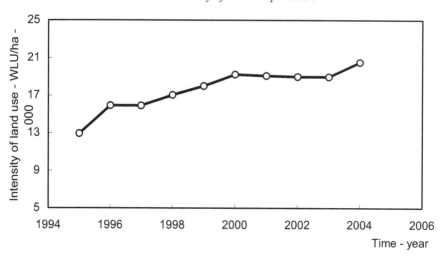

Figure 3.7   Intensity of land use at Amsterdam Schiphol Airport
Compiled from Schiphol Airport (2004)

For such purposes, information about the length of the incoming and outgoing atm (that is flights) and the number of passengers (and freight) on board should be provided.

3.2.6 Waste

Waste is generated at airports. The large airports generate relatively large quantities of waste. Regarding the micro location within the airport, waste can originate from the airport terminals, offices, food outlets, aircraft cleaning and apron activities, and other smaller collecting areas at the airport. Airport waste can be broadly classified into solid and liquid waste. In addition, it can be further segregated into non-industrial and industrial waste. The non-industrial waste originates from the passenger service onboard the aircraft, and the consumption of the airport employees and visitors (food, newspapers, cans, paper). The industrial waste originates from daily activities such as washing and cleaning the aircraft and ground vehicles, aircraft maintenance and repair work including painting and metal work, engine testing operations, the aircraft de/anti-icing operations, as well as ground vehicle maintenance. The industrial waste is further categorised into hazardous and non-hazardous waste. The former is managed according to the strict national and airport regulations governing collection, treatment, storage and disposal. The aim is to prevent contamination of the soil and drinking water at and around the airport (FAA, 2001).

The quantity of waste at an airport is usually expressed in absolute terms as the total quantities over a given period of time – a year, and in relative terms as the quantities of waste per unit of airport output (WLU – Workload Unit). Figure 3.8 shows an example of expressing waste in relative terms at Frankfurt main airport (Germany) during the period 1995–2005 (http://www.fraport.com).

As can be seen, the quantity of waste per WLU has been relatively stable during the observed period despite the traffic growth, implying a relatively sustainable airport development regarding this type of impact.

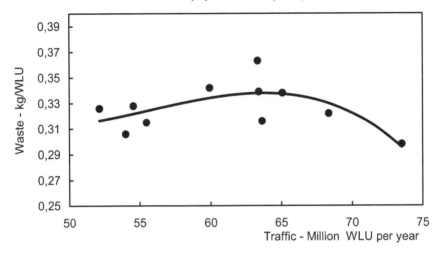

Figure 3.8 **The quantity of waste per unit of traffic vs the volume of traffic at Frankfurt Main Airport**

Compiled from http;//www.fraport.com

3.3 Evaluation of Impacts

3.3.1 Principles of Assessment

The theory of environmental economics has developed principles for the assessment and quantification of particular effects and impacts of the air transport systems on society and the environment (Button, 1993). In general, the effects − benefits − and the impacts as the costs of damages to people's health and the environment should be expressed in monetary terms in order to enable their consistent comparison within the same and/or across different transport modes. As far as the assessment and quantification of direct and indirect benefits of the air transport system is concerned, the method is relatively straightforward. However, assessment of the real costs of impacts appears to be much more complex. These costs are generally dependent on the product of two variables: the quantity of produced impact (externality), which correlates with the volume of transport output, and the economic cost per unit of impact (externality).

In addition to the volume of output, the quantity of a given externality depends on the technology in use and the amount of defensive and/or mitigating measures deployed. These externalities are physically measurable. The economic costs of externalities have been estimated using two approaches: the 'damage' based and the 'protection' based approach. The former approach implies that the damages from a given externality inherently lower the value of property, quality of life, landscape, and the health level, which can be expressed in monetary terms. The latter approach implies the costs of protecting against a given externality through mitigation, defence or abatement. These methods usually produce a range of different cost estimates even for the same externality. More systematic methods have therefore been proposed, such as: i) revealed preferences or hedonic pricing based on the observed conditions and behaviour of individuals under such conditions; ii) stated preference or contingency valuation based on collecting surveys from individuals in hypothetical situations; and iii) implied preferences based on the costs implied by legislation.

For the revealed preferences method, the cost of impacts is based on the assessment of damage and the scale by which it reduces the value of a given good. In addition, it can estimate the price people are willing to pay to avoid or mitigate the intensity and scale of the effect. In the present context, one example is the willingness of people to pay for insulation of their houses from the noisy nearby airport. With the stated preferences approach, hypothetical questions are used to determine the economic cost of a given externality. This is usually the amount that a group of individuals intends to pay in order to keep (or reduce) certain impact(s) at their current level after the expected (perceived) degradation of their neighbourhood takes place. Alternatively, this is the amount, which this group is willing to pay to prevent escalation of the impact of such (negative) development above the current level. The implied preference approach embraces the costs of externalities, which are set up by the governments based on the estimates of the costs of some protective and/or mitigating measures (Levison et al., 1996).

3.3.2 Fuel Consumption and Air Pollution as Externalities

Dealing with the air pollution from the air transport system as an externality appears to be a very complex task. On the one hand, quantification of the fuel consumption and the associated emissions of air pollutants is shown to be a straightforward task given the aircraft fleet, its operating regime, and fuel consumption and emission rates. On the other hand, while the contribution that fuel consumption makes to the depletion of the natural fuel stocks is quite clear, estimation of the marginal damages caused by the associated air pollution seems to be rather complex. What is certainly known is that air pollution affects society and the environment at both a local – airport – and global – troposphere and stratosphere – scale. Therefore, a polluter, in this case the air transport system, should be charged for such damages as compensation. Such a charge would generally cover four categories of air pollution costs: Photo-Chemical Smog, Acid Deposition, Ozone Depletion, and Global Warming (Levison et al., 1996). However, if one really intends to implement such a charging system, controversy emerges due to the lack of direct translation between the quantity of emitted pollutants and the damage they have already caused and/or are expected to cause in the future. Therefore, substantive research has been carried out aiming to estimate the cost of such short- and long-term damage. The first general research has dealt with the estimation of the physical damage to human health and the contribution of air pollutants such as CO_2 and NO_x to global warming. The following step has been evaluation of these damages by expressing them in monetary values. These figures have been applied to set up different taxes on the emissions of pollutants such as CO_2 and carbon C. The second part of the research relates to the estimation of the costs of protective, mitigating and defence measures from the impacts of air pollution. One example of mitigating/protective measures has consisted of planting trees in order to soak up the emitted CO_2. Finally, research on how much people would be willing to pay to avoid the air pollution up to a certain

Table 3.1   An example of setting up the tax on carbon emissions

Decade centred on the year	Carbon tax equivalent 1989 USD
1965	0.0
1975	0.0
1985	0.0
1995	5.29
2005	6.77
2025	10.03
2075	17.75

Source: Nordhaus (1991)

level has been carried out, mainly using stated preference surveys combined with analysis of the costs of preventing air pollution.

Some figures on the quantification of damage to people's health and the environment using a macro economic/global model has been shown to be relatively convenient. In this way the 'carbon-tax,' reflecting the cost of damages caused by the (air) transport system has been estimated. This tax is supposed to be applied at a given point in time to optimise the amount of air pollution on the one hand and trade-off the economic costs of damages by the greenhouse gases on the other. The values obtained from this model can be expressed in monetary units per tonne of carbon equivalent, as shown in Table 3.1.

In addition, some other proposals for carbon tax have been made. For example, in Europe suggestions are that it should be from about 53 to USD123/ton C; in the US it has been proposed to be between 83 and USD179/ton C (Levison et al., 1996). In some other cases, these taxes have been proposed to be equivalent to the costs of perceived damages – from about 31 to 171 EUR/ton of CO_2 depending on the social discount rate as the main governing factor (Tol, 1999). In addition, in some other studies these costs have varied between 30 and 50 EUR/ton of CO_2 (CE Delft, 2002).

The damages from the total human-made emissions of CO_2 equivalents have also been estimated in rather absolute global terms. The figures have ranged from about 1.0–2.7 per cent of GDP (Gross Domestic Product), depending on the rate of increase in the average global temperature – from about 2.5–3.0 °C respectively (Nordhaus, 1991; Tol, 1997). Consequently, if the air transport system contributed to the increasing of the average temperature (that is global warming) by about 0.052–0.096 °C, the corresponding costs of damage would be 0.017–0.032 per cent GNP in the former and 0.056,2–0.104 per cent in the latter case.

The cost of mitigation of man-made air pollution has been considered as an option for expressing the perceived damage indirectly. One example refers to the cost of planting trees to soak up the emitted CO_2. Some investigations have shown the very high diversity of these costs – from about 25 to USD120/ton C. Another example specifically related to the air transport system has consisted of considering the costs of developing technologies – more fuel efficient aircraft engines and airframes aiming at mitigating (or preventing) emissions of air pollutants – as the basis for setting up the carbon taxes (see Figures 3.4 and 3.5).

If it is assumed that the air pollution tax really reflects the cost of damage to people's health and the environment, the following equation could be used to estimate the cost of the air pollution damages per unit of output of a given flight (p-km):

$$C_{ap} = (t * FC * SE * p)/(n * d) \qquad (3.15)$$

where

t is the average flight duration (hours);

FC is the fuel consumption rate specified for the given aircraft type (tonne per hour);

SE is the emission rate of air pollutants specified for the given aircraft type (tonnes of pollutants per tonne of fuel burned);

p is the charge (that is tax) paid for the inflicted or perceived damage (monetary units per tonne of air pollutants);

n is the number of passengers per flight; and

d is the length of the flight (in miles or kilometres).

Similarly, the above-mentioned cost of damages could be estimated for the entire air transport system in a given region. Some estimates have shown that this cost is about USD0.0009/p-km (Levison et al., 1996). By multiplying Equation 3.15 by the total volume of output (p-km) carried out by a given aircraft fleet and then summing up the costs from all the aircraft fleets operating during a given period of time, the total cost of inflicted and prospective damages can be estimated.

3.3.3 Noise as Externality

3.3.3.1 Regulatory framework for mitigating noise

Different restrictive measures have been implemented to mitigate the noise burden on local populations in the vicinity of many busy airports. In the scope of the international efforts, the 33rd ICAO Assembly taking place in 2001 introduced the concept of a 'Balanced Approach' to noise management and control at airports. This implies identification of noise problems at an airport, and analysis and implementation of mitigation measures through the exploration of the following elements:

- Reduction of noise at source – allowing only operations of aircraft according to the above-mentioned Chapters 3 and 4;
- Restricting operations of particular aircraft types – forbidding operations of particular aircraft types during specific periods of the day;
- Noise abatement (operational) procedures – redistributing noise, which includes the use of preferential runways and approach/departure routes, and the noise abatement approach/landing procedures; any of these procedures need to satisfy the necessary safety standards (see also Chapter 2);
- Land use planning and management -introducing land-use zoning around an airport aiming to minimise the number of the population affected by aircraft noise; and

- Charging for excessive noise -introducing noise charges if the severe effects of noise exist at the airport (ICAO, 1991).

The above-mentioned measures have followed the already introduced measures where ICAO has prescribed the maximum levels of noise for given aircraft types (categories) as standards. Consequently those aircraft not fulfilling these standards have not been given the airworthiness certificate and have been unable to enter commercial operations (ICAO Annex 16 Chapter 2 and Chapter 3, and US FAR Part 36). According to these standards civil aircraft have been classified into two broad categories:

- Chapter 2: Subsonic jet airplanes certificated before 6 October 1977; and
- Chapter 3:
  - Subsonic jet airplanes certificated in or after October 1977;
  - Propeller-driven airplanes over 5,000 kg certificated after 1 January 1985 and before 17 November 1988; and
  - Propeller driven airplanes over 900 kg certificated on or after 17 November 1988.

The ICAO Member States had agreed that all Chapter 2 aircraft would be excluded from service from 1 April 2002 although some countries such as for example the US have applied this rule since the end of 1999. In the year 2002, a new Chapter 4 in the ICAO Annex 16 (Volume 1) was set up to be effective from the beginning of the year 2006. This new standard is more stringent by about 10 dB and is supposed to be applied to the newly certificated aircraft as well as to the re-certificated Chapter 3 aircraft.

The noise standards for particular aircraft types have been specified by taking into account the aircraft take-off weight (that is size) and phase of flight (take-off and flyover, approach and landing, and lateral flight. Table 3.2 summarises these standards (Smith, 2004).

The developments so far have shown that aircraft manufacturers and airlines have made a lot of effort to replace their fleets with aircraft fulfilling the standards of Chapter 3.

In addition, a lot of research has been devoted to investigating the possibilities for mitigating noise at airports. One of them has been Boeing's study making an inventory of the noise restrictions at 590 airports worldwide (http://boeing.com/commercial/noise/html). Intended primarily for Boeing's customers, the study has shown that the most frequent measures for mitigating noise have generally consisted of the implementation of noise abatement procedures with restrictions on runs up, the use of preferential runways, the implementation of curfews, and the use of noise monitoring systems. About 21 per cent of airports in 16 countries have opted for charging a noise fee and only a minority have charged for the emission of air pollutants. As a rule, the noise charging systems have not taken into account the actual noise nuisance caused by local aircraft operations (Morrell and Lu, 2000; Betancor and Carlos Martin, 2005).

Table 3.2   The noise standards for aircraft certification (US FAR36)

Phase of flight Take-off/Flyover[1]	Aircraft Maximum Take-Off Weight – MTOW (tones)
Noise level	
93 EPNdB	57 < MTOW < 34
93–108 EPNdB	34 < MTOW < 272
108 EPNdB	MTOW > 272
Phase of flight[2] **Approach/Landing/Lateral**	**Aircraft Maximum Take-Off Weight – MTOW** **(tones)**
Noise level	
102 EPNdB	57 < MTOW < 34
102–108 EPNdB	34 < MTOW < 272
108 EPNdB	MTOW > 272

[1]*ICAO Annex 16 is more restrictive for Δ = 3–5 EPNL*
[2]*ICAO Annex 16 is more restrictive for Δ = ½ EPNL for approach and for*
Δ = 3–4 EPNL for lateral phase.
Source: Smith (2004)

3.3.3.2 Technological framework for mitigating noise
Improvements in the noise performances of contemporary aircraft by investments in new technologies (aircraft engines) have been substantive and permanent. Some general trends are shown in Figures 3.9 and 3.10.

Figure 3.9 shows the dependability of INA (Integrated Noise Area) on the aircraft price.

In this case, INA corresponds to 100 EPNL pseudo contour noise area selected as an equivalent for the noise contours generated during the aircraft landings and take-offs at the airport (Equivalent Perceived Noise Level in EPNdB describes the noise of a single aircraft) (Smith, 2004).

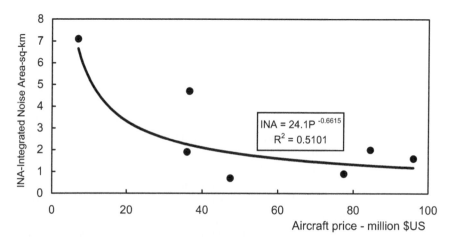

Figure 3.9   Dependence of Integrated Noise Area on the aircraft price
Compiled from Smith (2004)

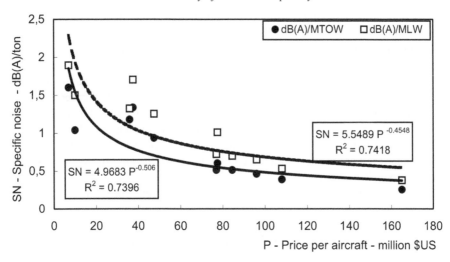

Figure 3.10 Dependence of aircraft specific noise and price
Compiled from Smith (2004)

As can be seen, the area of the pseudo contour decreases more than proportionally with the increasing aircraft price, indicating that investment in more expensive (larger) aircraft also contribute to reducing noise. However, it is easy to show that the marginal contribution of the unit of investment to the diminishing of noise has decreased with the increasing size of investment, that is the aircraft price, as follows: $\partial(INA)/\partial P = -2.524 P^{-1.506}$.

It becomes more obvious in Figure 3.10 where dependency of the aircraft specific noise during landings and take-offs on its price is shown.

The specific noise (SN) as the dependent variable is the ratio between noise in dB (A) generated at the aircraft noise certification points at airports (for landings it is at a distance of 2 km and for take-offs a distance of 6.5 km) and the corresponding aircraft maximum weight (Maximum Take-Off Weight – MTOW; Maximum Landing Weight – MLW) (Smith, 2004). The independent variable is the aircraft price generally increasing with the increasing aircraft MTOW. As can be seen, the specific noise SN decreases more than proportionally with the increasing aircraft price for both landings and take-offs, indicating that the higher investments in heavier-bigger aircraft have resulted in reducing the relative noise. A marginal contribution by the aircraft price-investments to decreasing the noise per unit of aircraft weight (that is the specific noise) has also decreased more than proportionally with the increasing aircraft price as follows: i) for take-off: $\partial SN / \partial P = -2.541 P^{-1.506}$; and ii) for landing: $\partial SN / \partial P = -2.524 P^{-1.455}$.

3.3.3.3 Operational framework for mitigating noise
The ATC/ATM has developed several innovative operational procedures for mitigating noise around airports. Some of them have been based on existing and others on innovative technologies. In addition, some have expected to increase the airport capacity. The most known procedures have been Low Drag/Low Power (LD/

LP), Continuous Descent Approach (CDA), Increased Glide Slope (IGS), Displaced Threshold (DT), and Curved Approach (CA) procedures (EC, 2005).

The LD/LP procedure implies minimising the drag and consequently power during the aircraft approach and landing. This is achieved by maintaining the aircraft's 'clean' configuration as long as possible, which reduces the aerodynamic drag and thus requires less engine power to keep the aircraft at the prescribed trajectory. This procedure can be combined with the CDA procedure described in Chapter 2.

The IGS procedure is based on the increased Glide Slope angle of the ILS (Instrumental Landing System) during the aircraft's approach and landing. Usually the standardised angle of 3° is increased to 4–6°. Thus, the aircraft trajectory closer to the airport is higher and consequently the noise beneath it is lower. Six regional airports in Europe have such a steeper instrumental approach procedure.

The DT procedure implies the higher final approach and landing trajectory. The vertical increase of the trajectory is $\Delta z = \Delta x * tg\theta$ where Δ_x is the distance of the displacement threshold and θ is the standardised ILS descent angle of 3°. Since the aircraft remains higher, the noise beneath it is lower (see also Chapter 4). This procedure could also be combined with the CDA procedure.

The CA procedure implies a curved multi-segment approach trajectory, which bypasses the noise sensitive areas. This procedure can also be combined with the CDA procedure.

Nevertheless, these procedures still have to be certified regarding safety under concrete circumstances (EC, 2005).

3.3.3.4 Scale of impact – The number of people exposed to aircraft noise
Theoretical estimation   The EU (European Union) Position Paper has developed a methodology for assessing the proportion of the human population annoyed by transport noise. The methodology takes into account air, rail and road transport (EC, 2002). The noise exposure indicators L_{den} and L_{night} in Equation 3.11 have been used to predict the air transport level of annoying noise. The descriptors have been defined in terms of the per cent of annoyed (%A) and the per cent of highly annoyed (%HA) people in the population. These descriptors have been derived by combining different noise scales and cutting values of 50 dB(A) for (%A) and 72 dB(A) for (%HA). Consequently, the following polynomial approximations have been proposed for the estimation of the noise annoyance (%A) and (%HA), dependent on the daily exposure indicator L_{den} in Europe (EC, 2002):

$$\%A = 8.5888 * 10^{-6} (L_{den} - 37)^3 + 1.777 * 10^{-2} (L_{den} - 37)^2 + 1.221(L_{den} - 37) \qquad (3.16a)$$

and

$$\%HA = -9.199 * 10^{-5} (L_{den} - 42)^3 + 3.932 * 10^{-2} (L_{den} - 42)^2 + 0.2939(L_{den} - 42) \qquad (3.16b)$$

Particular symbols are analogous to those in the previous equations. Figure 3.11 shows some results of the application of Equation 3.16.

As can be seen, at least in such calculations, %HA has been much lower than %A, and both have increased more than proportionally with increasing daily noise

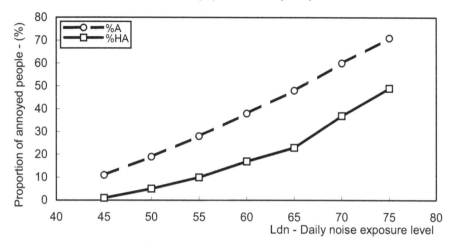

Figure 3.11   Dependence of the proportion of annoyed people on the level of exposure to air transport noise

Compiled from EC (2002)

exposure level L_{dn}. As such, the relationships (3.16) have been proposed for the strategic assessment of the air transport noise in the scope of noise management and the effects on the annoyed population (EC, 2002). In addition, as specifically applied to an airport – in this case London Heathrow – the above-mentioned methodology has produced the following results (Table 3.3).

Empirical estimation   Some other empirical research has dealt with estimations of the scale of noise impacts at airports. It is interesting to look at the situation in the matured well-developed air transport systems such as those in the US and Europe. Figures 3.12 and 3.13 illustrate these cases, respectively. In particular, Figure 3.12 shows the causal relationship between the total population exposed to noise around 250 US main airports and the total urban population. As can be seen, during the period 1975–1998, the proportion of the population exposed to air transport noise decreased more than proportionally with the increasing urban population, from about three to less than 0.5 per cent. Certainly, such a long-term trend was achieved by improvements to the airport land use planning; resettlement of annoyed populations who previously lived closer to the airports; improvements of the aircraft noise mitigating (operational) procedures; and modernisation of the aircraft fleet.

Table 3.3   Estimates of areas, population, and households within given L_{den} noise contours at London Heathrow Airport

Contour level dB(A)	Area (km²)	Population (thousands)	Household (thousands)
> 55	302.3	782.9	344.9
> 60	114.3	260.5	109.8
> 65	47.7	74.5	29.9
> 70	20.8	16.6	6.5
> 75	7.5	1.7	0.7

Source: CAA (2004)

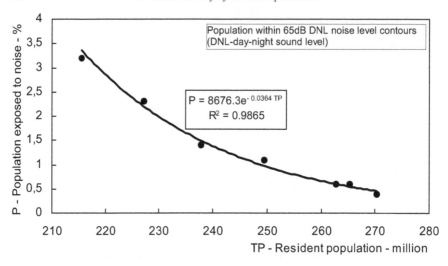

Figure 3.12 Population exposed to noise at 250 main US airports (1975–1998)
Compiled from BTS (2000)

Figure 3.13 shows the relationship between the number of the population exposed to transport noise and the level of noise in 17 European countries (INFRAS, 2005).

The information for rail and road transport is provided for comparative purposes. The dots indicate the population exposed to the noise level from the first neighbouring lower level to the dotted level, that is to noise from 50 to 55, 55 to 60, ..., dB(A). As can be seen, the population exposed to higher noise decreases more than proportionally with the increasing noise level for all transport modes. Specifically, the population exposed to air transport noise is the lowest, independent of the level of noise, which is seemingly due to the character and size of the air

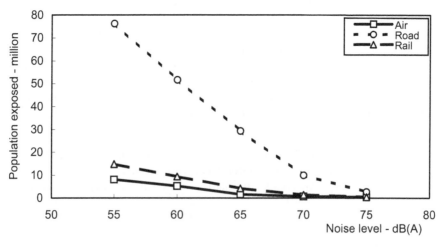

Figure 3.13 Population exposed to noise at 250 main US airports (1975–1998)
Compiled from BTS (2000)

transport infrastructure (airports) and its use as compared with other transport modes – road and rail.

3.3.3.5 Cost of noise nuisance

The most well known techniques for estimating the costs of noise damage are the hedonic and contingent valuation methods. Both belong to the direct valuation methods. The former measure is based on the revealed and the latter one on the stated behaviour. The hedonic price method has been the most widely used method for evaluation of the social cost of noise in the air transport system, at least within the academic community (Pearce and Markandya, 1989; Levison et al., 1996; Morrell and Lu, 2000; Betancor and Carlos Martin, 2005).

The hedonic price method usually uses the property values determined by location, attributes of the neighbourhood and community, and the environmental quality in the form of the property value function $P = f(S, N, Q)$ (S is the vector of location characteristics, N is the vector of the neighbourhood characteristics, and Q is the vector of environmental characteristics) (Morrell and Lu, 2000). Noise is considered as an attribute (i) of the vector of environmental characteristics and $\partial P / \partial Q_i$ is the hedonic price considered as the marginal implicit price of the noise social cost. Subsequently the so-called Noise Degradation Index (NDI), expressing the percentage of the reduction in the house price per A-weighted decibel (dB(A)) above the common background noise is empirically derived using the regression analysis. Dependent on the study and the time carried out, the values of NDI vary from 0.3 per cent to 0.7 per cent per additional dB. These values are also dependent on the noise scale in use (Levison et al., 1996). From the hedonic price method, the total annual social cost of noise C_n can be estimated as follows:

$$C_n = \sum_{i=1}^{M} \left(I_{NDI} P_v \right) \left(N_{ai} - N_i \right) H_i \tag{3.17}$$

where
M is the number of zones per noise contour;
I_{IND} is the NDI expressed as a percentage of the property value;
Pv is the annual average house rent in the vicinity of a given airport (monetary units);
N_{ai} is the average noise level for the *i*-th noise zone (dB(A);
N_0 is the background noise level (dB(A)); and
H_i is the number of residents within the *i*-th noise zone.

In Equation 3.17, the product $I_{NDI} P_v$ represents the annual noise social cost per resident per noise unit (dB(A)). In about 30 studies the average value of I_{NDI} has been estimated to be 0.81 with a standard deviation of 0.72 (Schipper, 2004). In many other studies the value of I_{NDI} has been assumed to be about 0.48. The factor $I_{NDI} P_v$ should take into account that the annoyance caused by noise increases more than proportionally with the increasing noise level above the ambient noise (see Equations 3.10 and 3.11). The annual house rent can be calculated as the product of the average house value in the vicinity of airport P and the capital recovery factor dependent on the mortgage interest rate r as follows (Levison et al., 1996; Morrell and Lu, 2000):

$$P_v = P\left(\frac{r(1+r)^n}{(1+r)^n - 1}\right) \tag{3.18}$$

The total social cost of noise determined by Equation 3.17 can be used for setting up the noise charging policies at the airport. In such a context the revenues collected by charging particular aircraft movements (atm) for noise should cover these costs. In such cases, different criteria can be used for charging for the noise from particular aircraft movements (atm). For example, one criterion can be the uniform allocation of charges to particular aircraft operations (atm). Another can be the selective allocation of charges regarding particular aircraft types in terms of their specific marginal contribution to the overall noise burden. Consequently, the noise charge for the specific flight i) has been set up as follows (Morrell and Lu, 2000):

$$c_i = \frac{R}{\sum_{i=1}^{N} L_i p_i} L_i \tag{3.19}$$

where
R	is the revenue collected by charging for noise at given airport (EUR);
p_i	is the proportion of aircraft of the engine characteristics (i);
L_i	is the noise impact index of the aircraft with engine characteristics (i); and
N	is the number of engines with different characteristics.

The revenues R in Equation 3.19 can also be determined using some other criteria. One of them is the cost of investments in noise mitigation measures such as for example the cost of isolating houses. In addition, knowing other characteristics of particular flight categories such as the number of passengers and length, the charge reflecting the noise social cost per unit of output (p-km) can be determined. As an example, according to Morrell and Lu (2000), the total cost of noise at Amsterdam Schiphol Airport (the Netherlands) estimated by Equation 3.17 amounted to about €124 million in the year 1999. Given the number of operations (atm) the cost of noise is about €312/operation. If the average number of passengers per atm is about 100 and the average route length about 1,000 km, the average charge per unit of output amounts to €0.00312/p-km.

In addition to the academic-based efforts, charging for aircraft noise at airports has also been the subject of national and international policies. There is one such policy in the European Commission's documents, which has proposed the following equation for charging for noise (EC, 1999):

$$c_n = c_a 10^{\frac{L_a - T_a}{10}} + c_d 10^{\frac{L_d - T_d}{10}} \tag{3.20}$$

where

c_a, c_d is the noise charged for an arrival and a departure, respectively, which theoretically can be equal to zero (monetary units per operation);

L_a, L_d is the noise level for an aircraft at the arrival and departure/flyover noise certificated locations, respectively (in $dB(A)$); and

T_a, T_d is the noise threshold during a departure and an arrival, respectively, corresponding to the category of relatively quiet aircraft (they are fixed around 13 dB below the upper threshold, corresponding to 95 per cent of the total noise energy emitted at an airport).

The main disadvantage of the method in Equation 3.20 is that the problem of choosing an appropriate technique to determine values c_a and c_d has not been completely resolved. Summarising, any method of charging for aircraft noise − from the cost of mitigation of noise burden to the marginal social cost based charging − can be used depending on local circumstances and specificity of the concrete airport.

3.3.4 Congestion as Externality

The total time, which an aircraft/flight spends between a given origin and destination airport can be divided into two parts: the first is the part where the aircraft does not experience congestion; another is the part where the aircraft experience congestion. The former part implies flying along the prescribed route without interference from other aircraft/flights, which might prolong the planned flight-time. The latter implies interference of the aircraft/flight with other flights, usually at the origin airport before departure and/or destination airport before landing and during the en route phase as well, which extends the planned flight-time. Extension of the aircraft/planned flight-time is considered as a delay, whose cost is considered as an externality. Such consideration is based on the assumption that delay is imposed by external factors − other aircraft/flights, airline, and users − passengers and freight shipments − on board. Dealing with the cost of congestion and delays includes estimating the relevant levels of delays to be charged, determination of the value of time for the affected actors − aircraft and users, and setting up the instruments for charging congestion based on the marginal costs of delays. These costs are those which the aircraft/flight imposes on others while being at the congestion facility − airport (see also Chapter 4) (Levison et al., 1996; Janić, 2005). The nature of congestion and delays in the air transport system has already been described above. This section deals with the value of time for the users-passengers as the necessary variable to determine the social cost of congestion. In general, the value of user-passenger time depends on factors such as mode of travel, time of the day, trip purpose (business, leisure), quality of service, and income. In addition, different phases of a trip may have different values of time: for example, the value of time might be different while waiting for a departure or while being onboard. In addition, the extended travel time in cases of disruption of regular (planned) services can be considered.

Different methods are developed to estimate the value of time of travelling people. In general, utility theory and theories of marginal productivity have been used. Over time, relatively consistent methods have been established: for business travellers assumed to travel during their working time, the value of time has been taken as equivalent to the wage rate. For non-business travellers the wage rate has also been used as a basis with a discounted coefficient compared with business travellers. Nevertheless, despite being based on similar principles in the methods used, the value of time has significantly varied in different research studies. Generally, this value has been calculated as the product of a given multiplier (factor) and the average wage rate. For business travellers, the multiplier has varied from 1 to 3 and for non-business travellers it has been equal to 1.0. Consequently the value of time for business travellers has varied from USD20 to 60 per hour and for non-business travellers from USD15 – 42 per hour (Levison et al., 1996). One should be aware of using the current wage rates while undertaking estimations of the value of passenger time for different categories of passengers.

The social cost of congestion per unit of the air transport system output (USD per p-km) can be calculated as follows (Janić, 2005):

$$C_c = (Q/\mu)\alpha/d \qquad\qquad (3.21)$$

where
  Q is the length of queue joined by an aircraft/flight (aircraft/flight);
  μ is the average service rate of a queue (minutes per aircraft/flight);
  α is the value of passenger time while onboard the aircraft/flight (USD per passenger-hour time); and
  d is the length of route of the flight (kilometres or miles).

At some very congested airports, the social cost of congestion can be very high. As these airports also have limited opportunities for expansion of their infrastructure capacity, policies of charging for congestion in order to prevent its further escalation and diminish the cost of lost time of particular actors are implemented. Charging for congestion at airports is discussed in Chapter 4.

3.3.5 Air Traffic Accidents as Externality

3.3.5.1 Assessment of the number of fatalities

The most important indicators of the severity of an air traffic accident are the number of deaths and the number of survivors. Solid statistics on these indicators exist. They can be used for different assessments. One of the examples relates to the distribution of the number of survivors and the number of people who perished dependent on the category of aircraft. The aircraft have been categorised into turbojets, turbo-props and piston-engine. The figures are presented in Table 3.4.

As can be seen, the turbo-prop aircraft has had the greatest number of accidents while the greatest number of people that have both perished and survived flew on turbojet aircraft. This can be explained by the fact that turbojets have always carried a greater number of passengers, which have inherently increased both the proportion of deaths and survivals. In the example, an average of 56 passengers died in turbojet aircraft accidents, and 42 and one in turbo-prop and piston-engine aircraft accidents, respectively. In a separate analysis, a sample of 258 air accidents was used (Janić, 2000).

Table 3.4   Characteristics of fatal accidents by aircraft category

Aircraft category	Fatal accidents per aircraft type (%)	People died per aircraft accident (%)
Turbojet	34.2	69.6
Turbo-prop	48.5	28.1
Piston-engine	17.3	3.3

Source: Janić (2000)

Of these 130 had no survivors. Based on the ratio between 'the number of accidents without survivors'/'the total number of accidents,' the probability of death was about 50 per cent (that is 130/258 = 0.504). By taking into account the number of people on board, the average number of deaths per accident is 76 and the standard deviation 81. In addition, the total number of passengers and crew per aircraft involved in an accident is 103 and the standard deviation 88. Comparison of the above-mentioned figures indicates that about 73 per cent of passengers are killed during an accident (the other 27 per cent survived, but some of them with severe injures) (Janić, 2000).

Statistics on air traffic accidents are being permanently updated, which enables assessment of the global trend in safety of the air transport system. The most recent statistics in Table 3.5 indicate that during the past six years, the system safety, expressed in absolute terms of the number of fatal accidents, fatalities, and the cost of lost aircraft has not considerably improved. However, since the system output has been permanently increasing, the accident and fatality rates have been diminishing.

3.3.5.2 Social cost of air traffic accidents
The principal approach in dealing with air traffic accidents as an externality involves estimating the costs of the damages and losses. Table 3.5 provides some information of such estimated costs per accident, which includes only the costs of loss of the aircraft. In general, this externality should also include the cost of damaged and/or lost property (in this case this is the aircraft and the third party properties), the cost of the loss of life, and the cost of the time needed for recovery from the injuries of those who survived. Numerous studies have dealt with estimations of the value of life. In most cases this value has been determined for different countries in relation to the individual's contribution to the national GDP during their working age. Obviously, the differences in the values of life have emerged due to the differences in the national GDP. For example, the value of life has been estimated to be between USD1.6–4.7mn (Levison et al., 1996). This raises questions, however, about the logic and ethics of the method used.

Table 3.5   Safety statistics for the air transport system

Year	1990s	2000	2001	2002	2003	2004	2005
Fatal accidents	42.5	35	34	36	25	29	32
Fatalities	1207	1,127	1,107	1,094	691	529	1,064
Cost major hull ($ million)	794.2	970	1,047	585	524	477	525

Source: AW (2006)

The costs of injuries have also been expressed as an externality by converting the duration of the injuries (time of non-working) into the equivalent years of life and then using the concept of the value of life. Research has indicated that the functional years lost due to injuries are mainly dependent on the degree of injury. Some examples have shown that they vary from about 0.07 years in cases of minor injury to about 42.7 years in cases of fatal injury. In addition, on top of the cost of losing functional years, the costs for hospitalisation, rehabilitation, and provision of the emergency services to the injured individuals need to be taken into account.

Using the concept of the value of life and injuries, the average cost of an air traffic accident per unit of the system output (p-km) can be estimated as:

$$C_a = a_r n \left[p_d \beta_d + (1 - p_d) \beta_i \right] \tag{3.22}$$

where

a_r is the air traffic accident rate (the number of accidents per unit of output (p-km));

n is the number of passengers on board the flight;

p_d is the probability of death during the accident;

β_d is the average value of life of a passenger killed in the accident (USD per passenger);

β_i is the cost of recovering from injuries (USD per passenger).

Equation 3.22 indicates that the social unit cost will increase in proportion with the overall accident rate, number of passengers onboard, that is the aircraft size, probability of death/survival, and the average value of life. As an example, the social cost of an air traffic accident expected to happen at the rate $a_r = 2.095x\ 10^{-16}$ (Equation 3.12), the number of passengers on board $n = 150$, the proportion of deaths $p_d = 0.7$, the proportion of survivors $(1p_d) = 0.3$ (Table 3.4), the average value of life $\beta_d = USD3.0mn$, and the cost of recovering injuries $\beta_i = USD10$ thousands, would be $C_a = 6.604x\ 10^{-8}$ USD/p-km (Janić, 2000).

3.3.6 Land Take (Use) as Externality

Considering the land taken (used) for building and/or expanding airports as an externality is generally ambiguous. The question is whether it is more socially feasible to take land for an airport or to use it for some other economical or non-economical purposes such as housing, agriculture, recreation, and the natural environment (green area with intact flora and fauna). In all these mutually exclusive cases, the land taken has a certain value, which in general can be the economic, non-economic, and market-based value. While assessing the social cost of such land, the economic value of land is relevant. However, there are certain conditions when such land take represents an externality that need to be specified. For example, let R_{ai} and C_{ai} be the total social revenues and costs, respectively, from operating an airport using the area of land S_i; let R_{ji} and C_{ji} be the social revenues and costs, respectively, from carrying

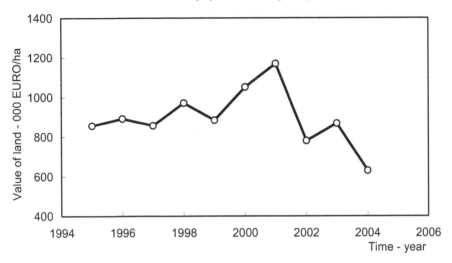

Figure 3.14 The value of land at Amsterdam Schiphol Airport
Compiled from Schiphol Airport (2004)

out some other economic or non-economic activity (j) on the same land S_i. The value (that is cost) of one unit of airport land can be determined as follows:

$$C_l = \frac{\left[(R_{ai} - R_{ji}) - (C_{ai} - C_j i)\right]}{S_i r} \qquad (3.23)$$

where

    r is the capitalisation rate converting the future monetary values into the present value.

In Equation 3.23 the capitalisation rate can be deduced for the annual rate of capital gain earned if the land was sold (Equation 3.18). The nominator of Equation 3.23 is often called the annual return to land. This equation also reflects the intensity of land use in monetary terms. Figure 3.14 shows an example of such land use at Amsterdam Schiphol Airport (the Netherlands).

The alternative economic activities, which could be carried out, instead are not taken into account. As can be seen, the value of land c_l (the vertical axis) has always been positive, which means that the land taken has always brought profits from the air transport operations. In such a case, the land taken should not be considered as an externality. However, in cases when the airport operations brought losses, the land taken could be considered as an externality. Multiplying the value of unit of land c_l by the total area S_i and then dividing the product by the output (p-km and t-km) expresses this externality per unit of the system output. If this value increases with increasing airport operations the economic value of the land will increase too, which may make further take of nearby land for the same activities increasingly expensive. This is an additional controversial element in the internalisation of land take around airports as an externality through considering the economic value of the land.

3.3.7 Waste as Externality

The airport waste can be considered as an externality under conditions when it causes damage to people's health and the environment. In particular, the incidental leakage of hazardous liquid industrial waste such as aviation fuels, oil, and vehicle washing and cleaning liquids can contaminate the soil and drinking water, and consequently endanger the health and possibly lives of people and other habitats. In these cases the externalities are counted as the costs of repairing the damage in the broadest sense. This usually implies cleaning up the contaminated areas and eventually strengthening the protective infrastructure and preventive measures. If the objective is to prevent incidents, the preventive measures are the most important. They are carried out through the so-called waste management system, which exists at almost all airports. This system usually includes identification of the sources, location, type and quantity of waste generated, then the infrastructure, facilities and equipment to deal with different types and quantities of waste, and finally the efficiency and effectiveness of the waste collection, storage, recycling and disposal.

An efficient and effective waste management system usually implies avoidance, minimisation, and recycling of waste. This involves segregation of waste into solid and liquid, hazardous and non-hazardous components at source, reduction of the quantities of waste generated, continuous increase in reuse, recycling and reprocessing of waste materials, and the continuous improvement of waste management practices. In particular, recycling implies conversion of waste into energy through thermal treatment (processing). For example, at Frankfurt Main airport (Germany), the rate of recycled waste has increased from 50 per cent in the year 1995 to about 85 per cent in the year 2005. Consequently, the cost of waste management can be reduced as the final objective (http://www.fraport.com).

Airport waste management systems are usually designed and operated respecting the valid national and local legislation. This particularly refers to the requirements for the storage and disposal of waste in dedicated areas, which cannot be used for some other more profitable activities (FAA, 2001).

3.4 Evaluation of Effects – Benefits

3.4.1 Principles of Evaluation

The air transport system has developed into a global industry. In addition to the impacts (externalities) described above, there have also been social effects-benefits to local communities around airports as well as to the national and international economies. The benefits have been broadly classified into internal and external benefits. The internal benefits relate to the direct employment by the air transport system operators – airlines, airports and ATC/ATM – as well as to the indirect employment by the system's supporting services. The external benefits include the system's contribution to developing tourism, trade, business, and productivity. Both categories of benefits can be expressed as the system's contributions to GDP. This enables quantification and comparison of the social benefits and costs (externalities) and thus assessment and quantification of the system's sustainability. However, in many cases, the precise assessment of both categories of effects appears to be complex due to the lack and in many cases relatively high costs of obtaining the relevant and consistent data, along with the absence of consistent methodology. Nevertheless, it is rather a question of 'must' than 'should' for everyone pretending to deal with sustainability of the air transport system to obtain consistent and relevant data for a fair assessment. In any case, understanding the global socio-economic environment in which the air transport system operates as a global transport mode as well as its contributions to the further globalisation of this environment appears to be of the greatest importance.

3.4.2 Globalisation

3.4.2.1 Globalisation of the world's economy and society
Globalisation is a process, which could be considered in either a narrower or broader sense (Yergin, Vietor and Evans, 2000). In the narrower sense, the process comprises integration and entry of the national economies into growing flows of international trade, investments, and capital exchange across national (historical) borders. In the broader sense, globalisation embraces the growing trend for exchanging organisational skills, technology, ideas, information, entertainment, people culture, and most recently the creation of monetary unions. The most important driving forces of globalisation have been:

- Promoting markets in terms of deregulation, liberalisation, and privatisation;
- Ending the ideological conflict of the Cold War;
- The growth of international trade and its further liberalisation;
- Growth of the capital markets and foreign trade investments;
- Technological change; and
- Regional integration.

Historically, the foundation for globalisation was set up just after the end of the Second World War, when it was realised that one of the main reasons for such severe conflict had been impoverishment, partly caused by the trade barriers in place at that time. Removing such barriers was seen as an opportunity for raising living standards, bringing nations closer to each other and even living together, creating economic and social interdependence, and consequently diminishing (or eliminating) the reasons for future conflicts on a large scale. The idea had materialised through the creation of international financial institutions such as the IMF (International Monetary Fund) and World Bank in the early 1950s, and the trade organisations such as General Agreements on Trade and Tariffs (GAAT) in the late 1950s. These institutions significantly contributed to the recovery of the world economy during the post-war period. For example, over the period 1949–1971 the annual economic growth in the so-called developed countries was impressive: 3.8 per cent in the US, 5 per cent in the Western Europe, and 10 per cent in Japan. Such development was generally stimulated by deregulation of the US economy, integration of Europe, regional integrations in other parts of the world (NAFTA-North America Free Trade Agreement), formation of multinational companies, the privatisation of industries and

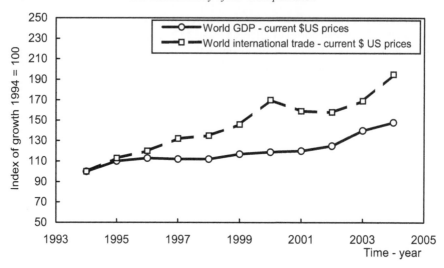

Figure 3.15   Growth of the world economy (GDP) and international trade
Compiled from WTO (2005)

other non-manufacturing companies, foreign investments, and strengthening of the
formal-institutional integration. For example, the share of foreign direct investments
as a proportion of the Gross Capital Foundation increased from about 2 per cent
in 1970 to about 14 per cent in the year 2000. In the developed countries, the net
investment flow was negative: USD–10.7bn from 1973 to 1978 and USD–86.6bn
in the period 1993–1996. At the same time, this investment flow in developing
countries was positive: USD4.6bn from 1973 to 1978 and about USD87.2bn from
1993 to 1996. Financial integration particularly stimulated increase in the net capital
flow towards developing countries: from about USD45bn in the 1990s to about
USD250bn in the year 2000. In parallel, the share of the export of goods and services
in particular regions such as the US and Europe increased from about 6.5 per cent
in 1971 to 11.5 per cent of GDP in the year 2000. The share of imports of goods and
services also increased from about 6.5 per cent to about 14 per cent, respectively.
Consequently, growth in international trade in relative terms has been higher than
that of the economy since the 1970s (Yergin, Vietor and Evans, 2000). Figure 3.15
illustrates developments over the past decade.

As can be seen, the gap between the growth in international trade and economic
growth has been widening, stimulated by further enhancement of the globalisation
forces making the national economies more global and consequently more dependent
on each other.

Technological developments and connections have also contributed to
globalisation. In particular these telecommunications, in combination with computer
technology and related software, have played a very important role. Together they
created the foundation for the Internet, which became fully effective in the year
1993 after the first commercial browser had been produced. The launch of Netscape
Navigator followed a year later. As a result, the global networks of distant encounters
and close connections of personal, governmental, entertainment and business users

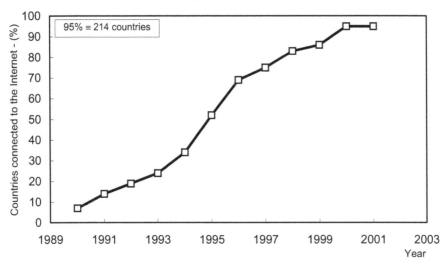

Figure 3.16   Countries connected to the Internet
Compiled from ITU (2002)

was established. Figure 3.16 shows the trend in countries connecting to the Internet. As can be seen, 95 per cent (214) countries across the world were connected to the Internet by the year 2001. Within this framework a global trade in the form of e-commerce has, in the US alone, reached the annual value of USD50–60bn billion.

Nevertheless, regardless of the above-mentioned advantages, globalisation has caused different reactions by three different groups. The first group, represented by economic and business leaders and managers, strongly support removal of all institutional barriers to economic and business efficiency. The second group, consisting mainly of labour unions, businesses facing competition from the imports and other 'economic nationalists,' prefers a restriction to free trade, investments and the strengthening of multinational companies. The environmentalists and human rights supporters complaining that globalisation offers only a lack of democracy and exploitation of the poor represent the last group. The confrontation between the opinions and views of the different groups continues.

3.4.2.2 Air transport system and globalisation
The air transport system is one of the most important driving forces of globalisation. On the one hand, the statistics on how it has facilitated the globalisation process are scarce. On the other, it is also interesting to learn how globalisation has influenced the transformation of the air transport system over time. In general, this transformation has been driven by internal forces such as the technological changes resulting in the development of more efficient aircraft with a relatively highly diversified capacity and performances, privatisation of the system components – airlines, airports and ATC/ATM, and other changes such as deregulation and liberalisation of the air transport markets. The last two changes in particular could be considered part of the response of the air transport system to globalisation. The other part comes from the airlines as the most flexible system component, which has formed global alliances. The airline

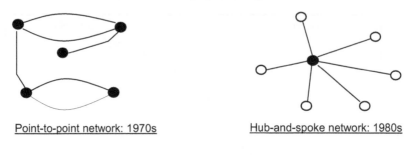

Point-to-point network: 1970s   Hub-and-spoke network: 1980s

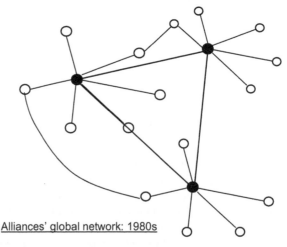

Alliances' global network: 1980s

Figure 3.17   Evolution of air transport networks
Compiled from Janić (2001); Yergin et al. (2000)

alliances represent an initiative by individual airlines to rationalise their operations, provide more efficient and effective market coverage, and offer a more seamless service compared with competitors under conditions where the current regulations restrict almost every airline acquiring or merging across national borders. In addition to the economic and business changes, the spatial and time configuration of airline route networks have also changed: from 'point-to-point' in the 1970s, 'hub-and-spoke' in 1980s, to global alliance networks linking the hub-and-spoke networks of particular alliance partners across the globe in 1990s. Figure 3.17 shows a simplified scheme of such evolutionary changes.

As mentioned in Chapter 2, at present three large alliances dominate the world's air transport market: OneWorld (British Airways), STAR (Lufthansa), and SkyTeam (Air France). How far the process of integration and consolidation will go will depend on how quickly the remaining barriers are removed – particularly those on cross border airline ownership and Open Sky agreements between Europe and the US. The process of deregulation and liberalisation of the air transport markets in Asia and between Asia and the rest of the world will also be of great influence. Nevertheless, the evidence indicates that the processes of liberalisation and deregulation of the air transport system have been slower and less intensive than globalisation processes in the rest of the world's economy. Evidently, the system has only followed these general economic processes. One of the main reasons is that the air transport system has not been a part of free trade agreements and integration processes that have taken place between the 1950s and the 1980s such as the EU, NAFTA and the agreement creating the common market of South (Latin) America. Anyway, despite constraints, the air transport system has played a fundamental role in globalisation although continuously facing barriers compromising its full consolidation as a real, affordable global entity to offer more efficient and effective services to its users – air passengers and freight (cargo) shippers.

3.4.3 The Social Benefits

3.4.3.1 Scale, scope and type
The most important social effects-benefits of the air transport system can be considered to be (EEC, 2005):

- Direct benefits from employment and other activities carried out within the system;
- Indirect benefits from employment and activities in the sectors supporting the air transport system in the widest sense;
- Induced benefits due to employment and activities of those people being employed due to direct and/or indirect relationship with the system; and
- Consumer welfare due to an increased opportunity to make a trip.

Some of the above-mentioned benefits, such as for example direct and indirect employment at airlines, airports and ATC/ATM have already been discussed in Chapter 2.

The effects – benefits can be considered at a local and global scale. In the latter case, effects-benefits come from supporting global socio-economic activities such as tourism, trade, productivity, and investments. These are all realised exclusively due to using the air transport system and contribute to the GDP of a given region. For example, GDP can be created by business investments realised partially by making the air trips (EEC, 2005). These aggregate contributions are usually expressed as a proportion of GDP.

Numerous studies and research have been carried out to estimate these benefits. Most of them have dealt with the direct and indirect benefits in terms of employment and consequently contributions to GDP. For example, research into the contribution of European airports has indicated that (ACI Europe, 1998):

- Airports currently create 1.2 million jobs, about 0.2 million of which are jobs directly related to the airport;
- Each of the above mentioned jobs supports about 2.1 additional jobs indirectly related to the airport or induced by spending by air transport employees; and
- Employment at airports represents about 25 of the total employment in Europe, which makes the overall contribution to GDP about 1.4–2.5 per cent.

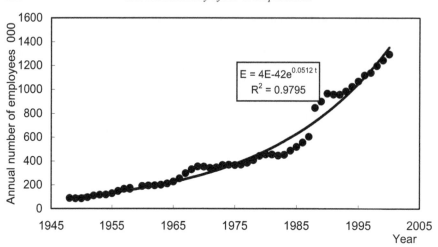

The plotted trend line shows:

$$E = 4E\text{-}42e^{0.0512\,t}$$
$$R^2 = 0.9795$$

Figure 3.18 Development of employment in the US air transport industry
Compiled from BEA (2001)

Some other research has estimated that the contribution of the European air transport system to GDP amounts to about 2.6 per cent of GDP (EEC, 2005). In the context of this discussion it should be mentioned that employment in the air transport system is dependent on the system size driven by growth of the air transport demand. Figure 3.18 shows an example for the development of employment in the US air transport system.

As can be seen, the long-term growth of employment over the past 60 years has been approximately exponential – starting with about 100,000 in the year 1945 and reaching about one million and 400,000 employees in the year 2001. This indicates an increase of about 14 times. The noticeable variations around the general trend indicate restructuring of the sector after deregulation of the airline industry in 1978 and the impacts of the global crisis before and after the Gulf War in 1991. One can easily correlate growth of the employment with increase in the system output (p-km or t-km) over the same period of time.

The assessment of the effects-benefits of the air transport system by its contribution to tourism, trade, productivity and business appears to be a relatively complex task. The most recent study carried out by the EEC (EEC, 2005) provides some assessments of these contributions for the European Union 25 Member States at present and in the year 2025, based on the scenario approach related to growth of the air transport system and GDP as well. The estimates have included the net effects-benefits from tourism, trade productivity and business for the area of EU. The so-called 'consumer surplus' of users whose airfares have been lower than the prices they would be willing to pay for an air trip has not been considered due to the complexity of measuring and the lack of clarity over the contribution to GDP.

Tourism benefits A large proportion of international and a smaller proportion of national-domestic tourists use the air transport system. A country, union of countries (EU), or a continent can be considered as regions where tourists arrive but also from where they depart to some other regions. Both groups spend while abroad, thus creating the outflow and inflow of money for a region. The net difference between this inflow and outflow over a period of time (one year) represents the net contribution of tourism by air transport to the regional GDP. This net contribution can be either positive or negative implying the direction of influence of the transport system.

Trade benefits   A relatively high proportion of high value goods is exported from and imported to a region using the air transport system. In addition to speed, the more distant markets also become accessible within a reasonable time. The net difference between the value of imports and exports of goods can be used as a measure of the contribution of export/import by the air transport system to the GDP of a region. This net value can also be positive or negative implying again a direction of influence of the air transport system.

Investment benefits   The availability of the air transport system links has always been considered supportive to the investments decisions. It implies the ability of a region to attract more inward investments and also invest more onward. In particular, the inward investments realised thanks to the air transport system may have added value through introducing new technologies and other skills leading to increased productivity, all contributing to an increase in GDP.

The labour supply benefits   The air transport system can make a region more attractive for a particular internal and external workforce as commuting with other distant regions becomes possible using the air transport system. This workforce can positively contribute to the GDP of the region in question.

Benefits to productivity   The air transport system expands potential markets for different companies from and outside a region. It therefore increases competition in a region, forcing companies to improve their internal efficiency and effectiveness, which in turn results in a larger contribution to the regional GDP.

The markets and investors benefits   Improved accessibility and service quality provided by the air transport system may stimulate and strengthen competitive pressure on the companies from and outside a region which intend to penetrate the internal (regional) and external markets.

3.4.3.2 Indirect assessment
Some aggregate figures of the effects-benefits of the air transport system obtained through its support of tourism, trade, productivity and business in the EU (25 Member States) are given in Table 3.6. Combined with the benefits from direct and indirect employment contributing about 2.5 per cent to GDP, these contributions of 4.3 per cent and 3.1 per cent give the total system's contribution to GDP of about 7 per cent and 6 per cent in the year 2003 and 2025, respectively.

Other research has identified four types of contributions of the air transport system to GDP (Han and Fang, 2000): i) the air transport system GDP; ii) the air transport system final demand; iii) the air transport system related GDP; and iv) the air transport system driven GDP.

Table 3.6   Benefits of air transport system in the EU (25 Member States)

Type of Net Effect	Contribution to GDP (%)	
	2003	**2025**
Tourism	−0.3	−0.2
Trade	+0.6	+1.5
Location and investment decisions	+2.0	+1.2
Productivity and business	+2.0	+0.6
Total:	+4.3	+3.1

Source: EEC (2005)

The air transport system GDP refers to the sum of all gross value-added created by performing the system activities and services. In this case, a measure of the contribution of the air transport system can be its share of the total GDP. For example, in the US this share was annually about 0.7−0.8 per cent of the transport related GDP during the period 1992−1997.

The air transport final demand as a part of GDP shows how much of the economy's net output is used for air transport purposes. This demand can be defined as the sum of the values of goods and services delivered to the final users. Its share of the GDP could be a useful indicator of the system contribution as a driving force of the economy. In the U. S this value was annually about 1.4 per cent of the transport related GDP during the period 1992−1997.

The air transport system-related GDP is the value-added (or the net value) generated while producing services to satisfy societal needs. The difference between the system GDP and the system-related GDP is that the system related GDP includes the value-added generated in production of the system components and energy for operation. In the US the share of this air transport related GDP in the total transport-related GDP was annually about 6.5 per cent over the period 1992−1997.

The air transport-driven GDP represents the sum of all the value added generated by activities providing the air transport services and the value-added by activities representing direct and indirect inputs for producing these services. Thus, the system driven GDP differs from the system-related GDP as it includes the value added from activities, which indirectly support the air transport services. In this context, it appears to be a rather complex measure of the contribution of the air transport system. For example, in the US, this annual contribution was relatively stable during the period 1992−1997, about 4.3 per cent of the total transport related GDP.

In this context, research on the assessment of the socio-economic effects-benefits of the air transport system in California (US) by using REMI (Regional Economic Models, Inc.) has been particularly interesting. The core of the approach consisted of the hypothesis of what would present if there were no air transport system in the region, as compared with the present situation (ERA, 2003). The effects − benefits of the air transport system have been assessed in terms of: i) direct employment in all activities that make up the air transport sector including the aircraft manufacturing and services, commercial space, governmental aviation and other airport businesses; ii) tourism to and within the region (California); and demand for goods and services induced by the air transport system in the region. The results have shown that the effects-benefits from the air transport system would be about 9 per cent of the regional GDP. For reference, the total contribution of the air transport system for the whole US to the national GDP has been estimated to be about 9.2 per cent for the year 2001 (DRI-WEFA, 2002).

3.4.3.4 Direct assessment

In general, the problem of assessing the benefits of the air transport system consists of two sub-problems: i) estimation of the shares of contribution of the air transport system to the regional GDP including estimation of the variations (stability) of these shares over time (that is calculating the air transport system-driven GDP in a region); and ii) linking such a contribution to causes – that is the system output, including investigation of the potential causal relationship by using regression analysis and relevant time series or cross-sectional data. The former sub-problem has been addressed as a complex and relatively expensive task usually requiring dedicated studies for the region in question. The studies already carried out have shown that the air transport system-driven GDP in a given region has been relatively stable over time. Under such conditions, tackling the latter sub-problem might be less complex. For such purposes the system-driven GDP needs to be presented in convenient form in order to enable easy and straightforward comparison with the system's social costs. For such purposes, it seems appropriate to interrelate the system driven GDP and the total system output (the volume of p-km or t-km) for a region over a given period of time. If such a relationship is shown to be relatively strong, the system's social benefits could be estimated as straightforward by forecasting the volume of the system output and assuming that the nature of the relationship will remain the same as it used to in the past. An example, of estimating this relationship for the European and US air transport system is given in Table 3.7.

In both cases, data for the period 1991–2003 are taken into account. For Europe, the data on GDP is for the countries who are members of the OECD (Organisation of Economic Cooperation and Development) multiplied by the assumed average contribution of the air transport system of 9.2–9.3 per cent, that is by the air transport-driven GDP, are considered as the dependent variable B_{EU} (in billion USD – current prices). The output of the air transport system in terms of the annual revenue ton-kilometres carried out by 30 AEA (Association of European Airlines) airlines is considered as the independent variable RTK (billion). The value of R^2 implies a relatively strong dependability. It can be seen that the marginal value-added by one revenue t-km is about USD16.9. Assuming that one revenue t-km is equivalent to 10 revenue p-km, the marginal value-added by one p-km will amount to about USD1.7. A similar but much stronger linear relationship (R^2) is obtained for the US

Table 3.7   **Dependence of the air transport system-driven GDP on the system's annual output (RTK – revenues tonne kilometres)**

Region	Relationship
Europe	$B_{EU} = 572.339 + 16.878RTK$; $R^2 = 0.632$; $N = 13$ for $RTK > 9.832$
US	$B_{US} = -253.809 + 8.795RTK$; $R^2 = 0.957$; $N = 13$ for $RTK > 100.591$

air transport system for the same period of time. The air transport-driven GDP as the dependent variable B_{US} (USDbn) is obtained by multiplying the national GDP with the share of the air transport system in it of about 11.2–11.3 per cent. The output in terms of the annual volume of t-km (billion) by all scheduled and non-scheduled airlines is considered as the independent variable *RTK*. Consequently, the marginal value-added is about USD8.8 from 1 t-km and/or about USD0.88 from 1 p-km.

3.5 Assessment of Sustainability

3.5.1 Principles of Sustainable Development

The air transport system develops in a sustainable way if its net social benefits, defined as the differences between the social effects and social costs, are constantly positive and preferably increase with the increasing system output. Such development can eventually be achieved by balancing (that is trading-off) between these benefits (effects) and costs (externalities). Generally, such a balance (trade-off) can be established at a global (world) level, regional (continental or country) level, and local (community) level (INFRAS, 2005). Another approach to dealing with the sustainability of the air transport system considers only the social costs (externalities). In such a context, the system is considered sustainable if its social costs remain within the prescribed limits (targets) while the system output over the medium to long-term period increases. This development can be achieved if the rate of decrease of the social cost per unit of output, thanks to the developments in technology, is higher than the rate of growth of the system output.

3.5.1.1 Global trading-off
The economic growth and the growth of air transport demand have been strongly driven by each other, producing evident consequences such as the increased energy (fossil fuels) consumption and emissions of air pollutants (greenhouse gases). A trade-off between the positive effects and negative impacts of such growth may be established by using one of the following policy scenarios:

- *Setting up a cap on impacts* This policy scenario implies setting up a cap on the total energy consumption, associated air pollution, and consequently growth of the air transport demand in absolute terms. However, implementation of such a restrictive scenario through the worldwide consensus of particular actors involved in the air transport operations and business seems unlikely to take place in the short to medium-term future. Nevertheless, the policy concept of emissions trading for aviation currently under discussion at a global ICAO's and local EU level could be an option to keep energy consumption and its associated emissions of air pollutants (mainly CO_2) within the prescribed limits, that is below the cap. Another option could be charging for emissions (Hewett and Foley, 2000; CE Delft, 2002; ICF Consulting, 2004; Trucost, 2004).
- *Decoupling air transport growth from the overall economic growth* This scenario implies weakening (decoupling) the strong dependence between

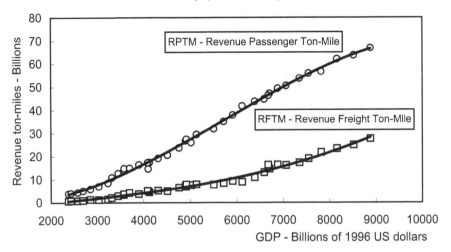

Figure 3.19   Air transport growth vs economic growth: US domestic traffic (1960–1999)

Compiled from BTS (2001)

the air transport demand and GDP as its main external driving force. Figure 3.19 shows an example of such dependence. As can be seen, the dependence is rather strong, which makes weakening it seemingly a very difficult task. Nevertheless, it seems that the long-term aim of encouraging people to change their habits of using air transport could be an option but with unpredictable success (EC, 1999). In addition, developing alternative options to air transport, particularly for long distance trips, are lacking at present.

- *Trading-off between global effects and impacts*   This policy scenario assumes that the long-term conditions guaranteeing faster growth of the system's positive effects than the negative impacts needs to be established at the global level. In general, this can be achieved through adequate technological improvements to the aircraft airframes and engines, and ATM/ATC procedures, as well as by more sophisticated use of land for expanding the system infrastructure. At present, this scenario seems to be the most acceptable.

3.5.1.2 Regional trading-off

At the regional (continental and/or country) level, particularly in the US and Western Europe, growth of air transport demand has been driven by liberalisation of the already matured air transport markets, higher productivity and diminishing airfares. At the same time, such growth has been confronted with the limited capacity of airports and ATM/ATC, which has increased congestion and affected the expected efficiency and effectiveness of services. This has given rise to the question of establishing an appropriate balance (trade-off) between demand and capacity at the regional level to maintain the prescribed effectiveness of services and hinder their further deterioration. Three policy scenarios seem to be reasonable:

- *Changing regional factors* This policy scenario assumes that regional factors are changed – liberalisation, market competition, productivity, and airfares – to discourage further growth in the demand for air transport. If this happened, previous positive progress and development would be annihilated. However, present trends indicate that this scenario is not likely to take place (Boeing, 2001).

- *Constraining the infrastructure expansion* This scenario is called the 'do nothing' scenario in terms of the further expansion of the air transport infrastructure. If such a scenario takes place in some developed markets, such as, for example, those in Western Europe and US, the system infrastructure will reach saturation due to the continuously growing demand, which in turn will cause widespread and severe deterioration of the efficiency and effectiveness of services and consequently deter prospective demand from using the system. Such a scenario has already partially taken place at some congested European airports and airspace, but without any noticeable evidence of a significant decrease in demand (EEC, 2001).

- *More efficient utilisation of the available resources* This policy scenario implies more efficient utilisation of the existing infrastructure – aircraft, airports, and ATM/ATC. This scenario could be made realistic by further improvement of the aircraft and ATM/ATC technologies and operational procedures, modification of the airline hub-and-spoke networks, and co-operation through complementarity with other transport modes (particularly High-Speed Railways). Some elements of this scenario have already taken place around particularly congested European airports included in the High-Speed-Rail network (see also Chapter 4) (Arthur, 2000).

3.5.1.3 Local trading-off

At the local level, the trade-off between the positive effects and negative impacts of growing airports may be established according to two policy scenarios:

- *Constraining the airport growth* This scenario assumes that the growth of demand at a particular airport should be constrained up to the capacity of the existing infrastructure in combination with constraints (caps) on noise and the emissions of air pollutants. On the one hand, such constrains would prevent further escalation of the airport's negative impacts on the local population and environment in terms of noise, air pollution, and land take, above the prescribed caps. On the other hand, they would also constrain the airport's direct and indirect positive effects, induced by airport growth, on the local economy. This scenario has already been partially applied to some of the very congested European airports. For example, for the first time, the Dutch Government has limited the maximum annual number of atm (aircraft movements) at Amsterdam Schiphol Airport, aiming to constrain the noise burden. Consequently, in 1998, the maximum annual number of atm was restricted to 380 thousand, with the possibility for an annual increase of 20 thousand until the year 2003. Currently, the annual limit is 460 thousand atm (Offerman and Bakker, 1998; Boeing, 2001).

• *Management of sustainable airport growth* This policy scenario seems to be the most reasonable and realistic for the development of most airports under present circumstances. The scenario assumes that the growth of an airport should be managed along a medium to long-term trajectory, focusing on enabling the airport to provide the higher rates of increase to the total local benefits rather than the rates of increase of costs of the associated negative impacts. How this scenario can work has already been discussed in Chapter 1 where the question of matching, in the long-term, capacity to demand at London Heathrow Airport (BAA, 2001).

3.5.2 Quantification of Sustainability

The quantification of the sustainability of the air transport system is far from easy. In theory, two approaches – the normative and incremental – have been developed (Nijkamp, Rienstra and Vleugel, 1998).

The normative approach implies comparison of the current and prospective system developments to an a priori and unambiguously defined target level. For example, this can be strict fulfilment of the Kyoto norms and/or the fulfilment of standards set up by national governments regarding the targeted volumes of emissions of greenhouse gases.

The incremental approach involves developing the indicator systems for assessing and quantifying the sustainability of the system. The idea behind the indicator systems – that is indicators and their measures – is to monitor the benefits and costs (impacts) of the system and its particular components over the medium (5–10 years) and long (20–50 years) term. Using the indicator systems leaves nevertheless some unanswered questions. For example, the question could be how much the current and prospective level of the system development contributes to the fulfilment of the objectives of the Kyoto Protocol, that is to the wider goals of environmental and social sustainability. Another question could be the assessment of the real potential of new technologies to the overall long-term system's sustainability.

In particular, the contribution of new technologies to the sustainability of the air transport system can be generally analysed by comparing the average and marginal costs on the one hand and the benefits of diminishing an environmental impact, dependent on the environmental improvements, that is reducing the quantity of given impact, on the other. Figure 3.20 shows the generic scheme (Button, 1993; Levison et al., 1996). The horizontal axis represents the environmental improvements. The vertical axis represents the marginal costs and benefits from these improvements. As can be seen, the marginal costs of improvements MC are likely to raise more than proportionally as the improvements are achieved. At the same time, the marginal benefits obtained by these improvements MB are likely to decrease as the improvements increase. Consequently, given the likely relationship between the marginal costs MC and marginal benefits MB, the optimal level of improvements will be OE_1. Beyond this level the marginal costs MC become higher than the marginal benefits MB, which makes further improvements economically infeasible. The area ABC indicates the losses of net welfare due to continuing environmental improvements beyond level E_1.

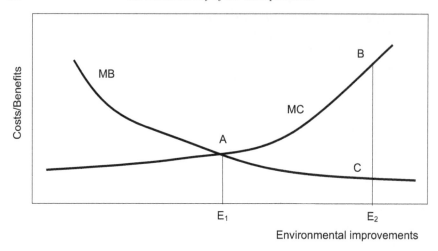

Figure 3.20 A scheme of the costs and benefits of environmental improvements
Compiled from Button (1993)

3.6 Incremental Approach – The Indicator Systems

3.6.1 Complexity and Requirements

The assessment and quantification of sustainability of the air transport system is seemingly a complex task due to the following reasons:

- Multidimensionality of the system performances, which implies a system with numerous and diverse components, actors involved, effects-benefits and costs of impacts including their (inherent) interdependency at the system (component), geographical (spatial) and institutional levels – as well as locally, nationally, and internationally;
- The complexity of setting up the sustainability targets due to the above-mentioned interdependency between the effects-benefits, costs of impacts, and the numerous influencing factors;
- The complexity of setting up the 'reference locations' for assessment and quantification of the system effects-benefits and costs of impacts; and
- The complexity of the assessment and quantification of the marginal and global contributions of particular policy measures and technologies on the system's sustainability.

In order to make this task easier, the comprehensive monitoring of sustainability of the air transport system needs to embrace: i) the multidimensionality and interrelationships of the system components and their performances, the related indicators and the sustainability targets; and ii) the interests and preferences of particular actors involved.

This monitoring can be carried out by the indicator systems. In the research to date, the recommended sustainability indicators and their measures have mainly focused on the system's economic, social and environmental performances at the community, regional, country (national) and global (international) level (Kelly, 1998; Janić, 2003). They have rarely, if at all, explicitly included the technical/ technological and operational dimensions of the system performances and related indicators, which have shown a strong if not crucial influence on the other three categories of performances. The institutional performances have only been implicitly addressed. In addition, many indicators have shown shortcomings in terms of meaning, appropriate measures, and the availability of relevant data. Consequently, in the scope of looking for good indicators, not pretending to be absolutely 'perfect' but always in accordance with the sustainability objectives, Kelly (1998) identified several criteria for such 'good' indicators, which are also applicable to the air transport system. Accordingly, a 'good' indicator should be:

- Calculated by using the already available or easily obtainable data;
- Easily understandable without ambiguity and exceptional overlapping;
- A measure of something important in its own right;
- Available with a relatively short lead time;
- Comparable in terms of different geographical (national and international) scales, and actors involved;
- User-driven, that is useful to the intended users, that is audience;
- Policy relevant, that is pertinent to policy concerns; and
- Highly aggregated, that is the final indexes should be few in number.

Regarding the above-mentioned criteria, it appears that a relatively large number of diverse indicators, inevitably increasing the complexity of assessment and quantification of sustainability of the air transport system, may be developed. In addition, the criteria do not explicitly state the need for considering possible dependability between particular indicators, external influencing factors, and their relevance for different actors (Basler and Partners, 1998; Morse et al., 2001; Litman, 2003).

According to the organisation of the air transport system, the various groups of actors may be involved in dealing with its sustainability. Sustainability may have different meanings and contents for particular actors, which in turn guides their current and prospective preferences, including the short- to long-term objectives (INFRAS, 2005; ATAG, 2003, 2003a). Figure 3.21 shows a simplified scheme for the organisation of the air transport system, focusing on the main actors directly involved in dealing with its sustainability:

- *Users* (consumers), the individual air passengers, freight and mail shippers, constitute the demand for the air transport system and prefer a smooth, frequent, easily accessible, relatively cheap, punctual, reliable, safe and secure service;
- *Airlines, airports and ATM/ATC*, as the air transport system operators, meet the passenger and freight demand by operating the fixed infrastructure, facilities and equipment, and mobile means – aircraft. Generally, they all prefer

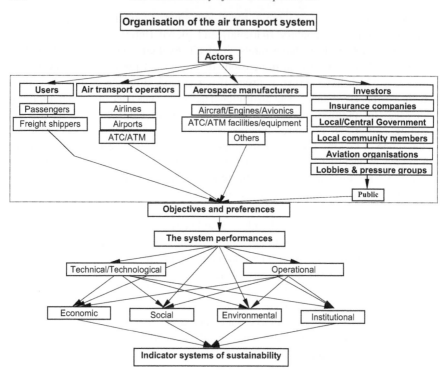

**Figure 3.21   Simplified scheme for assessment of sustainability of the air
           transport system**

profitability, safety and user satisfaction as attributes of the competitiveness
of their services;

• *Aerospace manufacturers* provide the airlines, airports and *ATC/ATM* with
  resources such as the aircraft, radio navigational facilities, equipment and
  other supporting systems and components. They usually prefer profitability of
  their qualitative, reliable, safe, and efficient products;

• *Private* and *public investors* (banks) at a local (national) and global
  (international) level mainly finance the investments in the air transport system
  infrastructure − airports, *ATM/ATC* facilities and equipment, and aircraft.
  They prefer the economic and social feasibility of their investments;

• *Insurance companies* deal with the financial protection of the air transport
  system and users of its services from damages. They always prefer low risk of
  air traffic accidents, low rate of loss of life and damage to properties caused
  by the operations of air transport operations;

• *Governmental authorities* and *policy makers* are usually involved in creating
  and implementing policies which regulate the air transport system operations
  at the local (community), regional, national and international scale. In some
  cases they also subsidise some unprofitable air transport services in order to
  protect and even increase the overall social-economic welfare. In addition,
  they control the externalities of the air transport system operations;

- *Community members* nearby airports benefit and suffer from the closeness to the air transport system. They usually tend to maximise the potential local benefits in terms of employment and availability of the system and minimise the costs of impacts mainly in terms of noise and local air pollution;
- *The national aviation organisations and international associations* coordinate the air transport system development at the local (national) and global (international) level by providing the framework and guidelines for its coordinated (and also sustainable) development;
- *Lobbies* and *pressure groups* are divided into two broad categories. The first category usually articulates the interests of those people opposing any expansion of the air transport system infrastructure by campaigning against the global harmful effects of the polluting systems on people's health and the environment. The aerospace manufacturers and related businesses as another category often tend to promote exclusively their commercial interests and projects and in many cases possibly postpone decisions on the system's regulation;
- *The general public* is mainly interested in the specific aspects of the air transport system such as severe disruptions of the system operations for any reason. Generally, the public prefers objective information.

3.6.2 Structure of the Indicator Systems

3.6.2.1 Assumptions
The indicator systems for the assessment and quantification of sustainability of the air transport system is based on the following assumptions:

- The indicator systems of sustainability are developed for particular groups of actors involved in dealing with sustainability of the air transport system. Thus, the number of indicator systems corresponds to the number of different actors involved;
- Each indicator system consists of six sub-systems corresponding to the six different dimensions of the system's performance (technical/technological, operational, economic, social, environmental, and institutional);
- Each sub-system consists of the individual indicators and their measures;
- The measure of particular indicators expresses quantitatively the system and/or components' effects-benefits and costs of impacts in either absolute or relative monetary or non-monetary terms, usually as functions of the system's output;
- If a 'threshold' or 'target' value for an indicator is set up, it will be used as the reference for comparison of the current with the required system development, and consequently of the current level of sustainability. Introducing 'thresholds' or 'targets' makes this incremental approach closer to the normative approach; and
- For all actors belonging to the same group of actors, the indicator systems are unique.

Different actors may use different indicator systems for the assessment and quantification of sustainability of the air transport system according to their objectives, preferences and interests. They relate to technical/technological, operational, economic, social, environmental, and institutional performances of the system's components.

3.6.3 Indicator Systems for Users – Air Travellers

The indicator system for users – air travellers – consists of eight individual indicators. Five indicators are defined for operational, and one each for economic, social and environmental performance. These indicators mainly relate to the airline and airport services and can be quantified for an individual airline, a route and/or given airport, as well as for the entire airline industry and/or airport network of a region.

3.6.3.1 Operational indicators
The indicators of operational performances are considered to be 'punctuality' and 'reliability' of service, 'lost and damaged baggage,' 'safety' and 'security'.

- *Punctuality* refers to the users' perception of the selected airline to carry out the flights-services on time. The quantification of punctuality can be carried out either by experience or by using airline data. In the former case, the individual's experience plays a crucial role. In this case, an airline is more punctual than another, and therefore has been selected. In the latter case, two measures appear convenient: i) the probability of an airline flight to be on time, which can be calculated as the ratio between the number of on-time flights and the total number of flights carried out during a given period of time (day, month, year); and ii) the average delay per flight, which may include the arrival delay, departure delay or both (as mentioned before, delays are usually categorised into those shorter and those longer than the threshold of 15 minutes (EEC, 2003; USDT, 2003). Both measures are relevant while making a decision off-line about the airline and route connecting given origin and destination. These measures are components of AQR (Airline Quality Rating) in the US (Headley and Bowen, 1992; BTS, 2001). Users usually prefer the former measure to be as high as possible and the latter one to be as low as possible when the number of flights is increasing.
- *Reliability* reflects the users' perception of a chosen airline to fulfil the schedule. Again, this indicator can be assessed either by experience or by using the airline data. In the latter case, the ratio between the number of cancelled (or diverted) and the total number of flights carried out by a chosen airline over a given period of time (day, month, year) can be used as a measure. This measure also belongs to AQR in the US (BTS, 2001a). Independent of the causes of cancellations or diversions of flights, the measure is always preferred to be as low as possible and to decrease with increasing output – the number of flights. Figure 3.22 illustrates an example of the reliability of two US airlines, American and Southwest. The measure of reliability is expressed

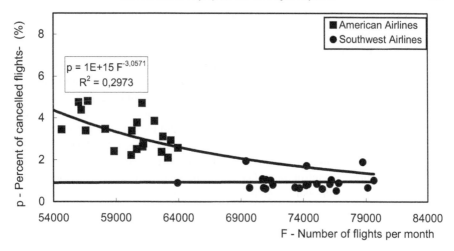

Figure 3.22 Reliability of some US airlines (1999–2000)
Compiled from USDT (2001)

as the proportion of cancelled flights dependent on the total number of flights carried out per month during the period of observation.

As can be seen, at American Airlines this proportion has varied between 2 and 6 per cent and has generally decreased with the increasing number of flights. At Southwest Airlines, it has varied between 0.5 and 2 per cent and has been nearly constant with the increasing number of flights. Southwest Airlines has also carried out a greater number of flights than American Airlines. The example indicates that the airlines with a greater number of flights have seemingly provided higher reliability for their services, which, according to the user's perception has made them more sustainable. An additional reason for the larger proportion of cancelled flights at American Airlines is in its operation of a hub-and-spoke network, which has been shown to be highly vulnerable to disruptions. Southwest operated a point-to-point network enabling them to contain the impacts of disruptions to individual (affected) flights. This is also analysed in Chapter 4.

- *Lost and damaged baggage* relates to the probability of losing or damaging users' baggage while it is within the air transport system. In addition to experience, the relevant data obtained from the chosen airline can be used to quantify this indicator. It also belongs to AQR in the US. The ratio between the number of lost (or damaged) bags and the total number of passengers served during a given period of time (year) should preferably be as low as possible and decrease with increasing output – the number of passengers served.

- *Safety* emerges also as a relevant indicator for users choosing between a few airlines. This indicator measures the perceived risk of death or injury while onboard. The number of deaths (or injuries) per unit of output – RPK (RPM) (RPK – Revenue Passenger Kilometre; RPM – Revenue Passenger Mile) – carried out over a given period of time usually expresses this measure.

The users prefer the measure to be as low as possible and to decrease with increasing output, or volume of RPK (RPM).

- *Security* relates to the perceived risk of exposure to the threat from illegally carried weapons or other dangerous devices (bombs, firearms, guns, and so on) while spending time at airports or onboard aircraft. The ratio between the number of detected illegal dangerous devices and the total number of passengers screened can be used as a measure for this indicator. The users prefer this measure to be as low as possible and to, independently of causes, decrease with increasing output, that is the number of screened passengers. In general, this ratio is out of the control of the airport security service, since it cannot directly control individuals bringing illegally dangerous devices to the airport. However, using sophisticated facilities and equipment for screening passengers may provide a high rate of detection.

3.6.3.2 Economic indicators
The air transport users usually consider the 'economic convenience' of air travel as an important indicator for choosing air transport from a few transport alternatives (Janić, 2001).

Economic convenience   This reflects the total generalised cost of the door-to-door trip, in which the air transport part has the largest proportion. However, some airfares charged by the low-cost carriers in Europe and the US represent exceptions to this general rule. Under such circumstances, for this indicator, the average airfare per passenger either individually or in combination with some other non-transport related indicators can be used as a measure. Users always prefer the airfares to be as low as possible and to decrease in time. Figure 3.23 shows the economic convenience of air transport services for users in the US. It is represented by the average airfares in combination with the Consumer Price Index (CPI).

As can be seen, two periods can be distinguished: the first period was between 1960 and 1982 when the index of airfares was above the index of CPI; the second period is the period from 1983 on, when the index of CPI has been below the index of airfares. The main driving forces of such changes have consisted of decreasing airfares after deregulation of the US aviation market (1978) on the one hand and of a permanent increase in the overall socio-economic progress, particularly from the year 1983, on the other.

3.6.3.3 Social indicators
An indicator of the air transport system social performances relevant for users is the 'spatial convenience' of air transport services.

- *Spatial convenience* reflects the users' opportunity to travel from a given airport by the selected airline(s) to places at medium and long distances. The number of destinations served from that airport (or a region) by a given airline can be used as a measure. In addition, connectivity by the non-stop, one-stop or multi-stop flights with respect to the trip purpose (business, leisure) can refine this measure. Recently, this measure has become one of the competitive tools

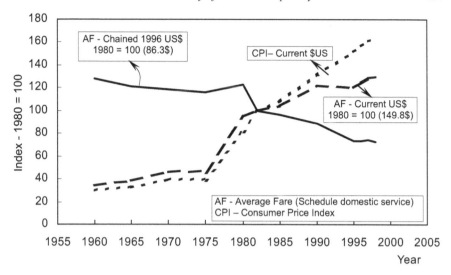

Figure 3.23   Economic convenience of the US air transport system
Compiled from BTS (2001)

of the large airline alliances – OneWorld, STAR, and SkyTeam and indirectly
of their European hub airport – London Heathrow, Frankfurt Main, and Paris
Charles de Gaulle and Amsterdam Schiphol Airport, respectively. In general,
users prefer this measure to be as high as possible and to increase over time.

3.6.3.4 Environmental indicators
Usually, air transport users consider the comfort and healthiness of the airport and
aircraft environment as the relevant attributes for assessing the quality of the travelling
environment. The indicator 'comfort and health' therefore reflects such attitudes.

- *Comfort and health* relates to the users' feeling of comfort while being at the
  airport terminal and onboard the aircraft. Different measures can be used.
  At an airport, in addition to the subjective judgement, density (that is the
  number of passengers per unit of space) and the average queuing time can
  be a measure of the passenger's comfort or discomfort, as a component of
  the airport's quality of service (Hooper and Hensher, 1997; Janić, 2001). The
  configuration and size of the economy class seats as well as the quantity of fresh
  air delivered to the aircraft cabin per unit of time can be the most appropriate
  measures of passenger comfort and the healthiness of the environment while
  onboard the aircraft. For example, the configuration of economy class seats
  on long haul flights has recently emerged as a matter of concern due to cases
  of passenger deaths caused by DVT (Deep Vein Thrombosis). The measures
  of airport discomfort are preferred to be as low as possible and to decline with
  the increasing number of passengers served during a given period of time.
  Both measures of comfort while onboard the aircraft are preferred to be as
  high as possible and to increase over time, which is achievable by renewing
  the airline fleet and setting up a more comfortable seat configuration.

3.6.4 Indicator Systems for Airlines

The indicator system for airlines consists of 11 indicators: five for operational performances, two for economic, none for social and institutional and four for environmental performances. These indicators can be quantified for an individual airline, an airline alliance or for the entire airline industry of a given region (country or a continent).

3.6.4.1 Operational indicators
The attributes 'size,' 'load factor,' 'punctuality' and 'reliability' of services, and 'safety' are considered as indicators of the airline operational performances.

- *Size* reflects the volume of airline output carried out during a given period of time (season, year). Several measures can be used to quantify this indicator: the total number of passengers, the total volume of freight and the total volume of r-tkm (revenue tonne kilometres) or r-tm (revenue tonne-mile) (Janić, 2001). In addition, r-pkm (r-pm) (revenue passenger kilometres (miles)) and freight tkm (tm) can be used separately instead of the aggregate r-tkm (r-tm). The number of aircraft and staff employed to carry out the output can also be used to measure the airline size. Some possible relationships between particular airline resources have already been discussed in Chapter 2. The preference is for all the above measures to be as high as possible and to increase over time.
- *Load factor* indicates the dynamic utilisation of an airline capacity over a given period of time (season, year). It is usually measured in the aggregate form as the ratio between the total r-tkm (r-tm) (revenue tonne-kilometre (-mile)) and the total a-tkm (a-tm) (available tonne-kilometre (mile)). The load factor can be determined separately for passengers and freight. In each case, this measure is preferred to be as high as possible and to increase with the increasing airline output – r-pkm (r-tm) or a-tkm (a-tm) (Janić, 2001).
- *Punctuality and reliability* of service, and *safety* are the indicators analogous to those of the users in terms of measurement and preferences. The airlines use them as competitive tools on the one hand and as indicators of the operational efficiency on the other (Janić, 2001).

3.6.4.2 Economic indicators
Two indicators express the economic performances of an airline, both of which have also been discussed in Chapter 2.

- *Profitability* relates to the airline's financial success. It is measured by the average profits (difference between the operating revenues and operating costs) per unit of output – r-tkm (r-tm). This indicator is preferred to be positive, as great as possible, and to increase with increasing airline output.
- *Labour productivity* reflects the airline efficiency in using its workforce. A measure of this indicator can be the volume of airline output (r-tkm) or (r-tm) per employee. The output and the number of employees are considered for

a given period of time (year). The preference for this measure is that it is as great as possible and to increase with the increasing number of employees.

3.6.4.3 Social indicators
The social indicators of airline performances, expressed by employment, have been already examined in Chapter 2 and earlier in Chapter 3, whilst discussing the social benefits of the air transport system.

3.6.4.4 Environmental indicators
Four indicators express the environmental performances of an airline. These are 'energy,' 'air pollution,' 'noise,' and 'waste' efficiency.

- *Energy* and *air pollution efficiency* relate to the rate of modernisation and efficiency of utilisation of the airline fleet in terms of the energy (fuel) consumption and associated emissions of air pollutants. These indicators can be measured by the average quantity of fuel and air pollution, respectively, per unit of output – r-tkm (r-tm), *d* (distance flown) or *fh* (flying hour). Both measures are preferred to be as low as possible and to decrease with increasing airline output (see Figure 2.11, which shows the efficiency of fuel consumption at British Airways during the period 1974–2000, expressed in grams of fuel consumed per r-pkm (revenue passenger kilometre) dependent on the total annual volume of r-pkm. This consumption has generally decreased more than proportionally with the increasing volume of airline annual output, implying a decrease in the associated air pollution. Such undoubtedly long-term sustainable development has been achieved thanks to the permanent modernisation of the airline fleet on the one hand and the provision of more effective services by the ATC/ATM on the other).
- *Noise efficiency* indicates the rate of modernisation of the airline fleet in terms of the use of noise Stage 3 and 4 aircraft (ICAO, 1993b; BA, 2001). This indicator can be measured by the proportion of Stage 3 and 4 aircraft in the airline fleet, which is preferred to be as great as possible and to increase as the airline fleet grows. Once an airline fleet is completely modernised by replacing all Stage 2 aircraft by Stage 3 and 4 aircraft, this indicator will become irrelevant.
- *Waste efficiency* indicates the generation of airline waste. This indicator can be measured by the average quantity of in-flight waste per unit of airline output – RTK (RTM) (BA, 2001). This measure is preferred to be as low as possible and to diminish with growing airline output.

3.6.5 Indicator Systems for Airports

The indicator system for airports consists of 11 indicators. Four indicators are defined for operational, two for economic, none for social, technological and institutional, and five for environmental performances. They can be quantified for an individual airport or the airport network in a region (Janić, 2003).

3.6.5.1 Operational indicators

'Demand,' 'capacity,' 'quality of service' and 'integrated service' can be regarded as the main indicators of an airport's operational performances. They have already been partially discussed in Chapter 2.

- *Demand* indicates a scale of airport operation. The number of passengers, air transport movements (atm) (atm is either an arrival or a departure), and the volume of freight accommodated during a given period of time (hour, day, year) can be measures of this indicator. Sometimes, it is more convenient to use Workload Unit (WLU) as an aggregate measure for both passenger and freight demand volumes. The airport operator prefers these measures to increase in time (Workload Unit is a physical equivalent for one passenger or 100 kg of freight) (Doganis, 1992).
- *Capacity* reflects the airport's maximum capability to accommodate a certain volume of demand under given conditions. Two measures can be used: the airside capacity in terms of the maximum number of atm, and the landside capacity in terms of the maximum number of WLU accommodated over a given period of time (hour, day, year). Both measures are preferred to be as high as possible and to increase in line with growing demand.
- *Quality of service* reflects the relationship between the airport demand and capacity. Generally, the average delay per atm or WLU, which occurs whenever the demand exceeds the capacity, can be used as a measure. This measure is preferred to be as low as possible and to decrease with increasing demand, such as the number of atm/WLU.
- *Integrated service* is an indicator, which may be relevant for the airports connected to the surface regional, national and international transport networks. Generally, these airports have the opportunity to improve utilisation of their capacity by substitution of some short-haul flights with adequate surface, usually High-Speed-Rail, services and by using such freed slots for more profitable long-haul services. For example, three European hubs – Frankfurt Main, Paris CDG and Amsterdam Schiphol Airport – are connected to the Trans-European High-Speed-Rail Network. There have been some indications that partial substitution of the short-haul flights with the HSR services has already taken place there (EC, 1997; HA, 1999; INFRAS, 2005). If the air-rail substitutions were carried out without filling in the freed slots with long-haul flights, congestion, associated local and global air pollution, and noise would eventually be diminished. Under such circumstances, this indicator could also be considered as an environmental indicator and a measure of this indicator could be the ratio between the number of substituted flights and the total number of viably substitutable flights carried out over a given period of time (year). This ratio is preferred to be as high as possible and to increase with the increasing number of substitutable flights.

3.6.5.2 Economic indicators

Airports as business enterprises also look after their economic performances. 'Profitability' and 'labour productivity' are defined as the most convenient indicators of these performances.

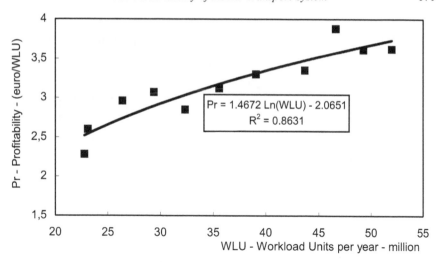

Figure 3.24 Profitability of Amsterdam Schiphol Airport (1990–2000)
Compiled from Schiphol Airport (2004)

- *Profitability* usually reflects the airport's financial-business success. It is usually measured by the operating profits (the difference between the operating revenues and operating costs) per unit of the airport's output – WLU (Doganis, 1992). This measure is preferred to be as high as possible and to increase the airport output increases. Figure 3.24 illustrates the profitability of Amsterdam Schiphol Airport (the Netherlands). Profitability in terms of €/WLU (Euro per Work Load Unit) has been related to the total annual number of WLU accommodated at the airport. As can be seen, this profitability has increased with the increasing number of WLU but at a decreasing rate. Thus, regarding this indicator, the long-term airport sustainability has been indicative.
- *Labour productivity* reflects the efficiency of labour use at an airport. The output in terms of the number of WLU (or ATM) per employee carried out over a given period of time (year) can be used as a measure (Doganis, 1992; Hooper and Hensher, 1997). It should be mentioned that only the airport's direct employees should be taken into account. This measure is preferred to be as high as possible and to increase with the increasing number of employees. Figure 3.25 illustrates an example of this indicator, again for Amsterdam Schiphol Airport (the Netherlands). The productivity indicator expressed as the number of WLU per employee has been related to the total number of WLU accommodated at the airport per year. As can be seen, during the observed period, this productivity has generally increased with the increasing number of WLU at a decreasing rate, which becomes zero after the annual number of WLU has exceeded 45 million. Such a development indicates how the sustainability of the airport with respect to this indicator could vanish with growth.

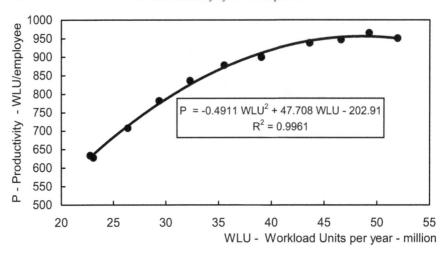

Figure 3.25 Labour productivity at Amsterdam Schiphol Airport
(1990–2000)

Compiled from Schiphol Airport (2004)

3.6.5.3 Social indicators
The indicators of the airport's social performances have not been explicitly
considered. Employment as an indicator of these performances has been elaborated
in the scope of the system's social performances relevant to the local community
around the airport and in Chapter 2 as well.

3.6.5.4 Environmental indicators
The indicators such as 'energy efficiency,' 'noise,' 'air pollution,' 'waste' and
'land use efficiency' have been defined as indicators of the airport's environmental
performances. These indicators, which appear relevant while undertaking the
mitigation measures, relate to impacts of the airport on the health of local people
and the environment.

- *Energy efficiency* relates to the total energy consumed by an airport over a
  given period of time (year). This energy, obtained from different sources,
  is used for lighting and heating. A measure for this indicator is the energy
  consumed per unit of airport output – WLU – over a given period of time
  (year). This measure is preferred to be as low as possible and to decrease with
  the increasing volume of airport output.
- *Noise efficiency* relates to the noise energy generated by atm (air transport
  movements) over a given period of time. This has been extensively discussed
  both in this chapter and Chapter 2. One measure for this indicator is the size
  of area exposed to the equivalent long-term noise level L_{eq} (in decibels –
  dB(A)). The size of this area is expressed in square kilometres. This indicator
  is preferred to be as low as possible and to diminish as the number of atm
  increases.

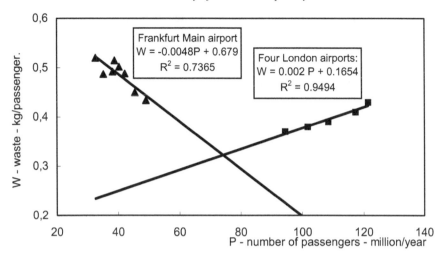

Figure 3.26   Waste efficiency at European airports
Compiled from Fraport (2001); BAA (2001)

- *Air pollution efficiency* relates to the total air pollution generated by the operations of an airport. The amount of all or only specific air pollutants can be taken into account. In addition to that of atm, the air pollution from the airport surface access (road) traffic, and the air pollution from the airport handling operations needs to be taken into account (EPA, 1999). Generally, the amount of air pollutants per polluting event – LTO cycle – can be used as a measure (ICAO has recommended the LTO cycle – Landing/Take-Off cycle – as the standard framework for quantifying the air pollution at airports; ICAO, 1993a) and the non-LTO cycle related pollution could be allocated to each of them. This measure is preferred to be as low as possible and to decrease with the increasing number of LTO cycles over a given period of time (year).
- *Waste efficiency* relates to the waste generated operations of an airport, excluding the airline waste (BA, 2001). A convenient measure of this indicator is the amount of waste generated per unit of the airport's output (WLU). This measure is preferred to be as low as possible and to decrease with the increasing airport's output over a given period of time (year). Figure 3.26 shows two examples of waste efficiency at Frankfurt Main airport (Germany) and three London airports – Heathrow, Stansted, and Gatwick (UK). The waste efficiency is expressed by the average amount of waste per passenger dependent on the annual number of passengers accommodated.

  As can be seen, this average amount decreases at Frankfurt Main airport and increases at the London airports as the annual number of passengers increases, which has indicated their sustainable and unsustainable development, respectively, with respect to this indicator.
- *Land use efficiency* relates to the utilisation of land acquired for building and operating an airport – both the airside and landside parts. Once the infrastructure has been constructed, the intensity of use of the acquired land becomes dependent on the volume of demand. However, this intensity is always limited

by the capacity of the infrastructure. In such context, a convenient measure for this indicator is the volume of workload units accommodated over a given period of time (year) per unit of land acquired. This measure is preferred to be as high as possible and to increase with the increasing amount of land taken by the airport. The case for this indicator has been partially considered when land use as an externality was discussed, earlier in this chapter.

3.6.6 Indicator Systems for the ATC/ATM

The indicator system for the Air Traffic Control/Management (ATC/ATM) consists of eight indicators, four for the operational, two for economic, none for the social and institutional, and two for the environmental performances. These indicators can be quantified for a part, that is for a given ATC/ATM sector or for the entire system covering the airspace of a given country or region (continent).

3.6.6.1 Operational indicators

'Demand,' 'capacity,' 'safety' and 'punctuality of service' are defined as the operational indicators.

- *Demand* is measured by the number of flights accommodated (that is controlled) in given ATC/ATM airspace over a given period of time (hour, day, year) (Janić, 2001). This measure is preferred to be as great as possible and to increase over time.
- *Capacity* expresses the operational capability of an ATC/ATM service unit to accommodate demand under certain conditions. This indicator can be measured by the maximum number of flights served in a given airspace per unit of time (Janić, 2001). This indicator is preferred to be as great as possible and to increase over time. It has already been discussed in Chapter 2.
- *Safety* expresses the risk of occurrence of air traffic accidents due to ATC/ATM operational error. Such accidents may take place at airports or in the airspace. Some convenient measures for this indicator are the number of individual aircraft accidents or the number of Near Midair Collisions (NMAC) per unit of the ATC/ATM output – the number of controlled flights. These measures are preferred to be as low as possible and to decrease with the increasing number of flights. Figure 3.27 shows examples of measures of safety in the European and US ATC/ATM. In this case the measure is the number of air proximities and level busts estimated dependent on the annual number of flights in European and US airspace. As can be seen, in both regions, this indicator has generally decreased with the increasing number of flights, but the rates of decrease have been different. Nevertheless, both systems have been developed in a sustainable way according to this indicator, that is flying has been less and less risky despite the increasing traffic density.
- *Punctuality of service* is a surrogate for the quality of service provided by the ATC/ATM to users – aircraft/flights. This indicator can be measured in two ways: the proportion of non-delayed flights due to ATC/ATM restrictions, and the average delay per delayed flight. The former measure is preferred to

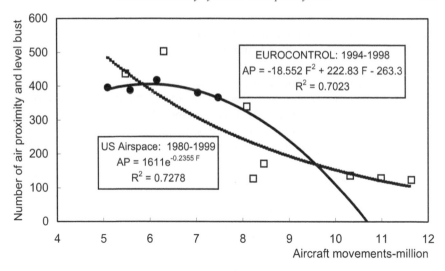

Figure 3.27 Indicators of safety in European and US airspace
Compiled from EC (1999b); DETR (2000); BTS (2001b)

be as high as possible and to increase with the increasing number of flights over a given period of time. The latter measure is preferred to be as low as possible and to decrease with air traffic growth. Both measures have already been discussed in Chapter 2.

3.6.6.2 Economic indicators

Two indicators are defined to express the economic performances of the ATC/ATM unit: 'cost efficiency' and 'labour productivity'.

- *Cost efficiency* relates to the ATC/ATM operating costs. 'Cost' is considered a more relevant indicator than 'profitability' because most ATC/ATM providers charge for their services on the cost-recovery principle. For example, the EUROCONTROL Member States and the ATM/ATC from Canada, Australia, New Zealand, and South Africa fully recover their costs by charges (INFRAS, 2005). This indicator can be measured by the average cost per unit of output – the controlled aircraft/flight. The measure is preferred to be as low as possible and to decrease with the increasing number of flights over a given period of time (year).
- *Labour productivity* reflects the efficiency of using labour by an ATC/ATM. The convenient measure for this indicator is the number of controlled flights per employee. This indicator is preferred to be as high as possible and to increase with the increasing number of employees.

3.6.6.3 Social indicators

The indicators of the social performances of ATC/ATM are not being considered in this context. Nevertheless, the measure of this indicator can be the number of employees discussed in Chapter 2.

3.6.6.4 Environmental indicators

Two indicators are defined to express the environmental performances of the ATC/ATM unit: 'energy efficiency' and 'air pollution efficiency'.

- *Energy efficiency* relates to the extra fuel consumption due to deviations of a flight/aircraft from the prescribed (fuel-optimal) trajectories dictated by the *ATC/ATM* safety requirements. This indicator can be measured by the average extra fuel consumption per flight (some examples are presented in Chapter 4). The measure is preferred to be as low as possible and to decrease with the increasing number of flights.
- *Air pollution efficiency* relates to the extra emission of air pollutants due to extra fuel consumption. The indicator is measured by the average quantity of the extra-emitted pollutants per flight. The measure is preferred to be as low as possible and to decrease with the increasing volume of traffic – the number of aircraft/flights.

3.6.7 Indicator Systems for Aerospace Manufacturers

The indicator systems for aerospace manufacturers consist of eight indicators as follows: three for technological/operational, two for economic, none for social and institutional, and three for environmental performances. These indicators can be quantified for an individual aerospace manufacturer or for a specific sector.

3.6.7.1 Technological/Operational indicators

The 'innovations of aircraft,' 'innovations of the ATC and airport facilities/equipment,' and 'reliability of the structures' are defined as indicators of the technological/operational performances.

- *Innovations of the aircraft* reflect the technological progress in the design of commercial aircraft including the increasing of aircraft speed, capacity and cost efficiency (RAS, 2001). The progress in speed and capacity can be measured by technical productivity as their product (tonne-kilometres (miles) per hour). The technical productivity of commercial aircraft has generally increased by introducing larger aircraft with higher subsonic speeds (Arthur, 2000). Figure 3.28 shows the historical development of technical productivity of the commercial aircraft expressed by the number of t-km/h (tonne-kilometres per hour). As can be seen, the average technical productivity has increased over time at a decreasing rate. After introducing the DC 3 aircraft, the increase in the technical productivity has been primarily achieved by development of larger aircraft and less by increasing the aircraft operating (cruising) speed. A culmination of such developments is certainly the introduction of the Airbus A380 into service in the year 2006/07. In addition, such a development has simultaneously included the development and upgrading of the aircraft engines (jet engines after DC3) in terms of their fuel and air pollution efficiency on the one hand and the development of sophisticated avionics on the other.

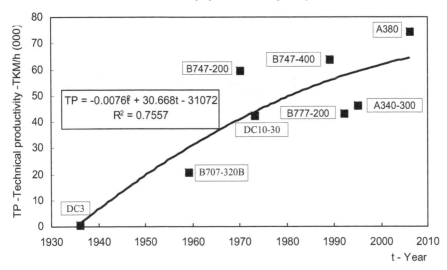

Figure 3.28 Development of aircraft technical productivity
Compiled from FI (2000; 2001)

Aircraft cost efficiency is usually measured by the average operating cost per unit of the aircraft capacity – a-tkm (a-tm) (a-tkm − available tonne kilometre; a-tm − available tonne mile). This cost generally decreases as the aircraft capacity increases, as discussed in Chapter 2 (Janić, 2001).

- *Innovations of ATC* and *airport facilities and equipment* relate to the technological progress in developing avionics, ATC/ATM, and airport facilities and equipment. Progress in developing avionics and ATC/ATM equipment can be measured by the cumulative navigational error of the aircraft position, which has been significantly reduced over time (Arthur, 2000). This has brought gains in the airspace capacity and safety. Progress in the development of airport facilities and equipment can be measured by increasing the efficiency (capacity) of the processing units in both the airport airside and landside area (Janić, 2001). This measure is preferred to be as high as possible and to increase over time.

- *Reliability of structures* reflects the feature of particular system components to operate without unexpected failures. This indicator can be separately measured for different components, but in any case, the average number of failures per unit of operating time (hour, day, year) can be used as a measure. For safety and operational reasons, this measure, independent of the system component, is preferred to be as high as possible and to improve over time.

3.6.7.2 Economic indicators
'The efficiency of using resources,' 'profitability' and 'labour productivity' are defined as indicators of the economic performances of aerospace manufacturers.

- *Efficiency of using resources* indicates the quantity of materials used for assembling an aircraft. In absolute terms this quantity increases as the aircraft

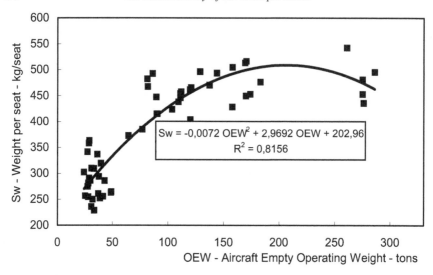

Figure 3.29   The intensity of resource-material use for aircraft manufacturing
Compiled from FI (2000; 2001)

size increases, that is seating capacity. Therefore, the convenient measure could be the quantity of material used per unit of aircraft capacity – seat. Figure 3.29 shows the relationship between the aircraft empty weight per seat and the total aircraft empty weight. As can be seen, the average weight per seat has generally increased at a decreasing rate as the aircraft empty weight increases. This illustrates the achievements in increasing the aircraft size using smaller amounts of lighter materials for each unit of its capacity (seat).

- *Profitability* expresses the financial success or failure of an airspace manufacturer. It can be measured by the average operating profits (the difference between operating revenues and costs) per unit sold. For any airspace manufacturer, this measure is preferred to be as great as possible and to increase the number of units sold increases.

- *Labour productivity* expresses the efficiency of the airspace manufacturers in using labour (workforce). Similarly as at airlines, airports and ATC/ATM providers, the average number of units produced per employee can be used as a measure. This measure is preferred to be as high as possible and to increase as the total number of employees increases.

3.6.7.3 Social indicators
These indicators have not been particularly elaborated. Nevertheless, the employment rate at the aerospace manufacturers could be an indicator. Some elements of an indicator defined so, and its relevance, have been discussed in the scope of the discussion on the social benefits of the air transport system earlier in this chapter.

3.6.7.4 Environmental indicators
Three indicators – the 'energy (fuel),' 'air pollution' and 'noise efficiency' – can be used to express the environmental performances of new aircraft and engines produced

by the aerospace manufacturers. These indicators have already been discussed in this chapter. Nevertheless, in this context they can be considered as follows:

- *Energy*, *air pollution* and *noise efficiency* reflect the reduction in fuel consumption, associated air pollution and noise energy generated by new aircraft and their engines, respectively, in both absolute and relative terms. They can be measured by an absolute or relative decrease in the amount of fuel consumption, air pollution and noise energy, respectively, per unit of the engine power or per unit of the aircraft operating weight as shown earlier in Chapter 2 and this chapter. These measures are preferred to be as low as possible and to decrease with increasing engine power and/or aircraft operating weight.

3.6.8 Indicator Systems for Local Community Members

People living permanently or temporarily near an airport represent a group of local community members. Logically, they are mostly interested in the social and environmental performances of the air transport system. Consequently, the indicator systems consist of four indicators: none for the operational, economic and institutional, one for the social, and three for the environmental system performances.

3.6.8.1 Social indicators
'Social welfare' can be considered as an indicator of the air transport system's social performances.

- *Social welfare* relates to the opportunity of local community people to get a job either directly or indirectly at the local air transport system (DETR, 1999). A convenient measure is the ratio between the number of people employed by the air transport system and the total number of employed community people. This measure is preferred to be as high as possible and to increase with increasing employment within local community.

3.6.8.2 Environmental indicators
'Noise disturbance,' 'air pollution' and 'safety' are defined as the indicators of the environmental performances.

- *Noise disturbance* reflects the annoyance of local people by noise from air transport operations at the neighbouring airport. The annoyance mostly depends on subjective and objective factors. The subjective factors reflect the individual's sensitivity to noise. In such a case, any noise equal to or exceeding a given (subjective) threshold is considered annoying. The most important objective factors include the amount of noise energy generated by aircraft flying over the affected area, the distance between the residential location and the aircraft flight paths, and the quality of insulation in the houses with respect to noise. Bearing in mind both types of factors, the number of complaints to noise by local people over a given period of time (day, month, year) can be

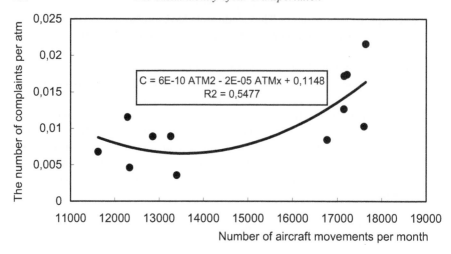

The number of complaints per atm

C = 6E-10 ATM2 - 2E-05 ATMx + 0,1148
R2 = 0,5477

Number of aircraft movements per month

Figure 3.30   Complaints to aircraft noise at Manchester Airport
Compiled from MA (1999)

defined as a measure. Figure 3.30 shows an example for Manchester airport
(UK), which is expressed by the average number of complaints per atm (atm
– air transport movement either landing or take-off) dependent on the total
number of atm carried out over a given period of time (a month). As can
be seen, up to about 13 thousands atm per month, the average number of
complaints decreased. However, after that it increases more than proportionally
with the increasing of the monthly number of atm. A possible explanation
is twofold: On the one hand, the number of complaints decreased after the
number of atm stabilised, thus reflecting the people's accommodation to a
given level of noise; on the other hand, after implementing a new (the second)
runway at the airport, the number of complaints increased due to increase in
the overall number of operations combined with the redistribution of the total
noise burden, also to the population in the vicinity of the new runway.

This population is experiencing the noise annoyance for the first time. Such
a development indicates that improvements in the operational performances
might sometimes compromise other, in this case the environmental,
performances at an airport. Anyway, this measure is preferred to be as low as
possible and to generally decrease with the increasing number of atm.

• *Air pollution* relates to the exposure of local people to the harmful impacts
of air pollution generated by the local air transport system. This indicator can
be measured as the ratio between the air pollution generated by the local air
transport system and the total air pollution generated by all local (air polluting)
sources. This indicator is preferred to be as low as possible and to decrease as
the total air pollution increases.

• *Safety* relates to the perceived risk of death or injury, and/or damage or loss of
local property due to an aircraft accident (crash) at an airport. This indicator
can be measured by the number of aircraft accidents per atm carried out at
the airport over a given period of time (year). This measure is preferred to

be as low as possible and to decrease with the increasing number of atm (air transport movements).

3.6.9 Indicator Systems for Local and Central Government

Usually, both local and central governments are not directly interested in the operational performances of the air transport system, except in cases of significant disruptions of services on both a local and global scale. A significant reduction in the punctuality and reliability of the air transport services may lead to deterioration in the overall system performances, other dependent socio-economic activities, and consequently the quality of life of the affected people. In addition, the governments at both levels are primarily interested in the system's economic, social and environmental performances. The indicator systems consist of seven indicators: none for the operational and institutional, three for the economic, one for the social and three for the environmental performances.

3.6.9.1 Economic indicators
'Economic welfare,' 'internalisation/globalisation,' and 'externalities' are defined as the indicators of the economic performances of the air transport system:

- *Economic welfare* relates to the contribution of the air transport system to local (community) and regional (national) wealth. An adequate measure can be the proportion of GDP realised by the air transport system in the total GDP of a given region. This measure is preferred to be as high as possible and to increase as the total GDP increases.
- *Internalisation/Globalisation* relates to the contribution of the air transport system to the internationalisation of local and regional (national) business – trade, investments, and tourism. Three measures can be used to quantify this indicator as follows:
  i.   The proportion of trade carried out by the air transport system in the total regional (or country's) trade. 'Trade' can be expressed by the volume and/ or value of the exported and imported goods and services. Figure 3.31 illustrates an example of the contribution of the air transport system to UK international trade during the period 1992–1998. As can be seen, the share of air transport by value in the country's import and export market has increased as the total value of international trade (import + export) has increased, thus reflecting the air transport system's ability to gain more expensive shipments.
  ii. The ratio between the number of long-distant business trips carried out by air and the total number of such trips carried out by all transport modes from/to a given region during a given period of time (year); and
  iii. The ratio between the number of long-distant tourist trips by air and the total number of long-distant tourist trips to/from a given region over a given period of time (year). All three measures are preferred to be as high as possible and to increase as the total value of international trade, and the number of business and tourist trips increases, respectively.

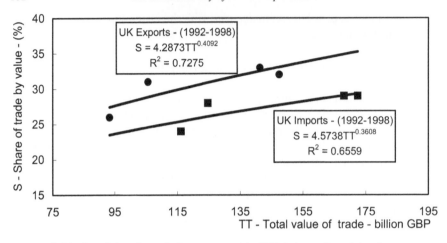

Figure 3.31   Participation of air transport in UK international trade
Compiled from DETR (2000b)

- *Externalities* relate to the costs of air transport noise, air pollution, and air incidents/accidents. Sometimes, congestion costs are also included (Levison et al., 1996). Local (community) and central (regional, national) governments are both interested in these costs due to their responsibility for creating policies for a healthier and more environmentally friendly society (EC, 1997; DETR, 2001). The measures of externalities for such purposes can be the average and marginal social costs of a given impact per unit of the system output – r-pkm (r-pm) (Ying-Lu, 2000). This measure is preferred to be as low as possible and to decrease as the system output increases.

3.6.9.2 Social indicators
'Overall social welfare' is defined as an indicator of the social performances.

- *Overall social welfare* represents the benefits gained by the total direct and indirect employment by the air transport system at a local (community) and regional (country) level, including the indirect system contribution to GDP. The measure can be the share of the system-realised GDP in the total GDP of given region. This measure is preferred to be as high as possible and to increase over time.

3.6.9.3 Environmental indicators
Four indicators are defined for the environmental dimension of performance: 'global energy efficiency,' 'global noise disturbance,' 'global air pollution,' and 'global land take (use)'.

- *Global energy efficiency* relates to the total energy (fuel) consumed by the air transport industry of a given country or region. This indicator is particularly important for central government while planning the energetic budget of the country. Nevertheless, for the purpose of assessing sustainability, a convenient measure for this indicator can be the average fuel consumed per unit of the system's output – r-tkm (r-tm) carried out over a given period of time (year). This measure is preferred to be as low as possible and to decrease as the system's output increases.

- *Global noise disturbance* relates to the global exposure of the local (community) and regional (country) population to noise generated by the air transport system. This indicator can be measured by the total number of people exposed to air transport noise over a given period of time (year). The measure is preferred to be as low as possible and to decrease over time (see Figures 3.11 and 3.12).
- *Global air pollution* relates to the global emission of air pollutants by the air transport system. This indicator can be measured by the total emissions of air pollutants over a given period of time (year). In this case, the total quantity of air pollutants comprises the air pollution during LTO cycle, climb, cruise, and descent phases of flight. This measure is preferred to be as low as possible and to diminish as the total system's output increases (see Figure 3.5).
- *Global land take* (use) relates to the total area of land taken (used) for the local (community) and regional (national) air transport infrastructure. An appropriate measure for this indicator is the intensity of land use expressed as the ratio between the total area of land taken and the total volume of the system output carried out over a given period of time (year). This measure is preferred to be as low as possible and to decrease as the system's output increases.

3.7 Perspectives of Sustainability of the Air Transport System

3.7.1 Basic Principles

The indicator systems developed with respect to the basic principles and rules regarding their generality, transparency and applicability have been able to measure the air transport system performances in both absolute and relative terms, and in relation to the volume of output assumed to generally increase over time. As such, it is possible to assess the sustainability of the entire system or of its particular components at a global, regional and local level. Fifty-eight individual indicators and 68 measures have been defined in the scope of the indicator systems, corresponding to seven groups of actors – users – air travellers, the system's operators – airlines, airports, ATC/ATM, aerospace manufacturers, local community members, and local and central governments.

Nevertheless, despite reflecting their particular effects and impacts, these indicators do not express the absolute benefits and costs but only trends in their development. Most indicators and their measures are expressed by the amount of their specific effect and/or impact per unit of the system's output, and in relation to the total volumes of this output are expected to generally increase over time.

In general, the measures of most indicators expressing the effects-benefits of the air transport system have increased as the output has increased. In most cases, this increase has been proportional or even more than proportional. The measures of indicators expressing the impacts-costs of the air transport system have decreased as the volume of system output has increased, usually more than proportionally. Consequently, the analytical form for any relative measure of the indicator systems expressing either the benefits-effects or impacts-costs could be as written follows:

$$I_{e/i} = aV^b \qquad (3.24)$$

where

V is the volume of system output over a given period of time (year);

a, b are the coefficients obtained usually by the last square regression technique applied to the empirical data.

In Equation 3.24, usually the coefficient a is positive for the measures of both effects and impacts. The coefficient b is generally positive for the effects ($b > 0$) and negative for the impacts ($-1 < b < 0$). The total effects and/or impacts in relation to the volume of the system output can be obtained using Equation 3.24 as follows:

$$IT_{e/i} = V * aV^b = aV^{b+1} \qquad (3.25)$$

Since, in Equation 3.24, the coefficient b is positive, it will definitely be greater than one in Equation 3.25, reflecting a more than proportional increase in the effects the system's output increases. At the same time, since the coefficient b for impacts is negative, but not less than (-1) in most cases, the total impacts will increase less than proportionally as the system's output increases. Consequently, the gap between the total effects and impacts will increase with the increasing system output, implying the system's sustainable development. An example of regulating noise at London Heathrow Airport (UK) shown in Figure 3.32 can be used to illustrate the above-mentioned principle.

The horizontal axis represents the annual number of atm. The vertical axis represents the number of people exposed to a given noise burden. As can be seen, the population exposed to noise (*35 NNI and 57 dB(A)L$_{eq}$*) has continuously decreased despite the increase in the annual number of aircraft movements (atm), which have been considered as obvious benefits for the airport. The number of exposed people has stabilised at about 250 thousand with the annual number of operations at about 440 thousand. It is expected to remain the same despite the expected continuous increase in the annual number of aircraft movements over the period 2002–2015, from about 460–610 thousand. The two parallel runways operating in the segregated and/or mixed mode will accommodate such traffic growth. This has and will be achieved mainly through further improvements in the aircraft's noise performance, selective access of noisy aircraft including a complete ban (Chapter 2 aircraft), and improvements of the ATC/ATM arrival and departures procedures around the airport by using innovative technologies and operational procedures. In addition, close co-operation between the airport and airlines will also be needed (BA, 2004). Nevertheless, changing the runway operation mode from the present segregated one into a mixed one and particularly building the new third parallel runway in the

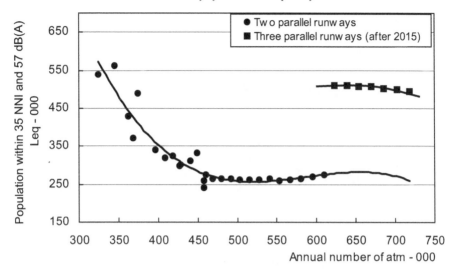

Figure 3.32 Dependence of the population exposed to noise of 35 NNI and 57 dB(A) L_{eq} on air transport movements (atm) at London Heathrow Airport (1988–2020)

Compiled from BA (2004); CAA (2002)

year 2015 (Figure 1.3 in Chapter 1) are expected to increase the number of people exposed to noise of 57 dB(A) L_{eq} again, to about 500 thousand. This number will slightly decrease afterwards despite a further increase in the annual number of atm mainly thanks to using new effective measures for reducing noise (CAA, 2002).

Another example refers to the air transport system's safety. In many cases, it is expressed by the number of aircraft accidents per unit of the system's output – the number of aircraft departures and/or the volume of p-km or t-km, over a given period of time (year). The development up to date has shown more than a proportional decrease in the number of aircraft accidents per unit of output over time (Boeing, 2005). However, these absolute numbers are dependent on each other. Figure 3.33 shows the relationship between the annual number of air traffic accidents and the number of aircraft departures over a given period of time in the US (ATA, 2006).

As can be seen, both the number of total and fatal accidents, despite high variations, has generally decreased as the number of aircraft departures has increased. This implies, according to the above-mentioned general principles, the long-term sustainable development of the system over the observed period. In addition, there is no correlation between the two variables indicating the actual nature of the aircraft accidents happening as random and truly unpredictable events.

Nevertheless, trends similar to the above-mentioned cannot be shown for the impacts of the major concern – fuel consumption and related air pollution. They continue to increase in absolute terms as the system's output increases, and thus do not offer any guarantee that the system will develop along a sustainable long-term trajectory. According to the normative approach to sustainability, the total impacts will continue to grow, threatening to compromise the prescribed targets (Figure 1.2 in Chapter 1). The possible slowing-down of such development could be carried

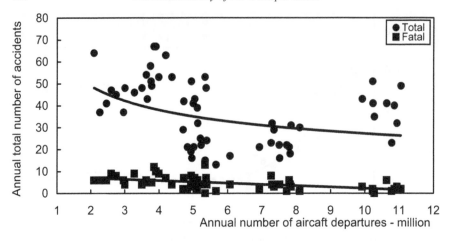

Figure 3.33   Dependence of the number of aircraft accidents on the number of departures in the US (1948–2005)

Compiled from ATA (2006)

out through the already discussed constraining of the system growth at different scales and scopes using the economic and/or institutional measures. However, such constraints would significantly compromise the system's benefits. Consequently, the solution for the problem seemingly lies in further technological development of the aircraft, which would be able to stabilise and maintain the total impacts below or just around the prescribed limits (targets) (see Equation 1.2 in Chapter 1). It would, of course, be even better if they would be able to turn around the trend of impacts – from increasing to decreasing – as the system's output increases. Equations 3.24 and 3.25 indicate the importance of such technological contributions. For example, referring to Equation 3.24, this actually means that the stabilisation of the trend of impacts or its eventual turning out was achievable if the coefficient b was just near or less than (-1). Under such circumstances, the total impacts would be nearly constant or even decreasing with an increasing system output.

3.7.2 Contribution of Aircraft Technologies

The innovative aircraft technologies, either as projects currently under way or as concepts, are mainly focused on the reduction of impacts such as fuel consumption and the related emissions of air pollutants (greenhouse gases). They can be roughly summarised as follows (Akerman, 2005).

- Further refining the conventional (present) turbofan aircraft/engine configurations as a continuation of the already achieved significant steps. This would include the use of lighter materials, refining the aerodynamic characteristics, and increasing the engine efficiency as discussed earlier in this and Chapter 2. The forthcoming aircraft A380, planned to enter commercial service in the year 2006/07, followed by the aircraft B787 and A350 to enter service in the year 2009/11 are the results of such efforts.

- Replacing the turbofan engines with a more fuel-efficient turboprop engine is the trajectory of development based on achieving the fuel savings through reducing the flying (cruising) speed. Current turboprops enable the cruising speeds of about 30 per cent and savings of the fuel consumption of about 20 per cent when compared with the turbofan engines;
- Reducing operationally the cruising speed of conventional turbofan aircraft is an easy option for implementation, which can diminish the fuel consumption. However, it does not have much potential since these aircraft have been specifically designed for the speeds at which they are flying (850–920 km/h);
- Developing the prop-fan engine as a hybrid between turboprop and turbofan engine has also been considered as a potential trajectory for decreasing fuel consumption. As a hybrid, it offers fuel consumption similar to a turboprop and nearly the cruising speed of a turbofan engine. Reduction in the fuel consumption has been estimated to be about 50–55 per cent as compared with the presently refined turbofan engines. The cruising speed would be 0.64–0.70 M as compared with about 0.80–0.85 M for the present turbofan aircraft (M – Mach Number). In addition to the lower cruising speed, the prop-fan engines have some other disadvantages. They generate more noise and vibration, and seemingly have higher maintenance costs, neither of which are particularly attractive for the airlines;
- Getting synthetic jet fuel (kerosene) has been considered as an option to slow down the depletion of the stock of fossil fuels currently in use. The sources for producing synthetic kerosene are hydrocarbons such as coal, fossil gas, and biomass – again exhaustive resources. In general, the technology for producing fuel from these sources is innovative; the refined conventional aircraft will use it;
- Gradually replacing the type of fuel – from kerosene to liquid hydrogen (LH_2) – is the trajectory of development, which can be combined with almost all current and prospective aircraft designs. In general, the quantity of fuel consumption (LH_2) would increase by about 10 per cent as compared with that of the conventional aircraft, emissions of CO_2 would be eliminated but emissions of H_2O and NO_x would increase (1 kg LH_2 = 8.94 kg H_2O) (Svensson, Hasselrot and Moldanova, 2004). The specific conditions are that the hydrogen needs to be produced in the CO_2 neutral way. In addition to adjusting the aircraft design, a completely innovative system of producing, distributing ad storing hydrogen fuel would be necessary.
- Gradual implementation of the new, as continuations to the forthcoming advanced aircraft designs such as A380, and B787 and A350 series, which would use one of two types of fuels – kerosene or liquid hydrogen. Some of these new designs are 'Flying Wing' and The Boeing Sonic Cruiser.

  ○ *Airbus A380* is the latest development of the large long-range commercial aircraft. At present, the information about the aircraft performances relevant for its contribution to the sustainability of the air transport system is relatively scarce. Nevertheless, some rough general structural (design) and operational characteristics are known. The MTOW is about 500 tons and the maximum payload 84–86 tons. The payload of about 55 tonnes is for passengers and the rest for freight. The seating capacity of the aircraft is about 550 passengers, each weighting about 100 kg together with his/her baggage. The aircraft trip fuel capacity is about 178 tonnes, which enables a typical full-payload range of about 12500 km with a cruising speed of M0.85 at flight level FL350 (RCEP, 2003). The above-mentioned performances provide an estimate of the average fuel consumption of about 25.9 g/p-km. If such an aircraft were fuelled by liquid hydrogen (LH_3) its fuel capacity would amount to about 68 tonnes, and consequently the MTOW to about 400 tonnes. This is about 100 tonnes (25 per cent) less than a kerosene-fuelled counterpart. The smaller quantity of fuel comes from the higher calorific value of liquid hydrogen compared to kerosene. Under such circumstances, the average fuel consumption per unit of output of a flight of range of 12500 km carrying 550 passengers would be about 9.9 g/p-km, which is about 2.65 times less than that of the kerosene driven version. It should be mentioned that the emissions of CO_2 would be completely eliminated while the emissions of H_2O would significantly increase, increasing the potential for formatting contrails at altitudes around FL350 (1kg of LH_3 = 8.94 kg H_2O). Therefore, the cruising altitude would need to be reduced to around FL270 where the conditions for the formation of contrails are negligible.

o   *Flying Wing* is considered as the future design for the commercial aircraft of the comparable size with the baseline model Airbus A380. This concept has been considered in Russia since the 1980s, aiming to continue the past and current trend of reducing fuel consumption by at least 30 per cent compared to the conventional aircraft configuration such as for example A380. The improvements would be achieved through reducing the aerodynamic drag (by about 8 per cent), increasing the overall engine efficiency (8 per cent), and reducing the aircraft empty weight (10–15 per cent) (RCEP, 2003). Flying wing would bring such improvements by its principal advantages – greater aerodynamic efficiency achieved by elimination of the fuselage and tail surfaces and consequently an increased L/D (Lift/Drag) ratio of about 15 per cent compared to the conventional design. The MTOW would also be reduced by about 14 per cent by using new materials. It would be between 380 and 430 tonnes depending on the configuration. The payload would be about 86 tonnes. The trip fuel capacity would vary from 83 to 137 tonnes, respectively. The cruising phase of the flight would take place at FL350 with a speed of about M0.85. Consequently, the range with the maximum payload would be 12500 km, similar to its conventional counterpart A380. Under the assumption of having 550 passengers onboard, the heavier version of Flying Wing would have an average fuel consumption of about 19.9 g/p-km, which is really about 30 per cent lower than the conventional design (25.9 g/p-km). The lighter version called 'Laminar Flying Wing' would have an average fuel consumption of about 12.1 g/p-km under the same operating conditions, which is about 2 times lower than that of the conventional baseline design (Bolsunovski et al., 2001).

Table 3.8   Fuel consumption of different aircraft designs

Aircraft configuration	Fuel consumption (*g/p*-km)	
	Kerosene	**Liquid hydrogen**
Baseline (A380)	25.9	9.9
Flying Wing	19.9	7.1
Laminar Flying Wing	12.1	4.6

Notes: Range 12,500 km, Seating capacity 550 passengers
Sources: http://www.airbus.com; Bolsunovski et al. (2001)

Flying wing using liquid hydrogen could be even more fuel-efficient. Their MTOW would be about 338 tons, the fuel weight for the heavier version about 49 tons and for the lighter version about 32 tons. Under the equivalent operating conditions (550 passengers and 12,500 range), the average fuel consumption would be about 7.1 g/p-km and 4.6 g/p-km, respectively, which is about 40 per cent and 215 per cent lower than the versions using kerosene, respectively. The emissions of CO_2 would be completely eliminated while the emissions of H_2O would significantly increase. This might require reducing of the cruising altitude due to the formation of contrails at certain cruising altitudes (around FL350). Table 3.8 summarises the main features of particular designs related to the average fuel consumption and type of fuel.

Information about the noise performances for any of these designs is still lacking. However, it is reasonable to conclude that the noise performances should fit the requirements for the ICAO Chapter 4 aircraft. Even more, an upgraded design of Flying Wing is under consideration in the project 'Silent Aircraft Initiative' carried out by The Cambridge-MIT Institute (UK–US) (http://silentaircraft.org). The project aims to design the aircraft and engines, which would not generate noise greater than the ambient urban noise outside the airport area during take-off and landing (about 60–65 dB(A)). This presumably might be achieved by designing completely new engines, which would have a much lower speed of gases exhausted and would consequently be less noisy. Since lowering the speed of the exhausted gases reduces the engine thrust, the size of the exhausted area needs to be increased in order to compensate the loss of thrust, particularly during take-off (see also Equation 2.2). A preliminary estimate shows that the size of this are would be about three times larger than that of existing turbofan engines of comparable size. However, benefits in terms of noise could be neutralised during the cruising phase of flight since larger engines increase the aerodynamic drag and consequently the fuel consumption and related emissions of air pollutants. During the approach and landing phase of flight, steeper descent in combination with a fast slow down could be a solution for diminishing noise significantly.

o *The Boeing Sonic Cruiser* has been a response to the launching of Airbus A380. This concept of new aircraft family is supposed to operate at higher cruising speeds than conventional aircraft, that is at M0.98 (1041 km/h

Table 3.9   **Some performances of advanced conventional and new generation aircraft**

Parameter	A340-300	A380	Flying Wing	Boeing Sonic Cruiser	Supersonic Transport (SST)
Capacity (Passengers)	295	550	550	200–220	100
Range (000 km)	13.5	12.5	12.5	11–14	6,1
Cruising altitude (000 ft)	35	35	35–40	40–50	50–60
Cruising speed (Mach)	0.86	0.86–0.89	0.86–0.89	0.95–0.98	1.4–2.0
Fuel Consumption (g/p.-km[1])	33.41	~ 25.9	~ 19.9	~ 29.45	707.06

[1] Average load factor – 100 per cent

Sources: http://www.boeing.com; http://www.airbus.com; Bolsunovski et al. (2001); Hepperle (2002)

at altitudes FL400-500) as compared to M0.82 (871 km/h at the altitudes FL350-390) (this is for about 20 per cent higher speed). In addition, the new aircraft is planned to have an extremely long range – about 18,000 km. Thanks to the higher speed, reductions in the flight time could be about 1 hour per 5500 km flown. The aircraft capacity is initially planned to be for between 100 and 300 passengers. The typical cruising altitudes are supposed to be between FL 400 and FL 500 (40,000–50,000 ft). However, later on, some of the above mentioned performances have been made more realistic, that is the range has been set up at about 11,000–14,000 km and the seating capacity between 200 and 220 passengers. These performances seemed to make sense for an aircraft expected to operate relatively frequent 'point-to-point' services between large airports (Hepperle, 2002). The aircraft fuel capacity is planned to be about 90 tons of kerosene, which would share about 39 per cent of the aircraft MTOW. This is lower than for other commercial long-range aircraft where the fuel weight shares about 47 per cent of the MTOW (Horonjeff and McKelvey, 1994). Equipped with jet engines derived from the current PW4082 and GE 90 models, these aircraft would have the average fuel consumption of 29.45 g/p-km, which is about 7 per cent and 9 per cent lower than that of types B767ER and B777-300LR, respectively. These estimates have been derived from the preliminary aircraft design. Emissions of CO_2 and H_2O would also be lower in the same proportions.

Ambitions regarding the reduction of noise have also been very high: the maximum noise during landings is expected to be about 86–86.5 EPNdB, and during take-offs about 93.5 EPNdB, both measured at the aircraft noise certification points around the airport. This would be a reduction of about 13 EPNdB compared to the noise of the conventional commercial aircraft of comparable size. In addition, the noise footprint during take-off would decrease due to steeper climbing. Some estimates show that climbing to FL400 would be carried out in about 24 minutes (http://www.boeing.com/news/feature/concept). Table 3.9 gives an indication of performances of this and other aircraft designs.

Some estimates based on the scenarios of gradual implementation of the above mentioned aircraft designs combined with other improvements have shown that moving along this trajectory would enable savings in fuel consumption of about 3 per cent per year until the year 2010 and about 30–47 per cent until the year 2050 (IPCC, 1999; Bolsunovski et al., 2001; Hepperle, 2002; http://www.boeing.com; http://www.airbus.com; RCEP, 2003).

The trajectories of developing new aircraft technologies mainly focused on the reduction of fuel consumption and related emissions of greenhouse gasses until 2050 and later until the year 2100 have also possessed potential disadvantages. These are mainly in terms of the uncertainty and complexity of the implementation of particular solutions:

- The aerospace manufacturers and airlines have managed to reduce the absolute number of aircraft accidents and fatalities significantly despite doubling the system output about every 15 years. Therefore, any departure from the current (convenient) aircraft technology trajectory will require timely and costly certification procedures, which may deter the airline industry from making radical changes. This particularly relates to the concept of the 'flying-wing';
- Turboprop aircraft increase the overall travel time due to flying at lower cruising speeds, which actually diminishes the 'speed advantage' and consequently the competitiveness of air transport compared with other transport modes. One such case is the High-Speed-Rail competing with the air transport mode on some short-haul routes in Europe. In further evaluation, the trading-off between both effects will certainly need to be established.
- The prop-fan engines save fuel but increase noise, vibration and maintenance costs. Consequently, they might not be acceptable to either the airline industry or the airports. Again, trading off between the long-term effects and impacts is needed.
- Using the synthetic kerosene is a quite realistic but limited option due to the limited sources of hydrocarbons for their production. Therefore, this option seemingly appears promising in the short- to medium but not in the long-term; and
- Replacing kerosene by hydrogen seems to be an unrealistic trajectory of technological development at least at the present time. This is primarily due to the requirements for completely changing the aircraft technology (cryoplane) and logistics of the fuel production and supply. In addition, it would require introducing operational constraints through limiting cruise altitudes within the range between FL240 and F310 (24,000–31,000 ft). Under such circumstances, consumption of the fuel-liquid oxygen (LH_2) and emissions of the associated pollutants – water vapor (H_2O) and nitrogen oxides (NO_x) – would increase but conditions for formation of contrails would vanish (Svensson, Hasselrot and Moldanova, 2004). Consequently, the main matter of concern seems to be the loss of the recent gains in the airspace capacity obtained by making all flight levels above FL290 usable for conventional aircraft, which would cause increasing congestion at the preferred flight levels used by the future cryoplanes.

The above-mentioned disadvantages may represent the main barriers to the implementation of particular new aircraft technologies. The main actors – the aerospace manufacturers and airline industry – seemingly prefer rather evolutionary (gradual) than revolutionary changes in moving towards more sustainable development while remaining aware of the following facts:

- Current medium to long-term development of the air transport system is on the right trajectory towards sustainability at least when compared with other transport modes and sectors of the contemporary economy. In general, the overall social benefits are increasing at a higher increasing rate than the overall social costs;
- Radical changes to the aircraft technologies will take place only if the threat of the depletion and disappearance of the present crude-oil reserves becomes realistic and certain in time. The current estimates of the available reserves and turbulent (generally increasing) prices of crude oil (and derivatives) are still not regarded as sufficiently convincing to undertake serious (revolutionary) steps in replacing current technology. Consequently, the main driving forces for technological changes seemingly remains the operational and economical viability within given (still acceptable and manageable) environmental constraints;
- Any technology trajectory always needs to be evaluated with respect to the interests, preferences and objectives of the actors involved, and then the air transport system's technical/technological, operational, economic, social, environmental, and institutional performances, as well as their interrelationship;
- The main actors are confident that the present and future air transport system is not actually a threat to the sustainability of the planet when compared with other transport systems and industries due to its relatively small contribution to global impacts as well as due to its successful dealing with the reduction of its own impacts gradually over time. They also count on the fact that development of the current and future global economy is unimaginable without an efficient, effective and environmentally friendly air transport system as its main driving force, and vice versa.

3.7.3 Contribution of the ATC/ATM

Contributions of the ATC/ATM to the future sustainability of the air transport system consist of contributions to reducing noise around airports and contributions to reducing fuel consumption and consequently emissions of air pollutants. Contributions to reducing noise around airports have already been discussed in Chapter 2. Continuous descent and even steeper approach and landing have been identified as the prospective solutions. Contributions to reducing fuel consumption embrace the benefit categories such as improving the efficiency of the aircraft flight paths, reduction of delays through increasing and better utilisation of existing airspace capacity, and contributing to better aircraft utilisation (http://www.boeing.com).

The efficiency of the aircraft flight path can be improved through direct and continuous climbing to cruising altitude, allocation of the wind-optimal routes, particularly for the long-haul flights, and enabling a continuous descent from cruising altitude to the altitude of final approach and landing. The targets to be achieved in the US air transport system until the year 2020 (as a vision) imply shortening the average flying time of each flight by about 4–8 min, which in turn would bring the annual savings of fuel cost of about USD1.5–3bn.

Reduction of delays through increasing and better utilisation of existing airspace and airport capacity can be achieved by increasing the precision of controlling air traffic, reduction of spacing between aircraft thanks to the advanced monitoring technologies both on the ground and onboard, provision of integrated services, and building new runways. In the US air transport system, these measures are expected to contribute to reducing delays by about 25–50 per cent (target in 2020), which in turn might result in savings of the fuel costs of about USD2–4bn.

Better utilisation of aircraft can be achieved through reducing Visual Meteorological Conditions (VMC) delays, increasing the aircraft flight path efficiency, and shortening the flight block pad. The expected cumulative effects are expressed in savings of the block time by about 10 min per each flight carried out in the US air transport system in the year 2020. In turn, this will result in annual monetary benefits of about USD0.5–1.0bn. Savings the fuel costs also contribute to savings of the cost of emissions of greenhouse gases, which is implicitly implied.

3.7.4 Contribution of Policies: Emission Trading

The most recent policy initiative is related to the policy of trading emissions of air pollutants as an economic option for keeping the emitted quantities under the prescribed cap. Indirectly, this is expected to reduce the emissions of air pollutants on the one hand and the cost of controlling air pollution on the other. The concept assumes that the central responsible authority at either a local, regional (country), or international (for example EU) level sets up the caps for each relevant air pollutant. Each actor involved is allocated the so-called designated limit or quota within the caps. The actors intending to exceed their designated limits may buy emission credits from those actors who are able to stay below their designated limits (quotas). This process is called trading.

In general, the central authority responsible for the emission trading grants the emission limits (quotas) or allowances to particular actors respecting their needs and previous air pollution performances. In such a case, the actor who does not need all their allocated allowances can sell part of them to those actors who might have insufficient allowances for their emission production.

In addition to the emission trading policies at a regional and national scale, the European Union Emission Trading Scheme (EUETS) is the largest multi-national greenhouse gases emissions trading scheme in the world. It started operation at the beginning of the year 2005 with participation of all 25 Member States (EC, 2003). In addition, the EU has requested from its Member states to consider the incorporation of the EU air transport system into the scheme since it has been exempted. This would start from the year 2012 in combination with setting up the caps on local emissions around airports through setting up the standards of local air quality. Initially, the air pollutant CO_2 would only be included in the emission allowances. The most convenient actors would be the airlines and their intra-EU flights. That means that flights to/from the EU area would be excluded.

If an efficient central (EU) authority is assumed capable of efficiently monitoring the scheme, two issues remain crucial: i) setting up a cap on the emissions for the air transport system; and ii) setting up an effective system of allocating the emission allowances.

Setting up a cap implies determining the total quota for emissions of CO_2 for the intra-EU flights carried out by the airlines involved in the scheme. This can be carried out in the scope of caps for the total transport sector and/or man-made emissions of CO_2 in the EU for the targeted year (for example, 1990).

The system of the allocation of allowances may be based on one of three methods: grandfathering allowances, auctioning allowances, and benchmarking allowances (Morrell, 2006).

The method of grandfathering allowances implies that the allowances are based on the past emissions performances. In such a context, the main issue appears to be setting up the length of the past period. Currently, it is suggested to be three years. In this case new entrants would have to buy their complete allowances.

The method of auctioning allowances applied by the responsible central authority, that is the European Commission assumes that the airlines would compete to buy the emissions allowances for the coming year or season. The main question is how to (re)-distribute the collected money.

The method of benchmarking allowances implies that the reference efficiency measure is determined in terms of either r-tkm (revenue tonne-kilometres), aircraft kms and flights per tonne of CO_2, setting up the cap on the overall emissions of CO_2, and then allocating the emission allowances to airlines in proportion to their share in the total r-tkm. One example of using r-tkm as a measure of the emission efficiency is:

$$A_i = [\sum_{i=1}^{N} E_i / \sum_{i=1}^{N} RTK_i]/* RTK_i \qquad (3.25)$$

where
  A_i is the emission allowance (quota) assigned to each airline;
  N is the number of airlines involved in the scheme;
  E_i is the emission assigned to the airline (i) in the base period; and
  RTK_i is the total volume of r-tkm assigned to the airline (i) in the base period.

Equation 3.25 indicates that more efficient airlines will be allocated more emission allowances, which will stimulate them to further improve their efficiency. In addition, less efficient airlines will be stimulated to improve efficiency in order to get more allowances.

The above-mentioned methods for allocating the emission allowances indicates that emission trading actually does not reduce pollution itself but stimulates the airlines to improve their efficiency below the allocated quotas and sell the differences to other less efficient airlines, which would need them in addition to their already assigned quotas.

References

ACI Europe (1998), *Creating Employment and Prosperity in Europe: A Study of the Social and Economic Impacts of Airports* (Brussels: Airport Council International Europe).

Akerman, J. (2005), 'Sustainable Air Transport – on Track in 2050', *Transportation Research D*, **10**, 111–126.

Archer, L.J. (1993), *Aircraft Emissions and the Environment, EV 17* (Oxford: Oxford Institute for Energy Studies).

Arthur D, Little Limited (2000), *Study into the Potential Impact of Changes in Technology on the Development of Air Transport in the UK,* (Cambridge: Arthur D, Little Limited).

ATA (2006), *Safety Record of US Air Carriers Operating under 14 CFR 121: Scheduled Passenger and Cargo Operations.* http://www.airlines.org.

ATAG (2003), *Aviation & the Environment* (Geneva: Air Transport Action Group).

ATAG (2003a), *Industry as a Partner for Sustainable Development* (Geneva: Air Transport Action Group).

AW (2006), 'Airline Safety in 2006', *Airliner World*, **9**, 66–69.

BA (2001), *From the Ground Up: Social and Environmental Report 2001* (Waterside, Middlesex: British Airways).

BA (2004), *Social and Environmental Report* (London: British Airways).

BAA (2001), *Waste Management* (London: British Airport Authority).

Basler, E. and Partners (1998), *Measuring the Sustainability of Transport, Materials of NRP 41,* Vol. M3, Transport and Environment Interaction, Switzerland/Europe, National Research Program 41 (Bern: Basler, E. and Partners).

Betancor, O. and Carlos Martin, J. (2005), 'Social Noise Cost at Madrid Barajas Airport', *8th NECTAR Conference, Las Palmas, Grand Canaria, Spain, June 2– 4, 2005*, 27.

Boeing (2001), *Current Market Outlook – 2001* (Seattle: Boeing Commercial Airplanes).

Boeing (2005), *Statistical Summary of Commercial Jet Airplane Accidents: Worldwide Operations 1959-2004* (Seattle: Boeing Commercial Airplanes).

Boeker, E. and Grondelle, R. (1999), *Environmental Physics, 2nd edn* (New York: John Wiley and Sons).

Bolsunovski, A.L., Buzoverya, N.P., Gurevich, B.I., Densiov, V.F., Dunaevsky, A.I., Shkadov, L.M., Sonin, O.V., Uzdhuhu, A.J. and Zhurihin, J.P. (2001), 'Flying Wing-Problems and Decisions', *Aircraft, Design*, **4**, 193–219.

BTS (2001), *National Transport Statistics 2000* (Washington, DC: US Bureau of Transport Statistics).

BTS (2001a), *Number of Pilot Reported Near Midair Collisions* (Washington, DC: US Bureau of Transport Statistics).

Button, K.J. (1993), *Transport Economics* (Aldershot: Edward Elgar).

CAA (2002), *Noise Trends and Controls at UK Airports* (London: UK Civil Aviation Authority).

CAA (2004), *Noise Mapping – Aircraft Traffic Noise Research Study on Aircraft Noise Mapping at Heathrow Airport Conducted on Behalf of DEFRA, ERCD Report 0306* (London: Civil Aviation Authority).

CE Delft (2002), *Economic Incentives to Mitigate Greenhouse Gas Emissions from Air Transport in Europe, Commissioned by European Commission, DG TREN* (Delft, the Netherlands: CE – Solutions for Environment, Economy and Technology).

DETR (1999), *Oxford Economic Forecasting: The Contribution of the Aviation Industry to the UK Economy* (London: Department of the Environment, Transport and Regions).

DETR (2000), '*Aircraft Proximity (AIRPROX): Number of Incidents 1988–1998*', *Transport Statistics Great Britain 2000 Edition* (London: Department of the Environment, Transport and Regions).

DETR (2000a), *UK Air Freight Study Report* (London: Department of the Environment, Transport and Regions).

DETR (2001), *Valuing the External Cost of Aviation* (London: Department of the Environment, Transport and Regions).

Doganis, R. (1992), *The Airport Business* (London: Routledge).

DRI-WEFA (2002), *The National Economic Impact of Civil Aviation* (Washington DC: DRI-WEFA).

EC (1997), *External Cost of Transport in ExternE* (Germany: European Commission, Non Nuclear Energy Programme, IER Germany).

EC (1998a), *Interaction between High-Speed Rail and Air Passenger Transport, Final Report on the Action COST 318, EUR 18165* (Luxembourg: European Commission).

EC (1999), *Air Transport and the Environment: Towards Meeting the Challenges of Sustainable Development* (Brussels: The Economic and Social Committee and the Committee of the Regions, The European Parliament).

EC (1999a), *Policy Scenarios of Sustainable Mobility – POSSUM*, 4th Framework Programme, EURES (Brussels: European Commission).

EC (1999b), *The Creation of the Single European Sky* (Brussels: European Commission).

EC (2002), '*Position Paper on Dose Response Relationships between Transport Noise and Annoyance,*' *The EU Future Noise Policy – WG2 – Dose/Effect* (Brussels: European Commission).

EC (2003), 'Directive 2003/87/EC of the European Parliament and of the Council of 13 October 2003 Establishing a Scheme for Greenhouse Gas Emission Allowance Trading within the Community and Amending Council Directive 96/61/EC (Text with EEA relevance)', *Official Journal*, **L275**, 0032–0046.

EC (2005), *Optimum Procedures and Techniques for the Improvement of Approach and Landing – OPTIMA, Deliverable D1.2, Sixth Framework Programme* (Brussels: European Commission).

ECAC (2004), *Methodology for Computing Noise Contours Around Civil Airports*, AIRMOD Group, ECAC.*CEAC Doc*29R, *Draft Version 6.0* (Brussels: European Civil Aviation Conference).

EEC (2001), *Forecasting Civil Aviation Fuel Burn and Emissions in Europe, Interim Report, EC Note N-8/2001* (Brussels: EUROCONTROL Experimental Centre).

EEC (2003), *ATFM Delays to Air Transport in Europe – Annual Report 2000* (Brussels: EUROCONTROL Experimental Centre).

EEC (2005), *The Economic Catalytic Effects of Air Transport in Europe, EEC/SEE/2005/004* (Brussels: EUROCONTROL Experimental Centre).

EPA (1999), *Evaluation of Air Pollutant Emissions from Subsonic Commercial Aircraft, EPA420-R-99-013* (Ann Arbor, MI: Environmental Protection Agency).

ERA (2003), *Aviation in California: Benefits to Our Economy and Way of Life* (Los Angeles: California Department of Transportation, Division of Aeronautics).

EU (2003), 'Guidelines on the Revised Interim Computation Methods for Industrial Noise Aircraft Noise, Road Traffic Noise, Railway Noise, and Related Emission Data', *The Official Journal of the European Union*, Annex, 50–64.

Evans, A.W. (1996), 'Risk Assessment by Transport Organisations', *Transport Reviews*, **17**, 145–163.

FAA (2001), *Management of Airport Industrial Waste, Report No. 150/5320-15* (Washington, DC: US Department of Transport).

FAA (2005), *FAA Aerospace Forecasts: Fiscal Years 2006–2017* (Washington, DC: US Department of Transport).

FI (2000), 'Commercial Aircraft Directory Part I, II', *Flight International*, Oct/Nov.

FI (2001), 'Regional Aircraft: World Airlines: Part I', *Flight International*, Sept.

Fraport (2001), *Frankfurt Airport: Environment – Noise Reduction* (Frankfurt: Fraport AG).

HA (1999), *Environmental Impact of Air Transport in Comparison with Railways and Transport by Rail in the European Union, Intermediate Report* (Madrid: Hera – Amalthea, S.L.).

Han, X. and Fang, B., (2000), 'Four Measures of Transportation's Economic Importance', *Journal of Transportation and Statistics*, April, 15–30.

Headley, E.D. and Bowen, D.B. (1992), *Airline Quality Issues: Proceedings of the International Forum on Airline Quality, NIAR Report 92-10* (Wichita, KS: NIAR – National Institute for Aviation Research, Wichita State University).

Hepperle, M. (2002), 'The Sonic Cruiser – A Concept Analysis', *International Symposium 'Aviation Technologies of the XXI Century: New Aircraft Concepts and Flight Simulation', 7-8 May 2002 Aviation Salon ILA-2002, Berlin, Germany*, 10.

Hewett, C. and Foley, J. (2000), *Plane Trading: Policies for Reducing the Climate Change Effects of International Aviation* (London: IPPR – The Institute for Public Policy Research).

Hooper, P.G. and Hensher, D.A. (1997), 'Measuring Total Factor Productivity at Airports', *Transportation Research E*, **33E**, 249–259.

Horonjeff, R. and McKelvey, F.R. (1994), *Planning and Design of Airports*, 3rd edn (New York, USA: McGraw-Hill).

Hunecke, K. (1997), *Jet Engines: Fundamentals of Theory, Design and Operation* (Shrewsbury: Airlife Publications).

ICAO (1988), *Review of the General Concept of Separation*, Sixth Meeting, Nov. 28, *Doc.* **9536**, RGCSP/G, Volumes I and II, Montréal, Canada.

ICAO (1991), *Airport Economics Manual* (Montreal: International Press/Civil Aviation Organisation).

ICAO (1993a), '*Aircraft Engine Emissions', Environmental Protection*, **2**, Annex 16.

ICAO (1993b), '*Aircraft Noise', Environmental Protection*, **1**, Annex 16.

ICF Consulting (2004), *Designing a Greenhouse Gas Emissions Trading System for International Aviation*, Study on behalf of the ICAO (London: ICF Consulting).

INFRAS (2005), *Sustainable Aviation, Study carried out for* Air Transport Research Group – ATAG (Zurich: INFRAS).

IPCC (1999), *Aviation and the Global Atmosphere, Intergovernmental Panel on Climate Change* (Cambridge: Cambridge University Press).

ITU (2002), *World Telecommunications Development Report 2002* (Geneva: International Telecommunications Union).

Janić, M. (1999), 'Aviation and Environment: Accomplishments and Problems', *Transportation Research D*, **4**, 159–180.

Janić, M. (2000), 'An Assessment of Risk and Safety in Civil Aviation', *Journal of Air Transport Management*, **6**, 43–50.

Janić, M. (2001), *Air Transport Systems Analysis and Modelling* (Amsterdam: Gordon & Breach).

Janić, M. (2003), '*An Assessment of the Sustainability of Air Transport System: Quantification of Indicators'*, 2003 ATRS, (Air Transport Research Society) Conference, 10–12 July, Toulouse, France, 35.

Janić, M. (2004), 'An Application of the Methodology for Assessment of the Sustainability of Air Transport System',, *Journal of Air Transportation*, **9**, 40–82.

Janić, M. (2005), 'Modelling Airport Congestion Charges', *Transportation Planning and Technology*, **28**, 1–26.

Jenkinson, L.R., Simpkin, P. and Rhodes, D. (1999), *Civil Jet Aircraft Design* (London: Arnold).

Kanafani, A. (1984), *The Analysis of Hazards and the Hazards of Analysis: Reflections on Air Traffic Safety Management, Working Paper, UCB-ITS-WP-84-1* (Berkeley, CA: Institute of Transportation Studies, University of California).

Kelly, K.L. (1998), 'A System Approach to Identifying Decisive Information for Sustainable Development', *European Journal of Operational Research*, **109**, 452–464.

Kuhlmann, A. (1981), *Introduction to Safety Science* (New York: Springer-Verlag).

Levison, D., Gillen, D., Kanafani, A. and Mathieu, J.M. (1996), *The Full Cost of Intercity Transportation – A Comparison of High-Speed Rail, Air and Highway Transportation in California, Research Report, UCB-ITS-RR-96-3* (Berkeley, CA: Institute of Transportation, University of California).

Litman, T. (2003), '*Issues in Sustainable Transportation', TDM Encyclopedia* (Victoria, Canada: Victoria Transport Policy Institute). http://www.vtpi.org.

MA (1999), *Manchester Airport Complaints and Community Disturbance: Report: 1998–1999* (Manchester: Manchester Airport).

Morrell, P. (2006), *'An Evaluation of Possible EU Air Transport Emissions Trading Scheme Allocation Methods'*, Air Transport Research Society Conference (ATRS) 26–28 May, Nagoya, Japan, 22.

Morrell, P. and Lu, C.H.-Y. (2000), 'Aircraft Noise Social Cost and Charge Mechanisms – A Case Study of Amsterdam Airport Schiphol', *Transportation Research D*, **5**, 305–320.

Morse, S., McNamara, N., Acholo, M. and Okwoli, B. (2001), 'Sustainability Indicators: The Problem of Integration', *Sustainable Development*, **9**, 1–15.

NASA (1996), 'Lingering Uncertainty about Aviation Impact Addressed by Growing Body of Scientific Data', *ICAO Journal*, **51**(January), 11–14.

Newell, G.F. (1982), *Application of Queuing Theory* (London: Chapman & Hall).

Nijkamp, P., Rienstra, S. and Vleugel, J. (1998), *Transportation Planning and the Future* (New York: John Wiley & Sons).

Nordhaus, W.D. (1991), 'To Slow or Not to Slow the Economics of the Greenhouse Gases', *Economic Journal*, **101**, 920–937.

Offerman, H. and Bakker, M. (1998), *'Growing Pains of Major European Airports'*, 2nd USA/Europe Air Traffic Management R&D Seminar, 1–4 December, Orlando, USA, 10.

Pearce, D.W. and Markandya, A. (1989), *Environmental Policy Benefits – Monetary Valuation* (Paris: Organisation for Economic Co-operation and Development).

Peeters, P.M., Middel, J. and Hoolhorst, A. (2005), *Fuel Efficiency of Commercial Aircraft: An Overview of Historical and Future Trends*, NLR-CR-2005-669 (Amsterdam: National Aerospace Laboratory – NLR).

RAS (2000), *Air Travel – Greener by Design*, Royal Aeronautical Society, Environmental Group, Report of the Technology Sub-Group (London: Royal Aeronautical Society).

RAS (2001), *Air Travel – Greener by Design – The Technology Challenge*, Royal Aeronautical Society, Report of Greener by Design Steering Group (London: Royal Aeronautical Society).

RCEP (2003), *The Environmental Effects of Civil Aircraft in Flight, Royal Commission on Environmental Pollution, Special Report* (London: Department for Environment, Food and Rural Affairs), www.rcep.org/uk.

Ruijgrok, C.J.J. (2000), *Elements of Aviation Acoustics* (Amsterdam: Het Spinhuis).

Sage, A.P. and White, E.B. (1980), 'Methodologies for Risk and Hazard Assessment: A Survey and Status Report', *IEEE Transaction on System, Man, and Cybernetics*, **SMC-10**, 425–441.

Schiphol Airport (2004), *Annual Statistical Review 1990–2004* (Amsterdam: ZG Schiphol).

Schipper, J. (2004), 'Environmental Costs in European Aviation', *Transport Policy*, **11**, 141–154.

Smith, M.J.T. (2004), *Aircraft Noise, Cambridge Aerospace Series* (New York: Cambridge University Press).

Svensson, F., Hasselrot, A. and Moldanova, J. (2004), 'Reduced Environmental Impact by Lowered Cruise Altitude For Liquid Hydrogen-Fuelled Aircraft', *Aerospace Science and Technology*, **3**, 7-32.

Tol, R.S.J. (1997), 'A Decision Analytic Treaties of the Enhanced Greenhouse Effect'. PhD thesis, Vrije Universiteit, Amsterdam.

Tol, R.S.J. (1999), 'The Marginal Cost of Greenhouse Emissions', *Energy Journal*, **11**, 61–81.

Trucost (2004), *Emission Trading and Aviation: The Effects of Incorporating Aviation into the EU Emission Trading Scheme, Trucost Sector Report – European Aviation* (London: Trucost).

USDT (2003), *Major Airport Flight Delay* (Washington, DC: US Department of Transportation).

Walder, R. (1993), 'Ageing Aircraft Programme Entails Major Effort and Expense', *ICAO Journal*, **48** (November), 6–8.

West, J. (2005), 'Flying the Boeing 777-300R', *Airliner World*, December, 62–64.

Williams, V., Noland, R.B. and Toumi, R. (2002), 'Reducing the Climate Change Impacts of Aviation by Restricting Cruise Altitudes,' *Transportation Research D*, **7**, 451–464.

WTO (2005), *International Trade Statistics 2005* (Geneva: World Trade Organisation).

Yergin, D., Vietor, R.H.K. and Evans, F.C. (2000), *Fettered Flight: Globalisation of the Airline Industry* (Cambridge, MA: Cambridge Energy Research Associates).

Ying Lu, C.H. (2000), 'Social Welfare Impacts of Environmental Charges on Commercial Flights', PhD thesis, College of Aeronautics, Cranfield University, UK .

Chapter 4

Modelling Sustainability of the Air Transport System

4.1 Introduction

This chapter presents cases of modelling sustainability of the air transport system and its main components – airlines, airports and ATC/ATM. Eight specific cases are presented.

The first section deals with the modelling of the operational, economic and environmental performances of an air transport network. This network consists of airports and air routes connecting them. The main performances of interest are the network capacity and profitability achieved, which could be achieved after setting up different environmental constraints in terms of noise and air pollution throughout the network after internalising their the externalities.

The second section deals with the qualitative evaluation of sustainability, that is the overall feasibility of an airline hub-and-spoke network. To this aim, the attributes of the network's technical/technological, operational, economic, environmental, and social performances are defined and modelled. In particular, preferences of the actors such as users-air passengers, airline, airports, the local community and policy makers at different levels (local, national, international) are considered in the process of the qualitative evaluation of the network.

The third section considers the vulnerability of an airline hub-and-spoke network to the large-scale disruptive events such as bad-weather, catastrophic failures of the network components, terrorist threats, industrial action by aviation staff, and so on. The aim is to quantify the airline costs caused by the disruption and to point out the non-sustainability (due to the high vulnerability) of the network configuration under specified circumstances.

The fourth section deals with the modelling of aircraft additional fuel consumption and associated emissions that occur due to flying at the fuel non-optimal flight levels (altitudes) in the en route airspace. The main causes of deviations of the actual from the optimal-planned aircraft vertical trajectories are an imbalance between the system capacity and demand and insufficient effectiveness of the ATC/ATM in handling the en route traffic. If this frequently happens, these deviations might cause a substantive impact on the environment due to the fuel consumption and associated emissions of air pollutants on the one hand and on the affected aircraft/flight and airline's operational costs on the other.

The fifth section relates to the options for increasing the airport runway capacity by introducing innovative operational procedures and technologies. In the shorter term, such an increase could diminish the burden on new land for building additional runway(s), which could prevent the spreading (re-distribution) of noise and air pollution, and thus contribute to the airport's sustainable development.

The sixth section deals with developing a tool for the optimal balancing of the airport demand and available capacity, aiming to minimise the aircraft delays and associated costs. Indirectly, this contributes to improving the utilisation of the airport runway capacity and reducing the loss of passenger time, additional fuel consumption in the arrival and departure queues, associated emissions of air pollutants, and noise. Better utilisation of the airport capacity relieves the pressure for building new infrastructure and taking additional land.

The seventh section elaborates congestion charging as an economic instrument of demand management at congested airports. The aim is to demonstrate the potential of a given instrument for balancing the airport-constrained capacity and demand by internalising the external congestion costs, which flights impose on each other during the congested period. The main idea is to set up a charge to prevent access of some aircraft/flights to an airport during the congested period and thus keep the time losses of users-passenger and airlines, fuel consumption and the associated emissions of air pollutants, noise, and pressure on the airport to expand under control.

The final section considers options for the substitution of air transport short-haul services (flight) by sufficiently attractive and commercially competitive surface transport alternatives, such as the High-Speed Rail. This is particularly considered for the markets-corridors connecting densely populated origins and destinations in Europe. For such purpose, the environmental performances of both alternatives are defined and estimated. The main material for each section is extracted from the author's previous work, which has been already published.

4.2 Performances of an Air Transport Network After Internalising Externalities

4.2.1 Background

An air transport network consists of airports and the air routes that connect them. It may have different spatial configurations depending on the physical size, that is the coverage of the geographical area, the number and location of airports included in the network, the pattern of the passenger and cargo demand, and the schedules and aircraft types the airlines use to serve the demand. This section describes the model to maximise the number of flights that can be carried out in an air transport network and consequently the network profits under specified operational, economic and environmental constraints. Most of the material is extracted from the author's already published research work (Janić, 2003).

The maximum number of flights, which can be carried out over particular network components such as airports and air routes under given conditions, represents its capacity. The conditions specify the flight pattern and constraints implemented to enable safe operations, and to control particular environmental impacts (Janić, 2003).

Operational constraints include the application of the air traffic control separation rules enabling safe flights. If the minimum separation rules are applied under conditions of constant flight demand, the number of flights served at the airports and on the air routes will be maximised. This represents the operational capacity and it is usually specified for a given period of time (Horonjeff and McKelvey, 1994; Janić, 2003).

Economic constraints imply the economic feasibility of carrying out particular flights in a given network, that is the difference between their revenues and costs on the one hand, and investments in the airport and air traffic control infrastructure providing sufficient capacity to appropriately handle realised flights on the other.

The environmental constraints limit the harmful impacts of the air transport network on people's health and the environment. In many cases, these constraints have an institutional or political character and are implemented as thresholds for particular types of impacts (IPCC, 1999). In order to keep the actual impacts below certain thresholds, the main generators – the number of flights serving the passenger and cargo demand – needs to be constrained or controlled. The maximum number of flights that generates the impacts meeting the thresholds can therefore be considered as the environmental capacity. At European airports, one of the earliest constraints on operations to protect local people from noise was at London Heathrow Airport in the early 1970s. The annual number of operations was limited to 270,000. However, under strong pressure from airlines this was lifted in 1991. A more recent case involves Amsterdam Schiphol Airport where legislation limited annual flights to 380,000 in 1998 with the possibility of an annual increase of 20,000 until the year 2003. At Zürich and Washington International airport, quotas on the number of flights have been introduced to control air pollution and noise, respectively (Offerman and Bakker, 1998). This capacity can be determined for a given network component, for example an airport, or with respect to different burdens, for example, noise or air pollution.

Network profits comprise the profits of the operators – airlines, airports and air traffic control, determined as the differences between their operational revenues and costs, including the internalised costs of particular burdens (Doganis, 1992).

4.2.2 Modelling Performances of Air Transport Networks

4.2.2.1 Problem formulation

Modelling performances of an air transport network aims to maximise its profits under given operational and environmental constraints. Usually, an objective function is defined to express profits. The constraints are expressed by the capacity of particular network components in terms of the operational (safety), noise, and air pollution thresholds, respectively. The objective function and constraints are assumed to be linear functions of the decision variables – the maximum number of flights allowed to be carried out in the network. These variables are the non-negative integers. That, together with linearity of the objective function and constraints enables the use of an Integer Programming (IP) technique to obtain the optimal solutions. This approach represents a continuation of the work of Ferrar (1974), which dealt with the allocation of an airport departure capacity under different types of constraints.

Modelling of the network performances is carried out for a given period of time (t) and following other notation:

N, K, L are the number of the airports, airlines and aircraft types in the network, respectively;

i, j are indexes of the flight origin and destination airport, respectively ($i \neq j$, i, $j \in N$);

k, l are indexes of the airline and aircraft types, respectively ($k \in K$; $l \in L$);

R_{ijkl}, C_{ijkl} are the revenue and costs per flight, respectively, carried out by the aircraft type (l) of airline (k) between airports (i) and (j);

X_{ijkl} is the maximum number of flights allowed to be carried out by aircraft type (l) of airline (k) between airports (i) and (j);

$b_{a/jikl}$, $b_{d/ijkl}$ are the amounts of air pollutants emitted by aircraft type (l) of airline (k) at airport (i), while coming from and going to airport (j), respectively;

c_{jikl}, c_{ijkl} are the amounts of air pollutants emitted by aircraft type (l) of airline (k) while cruising along the route between airports (j) and (i), and vice versa, respectively;

$a_{a/jikl}$, $a_{d/ijkl}$ are the noise levels generated by aircraft type (l) of airline (k) at airport (i), while coming from and going to airport (j), respectively;

A_{ai}, A_{di} are the operational arrival and departure capacity of airport (i), respectively;

P_{aji}, q_{dij} are the portions of the operational capacity of airport (i) allocated to the flights on the incoming and outgoing route (ji) and (ij), respectively;

B_i is the air pollution quota for airport (i);

B_{ij}, B_{ji} are the air pollution quotas for the en route airspace between airports (i) and (j), and vice versa, respectively;

A_{ai}, A_{di} are the noise quotas at airport (i) for the arriving and departing flights, respectively;

r_{ikl}, s_{ikl} are the portions of the available noise quota assigned to the aircraft type (l) of airline (k) during its arrival at and departure from airport (i), respectively;

Λ^*_i is the optimal capacity of airport (i)($i \in N$);

Λ^*_{ij} is the optimal capacity of a route between the airports (i) and (j), ($i,j \in N$); and

Λ is the capacity of the air transport network that maximises profits under given operational and environmental constraints.

The IP formulation of the problem can be stated as follows (Winston, 1994):

Maximise the objective function, that is the profits of a given air transport network:

$$Max \sum_{ijkl} (R_{ijkl} - C_{ijkl}) X_{ijkl} \qquad (4.2.1)$$

subject to:

Constraints on the operational capacity of airports
Allocation of the airport's operational capacity to the arrival and departure routes:

- *Incoming traffic – arrivals:*

$$\sum_{kl\,/\,j\neq i} X_{jikl} \leq p_{aji} * \Lambda_{ai} \qquad i,j \in N \tag{4.2.2a}$$

- *Outgoing traffic – departures:*

$$\sum_{kl\,/\,i\neq j} X_{ijkl} \leq q_{dij} * \Lambda_{di} \qquad i,j \in N \tag{4.2.2b}$$

where $\sum_{j} p_{aji} = \sum_{j} q_{dij} = 1$

Balancing the number of the incoming and outgoing flights:

$$\sum_{jkl\,/\,j\neq i} X_{jikl} - \sum_{jkl\,/\,i\neq j} X_{ijkl} \geq 0 \qquad i \in N \tag{4.2.2c}$$

Balancing the airline and aircraft types on the incoming and outgoing routes:

$$\sum_{kl\,/\,j\neq i} X_{jikl} - \sum_{kl\,/\,i\neq j} X_{ijkl} \geq 0 \qquad i,j \in N \tag{4.2.2d}$$

Constraints on the environmental thresholds at airports
Aircraft noise:
Policy 1   The noise quota is allocated in proportion to the noise from the expected flights on the incoming and outgoing routes:

- *Incoming flights – arrivals:*

$$\sum_{kl\,/\,j\neq i} a_{a/\,jikl} * X_{jikl} \leq p_{aji} * A_{ai} \qquad i,j \in N \tag{4.2.3a}$$

- *Outgoing flights – departures:*

$$\sum_{kl\,/\,i\neq j} a_{d/\,ijkl} * X_{ijkl} \leq q_{dij} * A_{di} \qquad i,j \in N \tag{4.2.3b}$$

Policy 2   The noise quota is allocated in proportion to flight and aircraft types:

- *Incoming flights – arrivals:*

$$\sum_{j\,/\,j\neq i} a_{a/\,jikl} * X_{jikl} \leq r_{ikl} * A_{ai} \qquad k \in K, l \in L, i \in N \tag{4.2.3c}$$

- *Outgoing flights – departures:*

$$\sum_{j\,/\,i\neq j} a_{d/\,ijkl} * X_{ijkl} \leq s_{ikl} * A_{di} \qquad k \in K, l \in L, i \in N \tag{4.2.3d}$$

where $\sum_{kl} r_{ikl} = \sum_{kl} s_{ikl} = 1$

Air pollution:
  • *Airports-local level:*

$$\sum_{jkl/i \neq j} (b_{a/jikl} * X_{jikl} + b_{d/ijkl} * X_{ijkl}) \leq B_i \quad i \in N \tag{4.2.4a}$$

  • *En-route airspace-global level:*

$$\sum_{kl/i \neq j} (c_{jikl} * X_{jikl/t} + c_{ijkl} * X_{ijkl}) \leq (B_{ji} + B_{ij}) \quad i, j \in N, \ i \neq j \tag{4.2.4b}$$

Constraints on the non-negativity of decision variables

$$X_{ijkl} \geq 0 \quad \text{(integers) for } i, j \in N, \ i \neq j, \ k \in K, \ l \in L \tag{4.2.5}$$

The optimal solution of Equation 4.2.1 is the maximum number of flights carried out at the particular airports and along the air routes under the constraints 4.2.2 to 4.2.5.

Constraints 4.2.2a and 4.2.2b limit the number of flights to the operational capacity of the airport's incoming and outgoing routes. The sub-constraints 4.2.2c and 4.2.2d provide a balance in the total number of arrivals and departures, the number of incoming and outgoing flights carried out by the different aircraft types operated by different airlines, and the number of aircraft on each airport arrival and departure route.

Constraints 4.2.3a–4.2.3d limit the maximum noise from the flights on the airport's incoming and outgoing routes to the noise quotas. The different, mutually exclusive policies can be applied to allocate these quotas to particular routes and flights. Similarly, the constraints 4a and 4b keep the maximum air pollution from flights at the airports below or at the allowed air pollution quotas. Finally, the constraints 4.2.5 ensure the decision variables – the number of flights – are the non-negative integers.

If $D^* = \left[X^*_{ijkl} \right]$ is the decision matrix, which maximises the objective function 4.2.1 subject to constraints 4.2.2–4.2.5, the optimal capacity of the air transport network and its components can be synthesised as follows ($i, j \in N; i \neq j; k \in K, l \in L; t \in T$):

  • For an airport (i):

$$\Lambda^*_i = \sum_{jkl} (X^*_{jikl} + X^*_{ijkl}) \tag{4.2.6a}$$

  • For a route between the airports (i) and (j):

$$\Delta^*_{ij} = \sum_{kl} (X^*_{jikl} + X^*_{ijkl}) \tag{4.2.6b}$$

- For the air transport network:

$$\Lambda^* = \sum_i \Lambda_i^* = \sum_{ij} \Lambda_{ij}^* \qquad (4.2.6c)$$

4.2.3 Determining Parameters of the Model

4.2.3.1 Coefficients of the objective function

Revenues   The revenues gained by an air transport network consist of the direct revenues of the network operators such as the airlines, airports and air traffic control. The benefits of air travel for the users and cargo consignors themselves and the indirect benefits for local communities are not taken into account (Button and Stough, 1998):

- The revenues of the network operators are expressed per flight. In such a context, the airline revenue per flight depends on the airfares and the number of passengers on board. It generally increases as these factors increase (Doganis, 2002). Assuming that the volume of passenger and freight demand is always at a level sufficient to make the maximum number of flights allowed in the network at least zero-profitable, the airlines may have an interest in carrying out as many as possible greater, fuller and more expensive flights in order to raise the revenues.
- The airport revenue per flight consists of the aeronautical and non-aeronautical part. In this case, the aeronautical part mostly consists of the aircraft and passenger fees and the non-aeronautical part consists of the revenues obtained from other 'not directly aviation' business. The revenue from both parts generally increases in line with the increasing number of flights and passengers. Consequently, airports may seek to raise their revenues by accommodating a larger number of larger aircraft and a greater number of passengers.
- Air traffic control mainly collects revenues by charging the navigation fees to the airline flights. The particular fee per flight increases as the flying distance and aircraft weight increases. Therefore, in order to recover its operating costs and gain profits more easily, air traffic control in a given area may seek to collect fees from as many longer flights as possible.

Costs   The costs of an air transport network include the direct cost of operators and the internalised externalities. Similarly as in the case of revenues, these costs are expressed per flight.

- The airline direct operational cost per flight embraces the costs of the resources (aircraft, labour, and energy) spent to carry it out. This cost generally increases as the aircraft size, number of passengers on board, and flying time increases (Doganis, 2002).
- The direct operational cost per flight at an airport consists of the costs of resources spent accommodating the passengers and aircraft. This cost generally increases as the aircraft size and number of passengers increases.

- The operational cost per flight of the air traffic control embraces the cost of the resources (facilities, equipment, energy, and labour) spent managing and controlling the flight within a given airspace. This cost is mainly proportional to the quality of the control, which depends on the sophistication of the control facilities and equipment (EEC, 1999).
- Externalities may include air pollution, noise, and air accidents (ATAG, 1996a; Levison et al., 1996). The cost of air pollution, noise and air accidents per flight depend on the pre-emptive and/or direct cost of damage caused by the related impacts. Each of the above externalities generally increases in the absolute sense as the aircraft size, number of passengers on board, and duration of flight increases. In addition, the congestion cost is considered as an additional externality. In this case, this cost exceptionally includes only the passenger and airline internal costs of flight delay and not the internalised cost of marginal delays, which this flight imposes on other flights during the congestion period (Daniel, 1995). This cost generally increases with the increasing value of passenger time, number of passengers on board, aircraft operating costs and delay.

4.2.3.2 Technological coefficients and the right-hand side of constraints
Constraints related to the operational capacity   For the sub-constraints 4.2.2a and 4.2.2b, the technological coefficients have a value of one for both types of decision variables – the incoming and outgoing flights at airports, X_{jikl} and X_{ijk}.

The right-hand sides, Λ_{ai} and Λ_{di}, as the airport's operational arrival and departure capacity, respectively, based on the minimum air traffic control separation rules, can be allocated to the routes and airlines according to the different 'slot-allocation' policies. For example, such policies can 'reserve' an adequate capacity in terms of slots, in proportion to the expected number of flights on particular routes, exclusively for the flights of particular airlines, or for a specific type of flights. At the capacity-constrained and busy airports such a policy may 'save' slots for the short hauls, new entrants, and so on (Coleman, 1999). The parameters p_{aji} and q_{dij} reflect the use of the first of the above policies.

Constraints related to the allowed noise quotas   Technological coefficients   In the constraints 4.2.3a and 4.2.3b, the technological coefficients $a_{a\,jikl}$ and $a_{d\,ijkl}$ represent the noise in terms of the sound energy $10^{L_{aji/kl}/10}$ and $10^{L_{dij/kl}/10}$ generated by an arrival and departure flight, respectively. The parameters $L_{aji/kl}$ and $L_{dij/kl}$ represent the aircraft noise in dB(A) (decibels), which is usually measured at the noise measurement locations in the vicinity of the airports (ICAO, 1993b).

The right-hand side   The noise quotas at airports, A_{ai} and A_{di} as the right-hand side of the sub-constraints 4.2.3a and 4.2.3b, comprise the total noise energy in terms of $(t *10^{L_{ai/eq}/10})$ and $(t *10^{L_{di/eq}/10})$ allowed to be generated by the incoming and outgoing flights during time (t) where the $L_{.i/eq}$s are the corresponding average energy sound levels, which accumulate all the sound energy for multiple flights. For example, the noise quota is zero during the night flying ban. During the day, it can be, for example 57 dB(A) at London Heathrow, 85 dB(A) at Birmingham, and 73 dB(A) at Frankfurt Airport (Horonjeff and McKelvey, 1994; DETR, 2000).

Different policies can be applied to allocate the noise quotas A_{ai} and A_{di} to the arriving and departing flights, airlines and aircraft types. For example, Policy 1 can exclusively allocate these quotas in proportion to the noise from the flights on the particular incoming and outgoing routes. In such a case, the parameters p_{aji} and q_{dij} in sub-constraints 4.2.3a and 4.2.3b will be identical to those in constraints 4.2.2a and 4.2.2b. Policy 2 can exclusively allocate the above noise quotas with respect to the noise performance of the particular aircraft types. The universal criterion to be applied should encourage the access of quieter and discourage the access of noisy aircraft. The parameters r_{ikl} and s_{ikl} in the constraints 4.2.3c and 4.2.3d reflect the possible application of this policy.

Constraints related to the air pollution quotas Technological coefficients In constraints 4.2.4a, the technological coefficients $b_{a\,jikl}$ and $b_{d\,ijkl}$, as the quantity of air pollutants emitted by a given flight at a given airport (kilograms/flight) depends on the time, that is the duration of the Landing/Take-Off (LTO) cycle, specific fuel consumption, and the emission index. The specific fuel consumption depends on the type of aircraft and engines. The emission index is the function of the type of air pollutant, aircraft engine, specific fuel consumption, and the LTO cycle mode. The particular modes of the LTO cycle are standardised. Here, the first standardised mode – the approach – is analogous to an arrival, the next two modes – take-off and climb – to a departure, and the last taxi/ground idle standardised mode to the sum of the flight taxiing-in and taxiing-out time. The air pollution from the airport passenger and aircraft ground-service vehicles, as well as the air pollution from the airport surface access systems, can also be appropriately added to the air pollution of a given flight (ICAO, 1993a).

In the constraints 4.2.4b, the technological coefficients c_{jikl} and c_{ijkl} represent the amount of the air pollutants emitted during the cruising phase of a flight. These amounts are proportional to the fuel consumption and emission index (both being dependent on the type of aircraft and engines), flight regime (speed/altitude), flying time, traffic, and the weather conditions.

The right-hand side The right-hand side of the constraints 4.2.4a, B_i is defined as the amount of air pollutants or quota (in kilograms) allowed to be emitted at an airport over a given period of time. This quota can be determined using either local or global criteria, or both (EPA, 1999). The quota is calculated as the product of the allowable concentration of given air pollutants and the volume of the airspace around given airport. If the allowed concentration is g_i (kilograms of pollutants per cubic kilometre) and the volume of the relevant airspace is V_i (cubic kilometres), their product will be $B_i = g_i V_i$. Alternatively, the air pollution quota B_i could be determined by trading-off with the air pollution quotas between airports in the network (IPCC, 1999).

The right-hand side of constraint 4.2.4b, the air pollution quotas B_{ji} and B_{ij}, can be expressed by the amount of air pollution allowed in the en route airspace between the corresponding airports. The quantifying procedure is similar to that at airports, but it should take into account the specifics of the airspace and the type and nature of the air pollution. At ground level the pollutants of relevance at airports are HCs (hydrocarbons), CO (carbon monoxide) and NO_x (nitrogen oxides) (ICAO, 1993a). In the en route airspace where the cruising phase of most sub-sonic flights takes place, the troposphere and the low stratosphere, the air pollutants are NO_x, CO_2 (carbon dioxide), and H_2O (water vapour).

4.2.4 An Application of Modelling the Network Performances

4.2.4.1 Input

Network configuration   The modelling of the air transport network performance is demonstrated with a simple network of one airport and a set of incoming and outgoing routes. Here, the case relates to the air transport network around London Heathrow Airport. At present, about a hundred routes and hundred airlines operating 40 different aircraft types connect the airport to the world's other airports. Such a high diversity of routes and airlines implies the existence of a large network optimisation problem of about $2 \times 100 \times 40$ decision variables and 10 constraints, which makes use of the IP technique complex. Therefore, the air routes are clustered and the aircraft types combined disregarding the differences between particular airlines, to reduce the size of the original problem. Consequently, the clusters of routes between Heathrow Airport and the domestic UK, European and US market areas are formed. These clusters comprise about 90 per cent of the traffic at Heathrow (Airline Business, 1999). However, due to the still very high diversity in the length of the particular clustered routes, the clusters are additionally grouped, this time into seven classes, and thus the original problem converted into a manageable surrogate consisting of 28 decision variables (seven routes and four aircraft categories) and 10 constraints (Groenewege, 1999). The classes of routes, as well as other relevant inputs such as aircraft types and categories, their seat capacity and the average route load factor are given in Table 4.2.1.

Constraints   Operational constraints   At present, the system of two parallel runways, one for arrivals and another for departures, is used at Heathrow Airport during the period $t = 19.5$ hours per day. The change of the two parallel runways at 3 p.m. every day and the night flying ban between 23.30 and 4.00 are already in place to relieve the noise burden. Under these circumstances, the airport declares an operational capacity of 78 operations per hour (39 arrivals and 39 departures) (DETR, 2000). The above-mentioned European and European/US flights use about 90 per cent of this capacity. Consequently, the airport's operational capacity allocated to this traffic is determined as: $\Lambda_0 = 78$ (flights/hour) x 19.5 (hours/day) x 0.90 $\cong 1,370$ (flights/day). Since the number of arrivals and departures is approximately balanced during the day, the corresponding operational arrival and departure capacity is determined as: $\Lambda_{a0} = \Lambda_{d0} = 1,370/2 = 685$ (aircraft/day) (the index '0' instead of 'i' is used to simplify the notation).

Each of the above capacities is allocated to seven route classes in proportion to the expected number of flights (that is according to the present slot-allocation policy). Since Heathrow Airport is mostly an origin/destination airport for flights, the portions of the operational capacity (slots) allocated to each incoming and outgoing route are adopted to be symmetrical and equal to: $p_{a10} + p_{a20} + p_{a30t} + p_{a40} = q_{d01} + q_{d02} + q_{d03} + q_{d04}$

Table 4.2.1   Characteristics of the classes of air routes, aircraft types and traffic in a given example

Route (index) (j)	Explanation	Average route length (km) (d)	Airline[1]	Typical aircraft type[2]	Aircraft category (index) (l)	Aircraft capacity (seats)	LF[3]
1	UK domestic and Closest EU (North)	402	European	B737/A320	1	140	0.70
2	Scandinavia and Central EU and Central Europe	816	European	B737/A320	1	140	0.70
3	South EU and Mediterranean	1461	European	B737/A320	1	140	0.70
4	Eastern Europe	2,380	European	B737/A320	1	140	0.70
5	North America (East Coast)	6,012	European/US: Virgin/BA/ American/United	B767/A330	2	250	0.70
6	North America (Central)	7,440	European/US: Virgin/BA/ American/United	B777/A340	3	350	0.70
7	North America (West Coast)	8,630	European/US: Virgin/BA/ American/United	B747	4	370	0.70

[1] Due to the institutional restrictions (Bermuda II), only four airlines fly between Heathrow and US airports
[2] These airlines operate more than 90% of this fleet news@airwise.com
[3] LF – Route Load Factor represents to the averages for the industry (Airline Business, 2000)

$= 0.68$ (European routes); $p_{a50} = q_{d05} = 0.15$ (European/US East Coast routes); $p_{a60} = q_{d06} = 0.10$ (European/US Central routes); $p_{a70} = q_{d07} = 0.07$ (European/US West Coast routes) (Airline Business, 1999). The operational capacity of the en route airspace between Heathrow and other airports is assumed to be unlimited.

Environmental constraints
*   *Noise*   All aircraft of the same category, independent of the route, are assumed to generate approximately the same level of arrival and departure noise. Table 4.2.2 contains the noise per individual aircraft category measured at points relevant for aircraft noise certification (Smith, 1989).
    The above-mentioned table is used to compute the coefficients $a_{aj0 \cdot l}$ and $a_{d0j \cdot l}$ of the noise constraints 4.2.3a and 4.2.3b. The noise quotas for the arrival and departure flights A_{a0} and A_{d0} are assumed to be identical and determined for the various values of the parameters $L_{ai/eq}$ and $L_{di/eq}$. These values are assumed as they are measured at the aircraft noise certification points around the airport. These quotas are allocated according to *Policy 1*, that is similar to the airport's operational capacity, and according to *Policy 2*, in proportion to the share of aircraft categories (types). In the latter case, the allocation

Table 4.2.2   The aircraft noise characteristics in a given example

| Aircraft category[1] | Noise dB(A)[1] | |
| | Arrival | Departure |
	$L_{a/•/l}$	$L_{d/•/l}$
1	80	74
2	90	75
3	90	78
4	94	93

[1] Adapted from Smith (1989)

looks like this: $r_{0•/l} = s_{0•/l} = 0.68$ (Category 1); $r_{0•/2} = s_{0•/2} = 0.11$ (Category 2); $r_{0•/3} = s_{0•/3} = 0.11$ (Category 3); and $r_{0•/4} = s_{0•/4} = 0.10$ (Category 4).

• *Air pollution*   The air pollution per LTO cycle, as the coefficients $b_{a/j0•/l}$ and $b_{d/0j•/l}$ of the sub-constraints 4.2.4a are determined by using data on the intensity of the emission of pollutants HCs, CO, and NO_x per aircraft engine (kg/s), the number of engines, the aircraft maximum take-off weight, and the duration of particular modes of the LTO cycle ($j = 1 - 7$, $l = 1 - 4$) (DERA, 1999). Regression technique is used to establish the relationships between the intensity of emissions per mode of the LTO cycle, EI_m (dependent variable) and the aircraft's maximum take-off weight W (independent variable). Half of the taxi-idle time of the LTO cycle is allocated to the taxiing-in and another half to the taxiing-out phase, which reflects the most common situation at Heathrow Airport. The results are given in Table 4.2.3.

Using the regression equations in Table 4.2.3 and the aircraft's typical take-off weight in Table 4.2.3, the quantity of air pollution for the arrival and departure part of the LTO cycle for the particular aircraft categories is calculated and given in Table 4.2.4.

In calculating B_o in constraints (4.2.4a), the relevant volume of airspace around the airport is assumed to be $V_o = 12(km^3)$ (the airspace is cylindrical with a base of *12 km²* and height of *1 km*). The total tolerable concentration of the relevant air pollutants HCs, CO, and NO_x for the period of $t = 1$ day is estimated to be $g_0 = 0.0246$ $(g/m^3) = 24,600$ (kg/km³), which gives the total

Table 4.2.3   Dependence of the intensity of emission per aircraft on the aircraft take-off weight during LTO cycle

LTO cycle (mode)	Code of LTO mode (m)	Duration of LTO mode (min)[1]	Intensity of emissions per aircraft (g/s)[2]
Take-off	1	0.7	$EI_1 = 0.832\ W^{1.026}$, $R^2 = 0.873$; $N = 33$
Climb	2	2.2	$EI_2 = 0.755\ W^{0.959}$, $R^2 = 0.907$; $N = 33$
Taxi-idle	3	26.0	$EI_3 = 0.225\ W^{0.754}$, $R^2 = 0.827$; $N = 33$
Approach/ Landing	4	4.0	$EI_4 = 0.367\ W^{0.759}$, $R^2 = 0.856$; $N = 33$

[1] ICAO (1993a); [2] Air pollutants – HCs, CO, NOx; $50 \leq W \leq 400$ tones (W – typical aircraft take-off weight)

Table 4.2.4   **Quantity of air pollutants emitted by particular aircraft categories during LTO cycle in a given example**

Aircraft category (l)	Typical aircraft take off weight (tonnes)[1]	The amount of the emitted pollutants (kg/aircraft)	
		Arrivals $b_{a/j0 \cdot l}$ (W)[2]	Departures $b_{d/0j \cdot l}$ (W)[3]
1	75	6.844	13.744
2	206	14.773	34.517
3	306	19.922	49.663
4	375	23.231	59.919

[1] FI, 1999a; b

[2] $b_{ad/0j \cdot l}$ (W) $= 88.08 \, W^{0.759} + 175.5 \, W^{0.754}$;

[3] $b_{d/0j \cdot l}$ (W) $= 175.5 \, W^{0.754} + 34.94 \, W^{1.026} + 99.66 \, W^{0.959}$

allowable amount of the air pollutants of $B_o = 24{,}600 (\text{tons/km}^3) \, x \, 12 \, (\text{km}^3) \, x$ $(19.5/24) = 239{,}850$ (tons/day) (DETR, 1999).

The air pollution quotas B_{j0} and B_{0j} in the constraints (4.2.4b) for the air-routes connecting Heathrow to other airports are assumed to be unlimited.

Coefficients of the objective function   The revenue and cost per flight as the coefficients of the objective function 1 are determined for each route class and each flight category. The revenue is calculated as the sum of the airport and airline revenue per flight. The cost per flight is calculated as the sum of the airport and airline operational costs and the internalised externalities per flight. The revenues and costs per flight of the air traffic control are assumed as already contained in the airline revenues. These calculations implicitly take into account the institutional restrictions (Bermuda II) of the access to Heathrow Airport for European non-EU and US airlines. Table 4.2.5 contains the calculated revenues and costs per route class and flight category ($j = 1-7$; $l = 1-4$). As can be seen, if given externalities are taken into account (internalised), some short haul flights become unprofitable. This is similar to the results of some previous studies that support the idea that as externalities are internalised, some flights might become unprofitable and are replaceable by surface transport alternatives (Levison et al., 1996).

4.2.4.2 Results

Using the inputs from Tables 4.2.1 to 4.2.5, two sets of experiments can be carried out by the software package Lindo aiming to investigate the impact of different environmental constraints on the network economic and operational performances. In the first set of experiments, the airport noise quotas A_{a0} and A_{d0} are varied. In the second set the airport air pollution quota B_0 is varied. In both cases, the operational capacities Λ_{a0} and Λ_{d0} are taken as constant. The results are shown in Figures 4.2.1, 4.2.2 and 4.2.3.

Impact of the noise constraints   Figure 4.2.1a shows the utilisation of the airport's operational capacity as affected by the noise quotas. The quotas of A_{a0} (65) and A_{d0} (65) based on the continuous noise of $L_{a0eq} = L_{d0eq} = 65$ dB(A) are allocated to the

Table 4.2.5 The revenues and costs per flight as coefficients of the objective function in a given example

Route/Aircraft category 0/j/l	Average cost ($/flight) $C_l/j \cdot /l \approx C_{j0-l}$					Average revenue ($/flight) $R_{0jl-l} \approx R_{j0-l}$[2]		Profit ($/flight)
	Airline[1]	Airport[2]	Internalised externalities (Noise/Air Pollution/Air Accidents)[3]	Congestion (Average delay)[4] Passengers[5]	Airline[6]	Airline[1]	Airport[2]	
1/1	7,970	2,842	158	163	453	8,027	3,528	−34
2/1	14,978	2,842	318	163	535	15,200	3,528	−108
3/1	23,786	2,842	570	163	537	24,419	3,528	49
4/1	32,659	2,842	928	163	484	34,089	3,528	541
5/2	76,863	5,075	4,187	583	484	96,637	6,300	15,614
5/3	107,525	7,105	5,891	817	677	131,091	8,820	17,896
5/4	113,669	7,511	6,227	863	715	138,538	9,324	18,877
6/2	95,046	5,075	5,182	583	488	115,878	6,300	15,804
6/3	133,065	7,105	7,254	817	683	162,229	8,820	22,125
6/4	140,669	7,511	7,669	863	722	171,449	9,324	23,359
7/2	110,249	5,075	6,011	583	490	134,412	6,300	18,324
7/3	154,348	7,105	8,415	817	686	188,177	8,820	26,226
7/4	163,168	7,511	8,896	863	725	198,930	9,324	27,091

[1] Airline operational cost: European routes (1,2,3,4): C(d) = $0.218e^{-0.000186d}$ ($/p.-km); $R^2 = 0.644$; N = 16; European/US routes (5,6,7): C(d) = 0.073 ($/p.-km) (AB, 2000); Airline revenue: European routes (1,2,3,4): R(d) = $0.218e^{-0.000168d}$ ($/p-km); $R^2 = 0.525$; N = 16; European/US routes (5,6,7): R(d) = 0.089 ($/p-km) (AB, 2000) where 'p.-km' is passenger-kilometre; 'd' is route length – (km)

[2] The average airport cost and revenue per passenger at Heathrow airport are estimated to be 29 and 36 $/pass, respectively (AB, 1999);

[3] Adapted from Levison et al. (1996): Cost of noise (UK case) – 0.002 USD/p-km; Cost of air pollution independently on type and scope – 0.00098 USD/p-km; Cost of air accidents – 0.001 USD/p-km

[4] Based on the average delay of five minutes per an arriving and departing flight – typical for Heathrow Airport (EEC/ECAC, 2001a); The average flying time on a route is: t (d) = 0.62 + 0.00021d; The value of passenger time is adopted to be 20 USD/pass/h for the European and 40 USD/pass/h for the European/US routes (FAA, 1995; Levison et al., 1996)

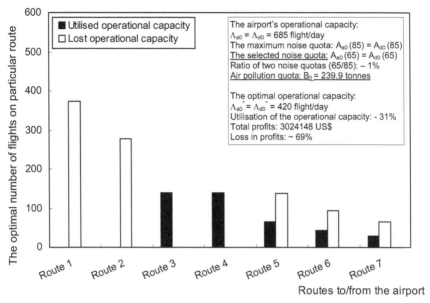

a) Operational capacity is allocated to all flights

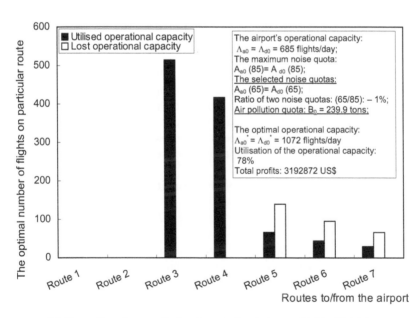

b) Operational capacity is allocated only to profitable flights)

Figure 4.2.1 Impact of noise constraints on the network performance: a) Operational capacity is allocated to all flights; b) Operational capacity is allocated only to the profitable flights

Compiled from Janić (2003)

incoming and outgoing routes in proportion to the expected traffic, that is Policy 1. The air pollution quota is fully used, that is B_0 = 239.9 tons/day. As can be seen, the maximum profits are realised by full utilisation of the operational capacity allocated to Routes 3 and 4 (medium-long haul European flights), and the partial utilisation of the capacity allocated to Routes 5, 6 and 7 (European/US flights). Routes 1 and 2 are excluded from the optimal solution because these handle short haul unprofitable European flights. Consequently, the optimal capacity Λ_0^* reaches only 420 flights/ day, that is about 30 per cent of the maximum, while profits are about 70 per cent lower when the noise and air pollution constraints are not included. Under these circumstances, Policy 2 for the allocation of the same noise quotas produces a similar impact.

Figure 4.2.1b shows the situation when unprofitable flights and their corresponding routes are excluded. Operational capacities are allocated to profitable flights. Other conditions remain unchanged. This change increases the utilisation of the airport's operational capacity to Λ_0^* = 1,072 flights/day – about 80 per cent of its maximum. However this increases profits by 5 per cent compared with the case in Figure 4.2.1a.

Impact of the air pollution constraints   Figure 4.2.2a shows the influence of the air pollution quota B_0 on the airport's profits and the utilisation of its operational capacity. The quota is set at 15.0 tonnes, that is at only 6 per cent of the maximum quota. The arrival and departure noise quotas are determined for the maximum continuous sound level of 85 dB (A), that is as A_{a0} (85) and A_{d0} (85), which means removal of these noise constraints. Nevertheless, these quotas are allocated to the routes and flights according to Policy 1.

The results appear different from Figure 4.2.1. The network profits are maximised by favouring more profitable, but also more polluting, long-distance intercontinental flights on Routes 5, 6 and 7 and by excluding both the unprofitable and profitable European flights on Routes 1 to 4. Excluding European flights and routes significantly reduces utilisation of the airport's operational capacity (to about 70 per cent) and consequently yields lower profits (for about 8 per cent) compared with the situation without air pollution quotas.

Figure 4.2.2b shows that allocating the operational capacity exclusively to the profitable flights has not improved the airport's overall operational and economic performance in comparison with those in Figure 4.2.2a.

Impact of the gradual relaxation of particular constraints   Figure 4.2.3 shows the relationships between the airport's profits and the utilisation of the operational capacity as influenced by the relaxation of the noise and air pollution constraints. The operational capacity is allocated only to the profitable flights. The noise capacity has been allocated in proportion to the expected flights on particular routes, that is according to Policy 1.

The lower curve shows the impact of gradual relaxation of the noise constraints $A_{a0}(\cdot)$ and $A_{d0}(\cdot)$h on the utilisation of the airport's operational capacity and corresponding profits. The continuous sound levels are varied in the range 55, 65, 75 and 85 dB(A). The air pollution constraint is essentially ineffective. Gradual

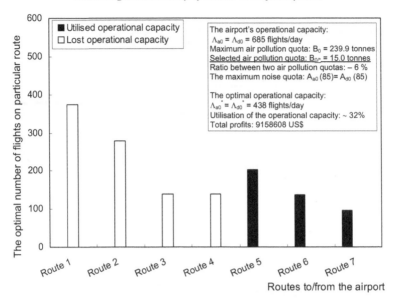

a) Operational capacity is allocated to all flights

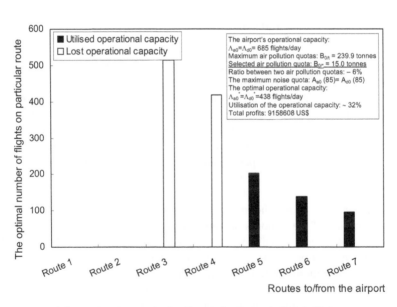

b) Operational capacity is allocated only to profitable flights

**Figure 4.2.2 Impact of air pollution constraints on the network performance:
a) Operational capacity is allocated to all flights; b) Operational
capacity is allocated only to the profitable flights**

Compiled from Janić (2003)

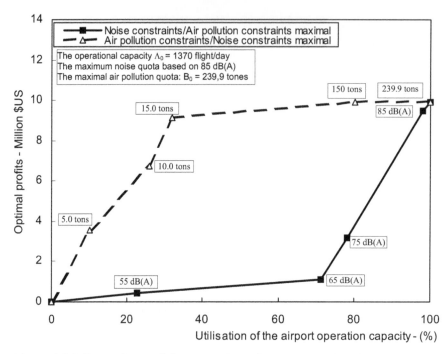

Figure 4.2.3 Dependence of the network performance on the relaxation of environmental constraints

Compiled from Janić (2003)

relaxation of the noise constraint increases utilisation of the airport's operational capacity. For example, the relaxation from 55 to 65 dB(A) increases utilisation of the operational capacity to about 70 per cent and profits to 32 per cent as compared with their maximums without any constraints. This implies that noise from the numerous, less noisy and less profitable European flights fills in this gap. The noise relaxation gaps from 65 to 75 dB(A) and from 75 to 85 dB(A) enable an additional increase in utilisation of the airport's operational capacity (30 per cent) and profits (70 per cent). This time, the noise gaps are filled by noisier more profitable long-distance European/US flights.

The upper curve shows the impact of relaxation of the air pollution constraints on the utilisation of the airport's operational capacity and consequently its profits. The noise quota is set at 85 dB(A), that is practically an ineffective level. The first relaxation gap of air pollution constraints to 15.0 tonnes/day significantly increases profits and only modestly utilisation of the airport's operational capacity, to the level of about 90 per cent and 32 per cent of the corresponding maximums, respectively. This implies that the more air polluting and more profitable European/US flights use this gap. The next relaxation gaps have increased the permissible air pollution quota from 15.0 to 150.0 and 239.9 tons/day. The less polluting and profitable medium and long haul European flights use these gaps. Consequently, this results in a substantial increase in the utilisation of the airport operational capacity by about 70 per cent, and modest increase in profits by about 8 per cent.

4.2.5 Concluding Remarks

This section has presented modelling performances of an air transport network consisting of airports and air routes connecting them. The network performance comprises the network profits maximised by maximising the number of flights allowed in the network under given operational, economic and environmental constraints. The constraints have been designed to internalise the environmental externalities. The profits and constraints are linear functions of the decision variables that allowed use of the Integer Programming techniques.

The model has been applied to a simplified air transport network around London Heathrow Airport. The noise and air pollution quotas have been taken into account. The results have shown that, after internalising externalities, some categories of flights have become unprofitable, and thus excluded from the optimal solution. Noise and air pollution quotas have significantly affected the air transport network's performances in terms of the utilisation of its operational capacity and consequently its profits. Different policies for allocating the noise quotas to flights and routes have not produced significantly different results. In this context, the influence of noise and air pollution quotas have differed during their gradual relaxation, but in general noise constraints have seemed to have a stronger relative impact. Both however, have lost influence after being set above given thresholds.

4.3 Evaluation of Sustainability of the Airline Hub-and-Spoke Networks

4.3.1 Background

Passenger and freight transport point-to-multipoint or hub-and-spoke networks have been developed after liberalisation of many national and international transport markets and privatisation of transport enterprises. The transport operators intend to run more profitable businesses, users ask for more effective, efficient and reliable services, and policy makers and society require the overall social feasibility of operations, that is sustainability. In most cases these objectives and preferences of particular actors have been in conflict.

The airline hub-and-spoke networks have emerged after liberalisation of the US air transport market in 1978 when the airlines were allowed free market entry/ leaving, setting up the airfares and supplying the capacity. Gradual liberalisation of the air transport market in the EU (European Union) over the past decade and a half has created similar conditions for European airlines, which have particularly used the opportunity for developing and/or consolidating their already existing hub-and-spoke networks – usually around the single domestic hub previously used as the airline base.

The main idea around the hub-and-spoke networks is the consolidation of traffic flows from a diverse range of origin (spoke) airports at the hub airport and directing them to a diverse range of destination (spoke) airports. The hub airport also appears to be the origin and destination of substantive traffic flows. Recently, debate has emerged about the overall social feasibility of airline hub-and-spoke network(s) aiming at considering the possible existence of a trade-off between the network advantages and disadvantages from the prospective of the different actors involved and not only the airline in charge (Button, 2002).

This section examines the development of a concept for evaluating sustainability, that is the overall social feasibility of an airline's hub-and-spoke network and the recently emerging modified concepts such as a 'rolling' and 'continuous' hub. The attributes of the technical/technological, operational, economic, environmental, social, and institutional performances of this network are defined, reflecting the interests and preferences of particular actors involved such as the users of air transport services – passengers, the airline as a provider of these services, airports and air traffic control as the infrastructure and traffic guidance/management service providers, respectively, policy makers, and local community subjects. In addition, simplified models enabling the quantification of particular attributes of performances are designed and applied to provide a qualitative evaluation of this network. Most of the material is taken from the author's previously published research work (Janić, 2005).

4.3.2 The Attributes of Performances of Airline Hub-and-Spoke Networks

Evaluation of sustainability of an airline hub-and-spoke network implies definition of the following attributes of the network performances:

- *Spatial layout and interconnectivity of nodes – airports* usually implies a star-shaped spatial-horizontal layout of a given hub-and-spoke network in which the airports are the network nodes, and the routes and flights connecting particular airport-nodes are the network links. Usually, the hub airport is located in the centre of an area divided into one or more concentric circles defining the spatial rings in which spoke airports are nearly uniformly distributed (Jeng, 1987).

- *Spatial and time concentration of the traffic flows/services* relates to the spatial concentration of traffic flows and services on each route-link connecting the hub and particular spoke airports over a given period of time and the time concentration of flights at the hub airport. The time concentration of flights at the hub airport implies their scheduling as 'waves' or 'banks' of incoming and outgoing services, that is complexes, very close in time. Each complex consists of an incoming and an outgoing bank of flights, which enables the effective and efficient exchange of traffic – passengers and freight – and other resources – aircraft and crews – at the hub airport. Several such complexes of flights can be scheduled by an airline at the hub airport over a given period of time (day) (Janić, 2005).

- *Economics of airline operations* implies economies of scale and economies of traffic density in a hub-and-spoke network. Generally, both occur when the average airline unit operational cost decreases as the volume of traffic – passengers – increases, served on a given link-route and the entire network as well (Button, 2002).

- *Size and utilisation of airline fleet* generally implies the structure and size of the aircraft fleet engaged to carry out flights in the network. In addition, this may relate to the diversity in utilisation of different aircraft types due to clustering flights into complexes at the hub airport (Jeng, 1987; Janić, 2001).
- *Utilisation of airport infrastructure* usually implies the development and use of infrastructure at the hub airport. Since the complexes of flights might be very large, handling them in terms of acceptable congestion and delay may require substantive airport infrastructure – runway(s), apron/gate parking stands and passenger/freight terminal(s). During the accommodation of particular complexes of flights, such an infrastructure might be over-utilised while at other times it might be under utilised. At the spokes, the available infrastructure might also be under-utilised.
- *Reliability and vulnerability* relates to the 'resistance' of an air transport hub-and-spoke network to disruptions. In particular, if the hub airport is affected by a disruptive event the consequences can be costly. Nowadays, the most common disruptive events are bad weather, failures of the aviation system facilities and equipment, industrial action by aviation staff, and terrorist threats. In general, disruptions are costly since many airlines tend to 'save integrity of schedule', 'avoid mixing of fleet' and 'maintain the scheduled order of complexes' at 'all costs' (Janić, 2005).
- *Environmental impacts* implies the harmfulness of excessive noise, air pollution, land use, waste, contamination of soil and water caused by the operation of a given hub-and-spoke network for people's health and the environment – the natural habitats. Most impacts appear particularly significant around the hub airport and in the airspace along the incoming and outgoing routes. They can be measured in absolute terms as the cumulative impacts and in relative terms as the quantity of impact per unit of the network output (flight-km of passenger-kilometre (p-km)) (Janić, 2003).
- *Social effects* mainly implies the contributions of an airline hub-and-spoke network to provide direct employment at airports, airline and air traffic control, and also indirect (supporting) local employment (ACI, 1998).

To date the qualitative and quantitative evaluation of airline hub-and-spoke network(s) has been fragmentary, that is based on the limited set of attributes – criteria of their performances and actors involved. The objectives of this section are to consider the evaluation of an airline hub-and-spoke network more integrally and generically through:

- Developing models for quantifying particular attributes of performances dependent on relevant factors and under given conditions; and
- Evaluating the network sustainability, that is the overall social feasibility qualitatively by using models of attributes of performances reflecting the preferences of particular actors involved.

4.3.3 Modelling Attributes of the Network Performances

4.3.3.1 State of the art and assumptions
In this section, extensive research into the performances of transport hub-and-spoke networks carried out up to date has been classified as follows:

- Analysis of configuration and design of transport hub-and-spoke networks including the problem of hub-location problems are addressed by Hall (1989, 1996); Aykin (1994, 1995, 1995a), O'Kelly (1998), O'Kelly and Bryan (1998), Kim et al. (1999), Marianov, Serra and ReVelle (1999), Cheung, Leung and Wong (2001), Kara and Tansel (2001) and Janić and Reggiani (2002).
- The hub operations within a given transport hub-and-spoke network are addressed in the work of Hall and Chong (1993), Hall (2001) and Bontekoning and Janić (2003).
- Some operational and economic characteristics of transport hub-and-spoke networks are considered by Economides (1996), Nero (1999), Pels, Nijkamp and Rietveld (2000), Wojhan (2000, 2001) and Button (2002).
- Operations of the hub-and-spoke networks under different disruptions are elaborated in the research of Shangyao and Chung-Gee (1997), Beatty et al. (1998), Lettovsky et al. (1999), Schavell (2000) and Janić (2005).
- The environmental characteristics of air transport hub-and-spoke networks are discussed by Button (2002) and elaborated by Janić (2003).

Using ideas from past research, models of the attributes of performances of an airline's idealised hub-and-spoke network are developed based on the following assumptions:

- The network serves a given volume of passenger flows during a specified period of time, say one day. These flows are approximately uniformly distributed during this period and are relatively inelastic to airfares. This enables conversion of the common airline objective of maximising revenues into the objectives of minimising costs and maximising quality of service for users (Jeng, 1987);
- The network consists of a few sub-networks, each characterised by a common hub airport, the specific set of spoke airports at approximately equal distances from the hub, equal volumes of passenger flows between particular origin and destination airports, and similar aircraft types serving these flows. Allocation of aircraft fleet to particular sub-networks is based on the aircraft's techno-operational and economic characteristics;
- The flights serving each sub-network are clustered into complexes of flights at the hub airport;
- The incoming and outgoing banks of flights serving particular sub-networks are hierarchically 'nested' within the same complex. The banks of the lower rank usually serve the shorter routes by the smaller aircraft, and vice versa;
- The passenger flows from particular sub-networks might mix at the hub airport, enabling each sub-network to carry, in addition to its 'internal' traffic, the traffic from other networks; and
- Particular sub-networks operate simultaneously over a given period of time.

4.3.3.2 Models of attributes

Following the notation used in developing models of particular attributes of performances of an airline hub-and-spoke network:

t_0 is the time period in which the network is analysed (one or part of the day);

M is the number of sub-networks in the given network;

i is the index, that is rank of the sub-network;

N_i is the number of nodes (hub plus spoke airports) in the sub network (i);

q_{i0} is the traffic flow between any origin and/or destination spoke and the hub airport in the sub-network (i);

q_i is the traffic flow between any pair of origin and destination spoke airports in the sub network (i);

q_{ij} is the traffic flow originating in the sub-network (i) and with a destination in the sub-network (j);

d_i is the length of route connecting the hub and any spoke airport in the sub-network (i);

$v_i(d_i)$ is the average aircraft speed on route d_i;

α_i, β_i is the value of unit of passenger time in the sub-network (i) while waiting for a flight, that is due to the schedule delay, and while being onboard, respectively;

f_i is the flight frequency on route d_i;

c_i is the cost per flight on route d_i;

t_{ri} is the aircraft 'rotation' time covering the circled itinerary 'spoke-hub-spoke' or vice versa;

W_i is the anticipated flight delay on route d_i;

$\tau_{0i}, \tau_i, \tau_{hi}$ is the prescribed, actual-spoke and actual-hub turnaround time, respectively, of an aircraft in the sub-network (i);

μ_{ai}, μ_{di} is the practical capacity (that is service rate) at the hub airport of a bank of incoming and a bank of outgoing flights, respectively, serving the sub-network (i);

y_i is the time gap at the hub airport between a bank of incoming and a bank of outgoing flights of a given complex serving the sub-network (i);

t_c is the duration of the scheduled complex;

K is the order of a flight among F complexes of flights taking place at the hub airport during time t_0;

v_{ai}, v_{di} is the practical capacity (that is service rate) of a bank of incoming and a bank of outgoing flights, respectively, serving the sub-network (i), compromised by disruption at the hub airport (usually $v_{ai} < \mu_{ai}$ and $v_{di} < \mu_{di}$);

$\chi_{k,k} + 1, \chi_{k,k} + 1^*$ is the scheduled and revised time between (k)-th and ($k+1$)-st complex at the hub airport, respectively;

b_k, b_{k+1} is the time buffer at the hub airport to neutralise the short foreseeable delays of two successive complexes of flights (k)th and ($k+1$)-st, respectively,

$c_i(D)$ is the cost of unit of delay of a flight in the sub-network (i);

$p_i(d)$ is the proportion of passengers given up from a flight serving the sub-network (i) due to disruption;

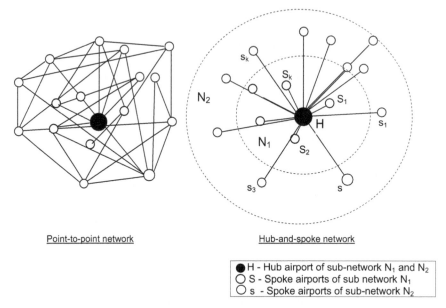

Point-to-point network   Hub-and-spoke network

● H - Hub airport of sub-network N_1 and N_2
○ S - Spoke airports of sub network N_1
○ s - Spoke airports of sub-network N_2

Figure 4.3.1 Simplified layout of the point-to-point and equivalent hub-and-spoke network

$r_i(d)$ is the revenue per passengers of a flight serving the sub-network (i);
C_{ai}, C_{di} is the cost of cancellation of an arriving and departing flight, respectively, serving the sub-network (i).

Spatial layout and interconnectivity of nodes The size and spatial coverage of an airline network implies the number of airport-nodes served and their spatial-geographical location. If airports are connected by direct flights the network spatial layout will be relatively dispersed. Conversely, if airports are connected indirectly by flights passing through the third centrally located airport, called the hub, the previously dispersed layout will transform into the characteristic star-shaped layout. Figure 4.3.1 shows both two-dimensional schemes.

If N_i airports-nodes, constituting a given network of type (i), are connected only by direct flights, the number of connections-links will be $1/2 N_i(N_i-1)$. This is a point-to-point network. If the same airports-nodes are connected only indirectly through the centrally located node-hub, the number of connections-links will be (N_i-1). Such configuration is called the hub-and-spoke network. The difference between the number of connections-links in two networks is $\Delta(N_i) = 1/2N_i(N_i-1) - (N_i-1) = 1/2 N_i^2 - 3/2N_i + 1$. Since this difference is always positive, the benefits of using indirect instead of direct connections for networks with a large number of nodes as well as for networks with expensive physical links such as rail and road network are obvious.

Spatial and time concentration of traffic flows/services In the point-to-point and hub-and spoke sub-network with N_i airports-nodes, the maximum number of origin-destination traffic-passenger flows is $N_i(N_i-1)$. Regarding (N_i-1) links in the hub-

and-spoke and $1/2N_i(N_i-1)$ links in the point-to-point network, the average traffic concentration per route will be of the order N_i in the former and *two* in the latter network. The total volume of passenger demand in sub-network (i) as a part of sub-network (i) can be estimated as follows:

$$Q_i = (N_i - 1)[2q_{i0} + \sum_{\substack{j=1 \\ j \neq i}}^{M}(q_{ij} + q_{ji})] + (N_i - 1)(N_i - 2)q_i \qquad (4.3.1a)$$

The first term on the right-hand side of Equation 4.3.1a represents the traffic flows originating at spoke airports and flying to the hub airport of sub-network (i), and vice versa. The second term represents the traffic flows originating in sub-network (i) and flying to other sub-networks, and vice versa. The last term denotes the traffic flows with both origin and destination in particular spoke airports of the sub-network (i).

The average traffic flow on the single link-route of the sub-network (i), based on (1a) is determined as follows:

$$\overline{Q}_i = Q_i / (N_i - 1) = 2q_{i0} + \sum_{\substack{j=1 \\ j \neq i}}^{M}(q_{ij} + q_{ji}) + (N_i - 2)q_i \qquad (4.3.1b)$$

The average traffic flow in the single direction of a route of the sub-network (i) is determined as:

$$\overline{Q}_i^* = \overline{Q}_i / 2 \qquad (4.3.1c)$$

Traffic flows carried out by flights arrive from (N_i-1) spoke airports at the hub airport close together in time (1 h or less), as a bank of incoming flights. After exchanging passengers, aircraft and crew, a bank of (N_i-1) outgoing flights departs towards the spoke airports, again during a short period of time. Both banks constitute a complex of flights. Figure 4.3.2 shows a simplified scheme of a single complex serving three sub-networks ($i = 1,2,3$). The number of complexes depends on the frequency of connecting hub and spoke airports during time t_0.

The handling time of an incoming and outgoing bank of flights serving the sub-network (i) at the hub airport depends on the number of flights and the corresponding airport service rates, that is the arrival and departure capacities μ_{ai} and μ_{di} as follows: $w_{ai} = (N_i-1)/\mu_{ai}$ and $w_{di} = (N_i-1)/\mu_{di}$, respectively. There is usually a time interval y_i enabling the exchange of traffic between the last incoming and the first outgoing flight of a given complex serving the sub-network (i), which makes its total handling time equal to: $t_{ci} = w_{ai} + y_i + w_{di}$ (Janić, 2005).

The average traffic flow in a single direction on a route of the point-to-point network is q_i. In this network, there might be some concentration of the incoming and outgoing flights at the airline base airport but it is mainly due to the volume and time pattern of passenger flows and associated flights-services and not due to intentional airline scheduling. Some airlines operate a relatively large number of non-interconnected flights at particular airports, which is called the concept of a 'continuous' hub.

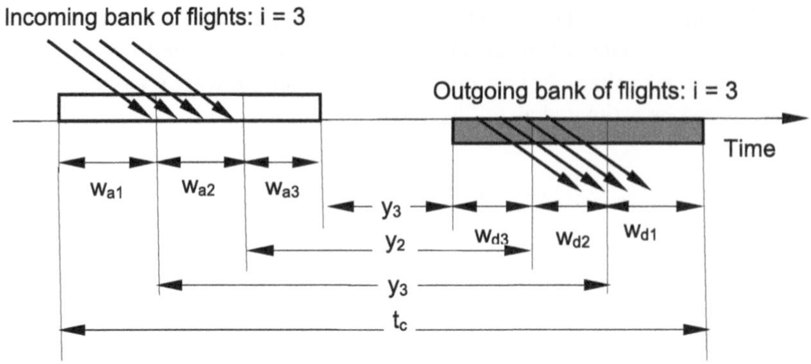

Incoming bank of flights: i = 3

Figure 4.3.2 Time scheme of a complex of flights at the hub

Economics of airline operations   The economics of airline operations in a hub-and-spoke network implies determining the flight frequencies on particular routes to minimise the total generalised costs consisting of the passenger time costs and airline operating costs. The airline costs are supposed to reflect airfares. The total cost on a route of the sub-network (i) with traffic flow $\overline{Q}_i / 2$ during time t_0 can be determined as (Jeng, 1987; Janić, 2001):

$$C_i = (\alpha_i \overline{Q}_i t_0 / 4f) + (\beta_i \overline{Q}_i / 2)[d_i / v_i(d_i) + W_i] + c_i f_i \qquad (4.3.2)$$

In Equation 4.3.2, the first and second term on the right-hand side represents the cost of passenger time due to the schedule delay and time onboard. The last term represents the airline's operating cost. By deriving the function C_i with respect to the flight frequency f_i, and after equalising the obtained first derivative with zero, the optimal flight frequency on a given route of the sub-network (i) in a single direction can be determined as follows:

$$f_i^* = 1/2(\alpha_i \overline{Q}_i t_0 / c_i)^{0.5} \qquad (4.3.3)$$

In Equation 4.3.3, the cost per flight c_i is generally proportional to the product of the average unit cost per aircraft seat-mile $c_{i0}(d_i)$, route length d_i and aircraft seat capacity S_i, that is $c_i = c_{i0}(d_i).d_i.S_i$. For example, an analysis for the US airline industry shows that the average unit cost per aircraft seat mile decreases with increasing distance as follows: $c(d) = 95.092d^{-0.2916}$; $R^2 = 0.959$; $n = 36$; $100 \leq d \leq 3,500$ miles (GAO, 2004).

After inserting Equation 4.3.3 into 4.3.2 and dividing the total cost by the route traffic $\overline{Q}_i / 2$, the airline minimum unit cost per passenger can be determined as:

$$_{min}C_{ai} = 2c_i f_i^* / \overline{Q}_i = (\alpha_i c_i t_0 / \overline{Q})^{0.5} \qquad (4.3.4)$$

In Equation 4.3.4, the cost per passenger generally decreases as the volumes of traffic increases indicating economies of scale (Button, 2002), which can generally be achieved by increasing the size and number of sub-networks around a single hub airport. For example, an investigation of the prospective relationships between the average airfare per passenger and the volume of passengers on a route is carried out for the US domestic market as follows: $r(Q) = 312.8e^{-0.0347\,Q}$; $R^2 = 0.878$; $n = 52$; $0.10 \leq Q \leq 38$ millions. The decreasing airfare in \$ US per passenger as the passenger volumes increase indirectly reflects economies of scale. This cost increases with the increasing aircraft size and route length. An example is the relationship between the average airfare per passenger and route length in the US domestic market shown to be: $r(d) = 6.876d^{0.4653}$; $R^2 = 0.964$; $n = 52$; $100 \leq d \leq 5,500$ (statute) miles. Data compiled from Huang (2000).

The airline minimum operating cost (4) can be the basis for setting airfares on the route connecting the hub and each spoke airport. Two times this amount can be the basis for an airfare on the route between any two spoke airports via the hub. In many cases, the airfare based on the costs of two (longer) indirect connections might be lower than the airfare of a single equivalent direct connection due to economies of scale and density (Jeng, 1987; Button, 2002).

Size and utilisation of airline fleet Size – the number of aircraft The aircraft fleet is one of the main physical resources of an airline (Janić, 2001). On a single route of a given sub-network (i), the number of aircraft of similar type carrying out flights during time t_0 can be determined using Equation 4.4.3 as follows:

$$n_i = (f_i^* t_{ri})/t_0 \qquad (4.3.5)$$

The number of aircraft of a given type needed to serve the entire sub-network (i) during time t_0 can be determined as:

$$n_{ti} = (N_i - 1)n_i = (N_i - 1)(f_i^* t_{ri})/t_0 = 1/2(N_i - 1)t_{ri}(\alpha_i \overline{Q}_i / c_i t_0)^{0.5} \qquad (4.3.6)$$

In the equivalent point-to-point network, the number of aircraft in the fleet might be smaller because: lower volumes of route traffic, generally shorter aircraft turnaround time due to the lack of the need for schedule synchronisation, and seemingly carrying out flights more sequentially than simultaneously. The factors contributing to deployment of the larger number of aircraft are longer flying times along longer-direct routes.

Utilisation – aircraft rotation time and apron/gate occupancy time In Equation 4.3.6, the aircraft rotation time t_{ri} consists of: i) turnaround time at a spoke airport, ii) en route flying time, iii) turnaround time at the hub airport. For an aircraft on a route of the sub-network (i) this time can be estimated as:

$$t_{ri} = \tau_{hi} + 2d_i / v_i(d_i) + W_i + \tau_i \qquad (4.3.7a)$$

In Equation 4.3.7a, the hub turnaround time of an aircraft serving the sub-network (i), $\tau_{hi,}$ depends on i) the hierarchical rank of the bank of flights it belongs to, that is the hierarchical rank of the sub-network it serves; ii) the size of bank it belongs to and the size of the higher ranked banks of flights; and iii) the service rate of

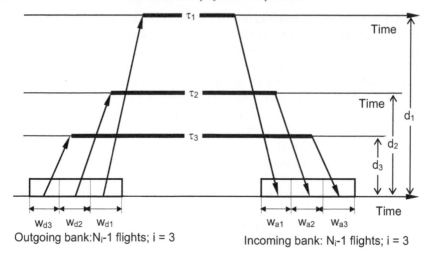

Figure 4.3.3 Dependence of the aircraft turnaround times at spoke airports

particular incoming and outgoing banks of flights at the hub airport. For example, at European airlines operating hub-and-spoke networks, the aircraft of higher rank, that is with the lower index (i) arrive at the hub airport earlier and depart later, that is they serve the sub-networks with the longer routes.

Based on the scheme in Figure 4.3.3, this average turnaround time of an aircraft operating within the sub-network (i) at the hub airport can be determined as:

$$\tau_{hi} = \max\left[\tau_{0i}; 1/2(N_i - 1)(1/\mu_{ai} + 1/\mu_{di}) + y_M + \sum_{j=i+1}^{M}(N_j - 1)(1/\mu_{aj} + 1/\mu_{dj})\right] \text{ for } i = 1, 2, ...M \quad (4.3.7b)$$

In Equation 4.3.7b, the aircraft serving the higher-ranked sub-network(s), that is those with the lower index (i), stay longer at the hub airport. This might be considered as a paradox of the nub and spoke-operations because, at least at European airlines, these are usually larger aircraft flying at longer routes.

In Equation 4.3.7b, the spoke-turnaround time of an aircraft serving the sub-network (j), τ_j in dependence on the spoke-turnaround time of an aircraft serving the higher-ranked sub-network (i) (that is $i < j$), τ_i can be determined follows:

$$\tau_j = \max\left[\tau_{0j}; \tau_i + 2[d_i/v_i(d_i) + W_i - d_j/v_j(d_j) - W_j] + \frac{1}{2}\sum_{k=i}^{j}(N_k - 1)[1/\mu_{ak} + 1/\mu_{dk}]\right] \quad (4.3.7c)$$

$$\text{for } i = 2, 3, ...M$$

Figure 4.3.3 provides a scheme. If $i = j$, τ_j handles position of the two aircraft serving the same sub-network of given complex. If $j > i$, an aircraft serving the lower-ranked sub-network will always stay longer at any spoke airport in order to fit the schedule of an aircraft serving the higher-ranked sub-network of the same complex, that is the smaller aircraft flying on the shorter routes will usually wait for 'nesting' with the larger aircraft flying on the longer routes. If stays all the time at spoke airports, the smaller aircraft will be lower utilised. Such dependability of the aircraft turnaround times appears to be an inherent characteristic of the hub-and-spoke networks as compared with the equivalent point-to-point networks.

Utilisation of the airport infrastructure If the demand rate, consisting of an incoming and an outgoing bank of flights of the sub-network (i), frequently exceeds the corresponding service rate of the runway system of the hub airport, that is if (N_i–1) $/w_{ai} > \mu_{ai}$ and (N_i–1) $/w_{di} > \mu_{di}$, the utilisation of the runway system is frequently equal to one and the traffic load is frequently greater than one. Under such conditions, serious congestion and delays might be expected, which could be considered as a 'sign' for planning a new runway (Janić, 2004).

If an incoming bank of (N_i–1) flights serving the sub-network (i) arrives at the apron/gate complex over a relatively short period of time, the number of parking stands-gates to accommodate them can be determined as (Horonjeff and McKelvey, 1994):

$$G_{hi} = \max[(N_i - 1)/w_{ai}; \mu_{ai}] * [\max(\tau_{0i}; \tau_{hi})] \tag{4.3.8a}$$

The total number of parking stands at the hub airport to accommodate the incoming banks of flights serving all sub-networks M of the given hub-and-spoke network can be determined as:

$$G_h = \sum_{i=1}^{M} G_{hi} \tag{4.3.8b}$$

The prolonged gate occupancy time caused by scheduling complexes of flights at the hub airport ultimately increases the occupancy time and utilisation of the airport apron/gate complex. The utilisation of parking stands-gates can be determined as the ratio between the total gate occupancy time by all complexes and the total gate available time as follows:

$$U = \sum_{i=1}^{M} [(N_i - 1)f_i^* \tau_{hi}]/G_{hi}t_0 = \sum_{i=1}^{M} [(N_i - 1)(1/2)(\overline{Q}_i \alpha_i t_0 / c_i)^{0.5} \tau_{hi}]/(G_{hi}t_0) \leq 1 \tag{4.3.8c}$$

The analogous reasoning can be applied to the spoke airports as well as to the airports of a point-to-point network, with necessary modifications of particular parameters in Equation 4.3.8.

Reliability and vulnerability Reliability and vulnerability of an air transport network can be expressed by the cost of delayed and/or cancelled flights caused by disruptive events (Janić, 2005). For an estimation of these costs for a given hub-and-spoke network, scenarios of occurrence of the disruptive event in terms of scope, time, duration and intensity of impact should be predictable. In particular, when a disruptive event affects the capacity of the hub airport, it prolongs the scheduled time of particular complexes and causes delays and cancellations of particular flights. For example, if a disruptive event takes place at the hub airport, lasts longer than the scheduled time of the given complex, and diminishes the airport's practical capacity,

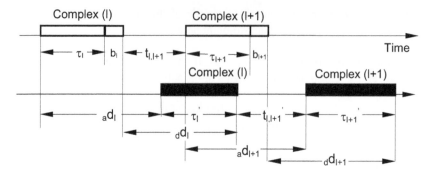

Figure 4.3.4 Propagation of delays between two successive complexes of flights at hub airport

that is service rate, the actual time for realising the (k)-th affected complex consisting of the flights serving all sub-networks of the given hub-and-spoke network can be determined as (Janić, 2005):

$$t_{ck}^* = \sum_{i=1}^{M} (N_i - 1)(\frac{1}{v_{ai}} + \frac{1}{v_{di}}) + y_k \tag{4.3.9}$$

The scheduled time of realising the (k)-th complex, t_{ck}, can be determined by replacing v_{ai} and v_{di} with μ_{ai} and μ_{ai}, respectively, in Equation 4.3.9.

The arrival and departure delay of the $(k + 1)$-st complex due to the delay of the preceding (k)-th complex can be determined from Figure 4.3.4 as follows (Janić, 2005):

$$_aD_{l+1} = \max\left[0;\ _dD_l + (t_{l,l+1}' - t_{l,l+1})\right] = \max\left[0;\ _aD_l + (t_{l,l+1}' - t_{l,l+1}) + (\tau_l' - \tau_l - b_l)\right] \tag{4.3.10a}$$
for $l = 1, 2, . . , L - 1$

and

$$_dD_{l+1} = \max\left[0;\ _aD_{l+1} + (\tau_{l+1}' - \tau_{l+1} - b_{l+1})\right] \tag{4.3.10b}$$
for $l = 1, 2, . . , L - 1$

If $k = 1,2,3, ...\ F^*$ complexes are delayed, the airline cost can be estimated as (Janić, 2005):

$$C_D = \sum_{k=1}^{F^*}[D_{ak}\sum_{i=1}^{M}(N_i - 1)[c_i(D_{aki}) + p_{aki}(D_{ak})(\overline{Q}_i c_i / t_o \alpha_i)^{0.5}] + D_{dk}\sum_{i=1}^{M}(N_i - 1)[c_i(D_{dk}) + p_{dki}(D_{dk})(\overline{Q}_i c_i / t_o \alpha_i)^{0.5}] \tag{4.3.10c}$$

If F^* complexes are cancelled, the airline cost of cancellations can be estimated as (Janić, 2005):

$$C_C = \sum_{k=1}^{F^*}\sum_{i=1}^{M}(N_i - 1)(C_{aki} + C_{dki}) \tag{4.3.10d}$$

In Equations 4.3.10c and 4.3.10d, the airline cost will generally increase with the increasing number of affected complexes (and flights), their size, cost per affected flight, and proportion of passengers giving up.

Some calculations have shown that the cost of delayed complexes is greater than the cost of cancelled complexes. In general, a flight is considered delayed if it departs or arrives 15 minutes or more behind the published schedule. For example, in the US during the past decade and a half the proportion of cancelled and diverted flights as compared with the proportion of delayed flights has been relatively modest. The proportion of cancelled flights has been between three and 10 times greater than the proportion of diverted flights. The number of cancellations has generally increased while the number of diversions has remained relatively stable, illustrating the airlines' preferential strategies. Some figures have shown that the total average direct annual cost of irregular operations of 10 major US airlines (the period 1996–1999) has been about USD1.9bn, of which delays have accounted for approximately USD900mn, cancellations between USD800 and USD900mn and diversions between USD50 and USD60mn. Both costs are generally higher if the disruptive event affects the capacity of the hub airport earlier in the day and lasts for a longer time. Consequently, the hub-and-spoke network appears to be very vulnerable to disruptions compared with the point-to-point network (Janić, 2005).

The environmental impacts Noise The aircraft noise around airports is usually measured by A-weighted sound level – dB(A) scale and the Effective Perceived Noise Level (EPNdB) scale. The approximate relationship between the two scales is: EPNdB $\approx$ d$B(A)$ + 12. The former scale is commonly used for land use planning and the latter one for aircraft noise certification (Horonjeff and McKelvey, 1994; Janić, 1999).

In an airline's hub-and-spoke network, both the hub and spoke airports are relevant for monitoring and measuring the aircraft noise. When the equivalent A-weighted sound level is applied and if $L_A(j)$ is the instantaneous A-weighted noise generated by a single event (j), the total noise in dBA generated by $\overline{N}$ events taking place at the given airline hub airport during time t_0 can be estimated as follows (Ruijgrok, 2000):

$$L_{Aeq,t_0} = 10\log\left[1/\overline{N}\sum_{j=1}^{\overline{N}} 10^{\frac{L_A(j)}{10}}\right] \qquad (4.3.11a)$$

Similarly, if $L_{AE}(j)$ is the sound exposure level due to the noise event (j) and t_1 the period of 1 s, the total noise exposure during time t_0 can be estimated as (Ruijgrok, 2000):

$$L_{Aeq,t_0} = 10\log\left[\sum_{j=1}^{\overline{N}} 10^{\frac{L_{AE}(j)}{10}}\right] - 10\log(t_0/t_1) \qquad (4.3.11b)$$

Some other methods for estimating the noise exposure level based on the perception of noise by the local population around the airport can also be applied. For example, the noise and number index (NNI) was developed in the UK for community noise assessment as follows: $NNI = 10 \log \left[(1/\overline{N}) \sum\limits_{j=1}^{\overline{N}} 10^{\frac{L\max(j)}{10}} \right] + 15 \log N - 80$ where $L_{max}(j)$ is the maximum perceived noise level of the noise event – flight (j) – and N is the number of events-flights during time t_0. Another measure based on the total noise load in Kosten units has been developed for the Dutch airports. It takes the perception of noise by the affected population during different periods of the day as follows: $B = 20 \log \left[\sum\limits_{j=1}^{\overline{N}} w(j) 10^{\frac{LA\max(j)}{15}} \right] - 157$ where $L_{amax}(j)$ is the maximum weighted sound and $w(j)$ is the time of day weighting factor due to the noise event-flight (j) (Ruijgrok, 2000). The number of noise events at the hub airport during time t_0 is $\overline{N} = 2 \sum\limits_{j=1}^{M} f_i^* (N_i - 1)$, of which half belong to the incoming and half to the outgoing banks of flights. Differences in noise of the incoming and outgoing flights should be anticipated by modifying $L_a(j)$ and $L_{ae}(j)$ in Equation 4.3.11. At each spoke airport the number of events is $\overline{N} = 2f_i^*$, of which half is for the incoming and another half for the outgoing flights. The total number of noise events at all the spoke airports is the same as that at the hub airport.

The number of noise events in the point-to-point network is $\overline{N} = \sum\limits_{j=1}^{M} f_i N_i (N_i - 1)$ of which, again, half are the incoming and another half the outgoing events-flights. Consequently, the ratio between the number of noise events in the point-to-point and hub-and-spoke network can be of the order of $(f_i N_i / 2 f_i^*)$. Nevertheless, the cumulative noise burden is the same as in the hub-and-spoke network but with a different spatial distribution and consequently a different affect on the local population around particular airports.

Air pollution   As mentioned in Chapters 1–4, the air pollution from traffic using mostly JP1 jet fuel consists of pollutants such as non-burned hydrocarbons (UHC), carbon monoxide (CO), carbon dioxide (CO_2), the nitrogen oxides (NO_x) and water vapour. As already mentioned, the average rate of emissions per kg of JP1 burned fuel is about 0.011 kg of NO_x, 3.16 kg of CO_2 and 1.25 kg of water vapour (Janić, 1999). For an airline's hub-and-spoke network, air pollution is particularly concentrated around the hub airport and along the incoming and outgoing routes. The reference time unit for measuring emissions at airports is LTO (Landing and Take-off Cycle) (Janić, 2003). For a hub-and-spoke network, the number of LTO cycles carried out at the hub airport during time t_0 is of the order of $\sum\limits_{j=1}^{M} f_i^* (N_i - 1)$. The same number of LTO cycles takes place at all the spoke airports. If $\tau_{LTO/i}$ is the duration of a single LTO cycle, e_i and E_i the average rate of fuel consumption during the LTO cycle and en route (units of fuel burned per unit of time), and r_i and R_i the corresponding emission rates (kg of air pollutant per kg of fuel burned), respectively, of an aircraft serving the sub-network (i), the total air pollution by the hub-and-spoke network during time t_0 can be determined as:

$$A_p = \sum_{i=1}^{M} f_i^* (N_i - 1) \left[\tau_{LTO/i} e_i r_i + 2[d_i / v_i(d_i) + W_i] E_i R_i \right] \qquad (4.3.12)$$

In the equivalent point-to-point network, the number flights on particular routes is lower due to the lower concentration of traffic flows, although in the case of full connectivity the number of routes is higher. Therefore, apart from the differences in the spatial and time concentration of air pollutants, the judgement of which network pollutes more remains inconclusive without concrete calculations.

Land use   The most land-demanding infrastructures at an airport are the runways, taxiways and apron/gate complex. For example, each new runway requires about 250 ha of land (Horonjeff and McKelvey, 1994). In addition, these might be the passenger terminals and the airport surface access systems. In an airline hub-and-spoke network, the 'pressure' for capacity is the highest at the hub airport due to the very sharp peak demand caused by clustering flights into complexes over very short time periods. The 'pressure' for the landside capacity can also be high. As an example, the public enquiry and governmental approval of London Heathrow Terminal 5 has taken a decade due to the environmental concerns, one of which was the impact of the surface access systems on the environment (Janić, 2004). On the other hand, the spoke airports may lose traffic, which, due to the lack of compensation, can result in underutilisation of the available airside and landside infrastructure.

Social effects   The social effects of an airline hub-and-spoke network can be measured by the total employment at the airline and airports. At the airlines, the employment is generally related to the fleet size. For example, some investigation has shown the following relationships between the number of airline employees and the fleet size: $E_a = -2,769.64 + 23.72(s/n) + 148.44n$, $R_2 = 0.856$; $n = 251$, where n is the total number of aircraft and s is the total number of seats in the airline fleet (Janić, 2001). At each airport, employment can be direct, indirect, induced and catalytic (ACI, 1998). The first type includes the number of employees directly engaged in carrying out direct aviation-related activities at the airports. The second type embraces jobs providing goods and services supporting these direct activities. The third type includes jobs providing goods and services for the direct and indirect employees living in the region. The final type of employment includes jobs generated by the general attraction, retention and/or expansion of the economic activities in the region around the airport (Button and Stough, 1998). In general, employment appears to be in some kind of relation to the annual number of passengers accommodated at an airport. For instance, compiling data from ACI (1998), the relationships between the airport employment and the annual volume of traffic at European airports is obtained as: I) direct employment: $E_d(Q) = 1.470,2Q - 4.209$; $R^2 = 0.901$; $N = 22$; and ii) total employment: $E_t(Q) = 0.557Q^{1.493}$; $R^2 = 0.930$; $N = 22$, where Q is the annual number of passengers at an airport (million). Consequently, an airline hub-and-spoke network contributes to the employment at the hub airport in proportion to the factor $365 * \sum_{i=1}^{M} (N_i - 1)\overline{Q}_i$ and at all spoke airports to the factor $365\overline{Q}_i$. The employment at each airport of the equivalent point-to-point network is proportional to the factor $365*2*\sum_{i=1}^{M}(N_i - 1)q_i$.

4.3.5 Modifications of the Airline Hub-and-Spoke Networks

4.3.5.1 The 'rolling' hub

Equations 4.3.7a–4.3.7c show that the turnaround times of particular aircraft at the hub-and-spoke airports are dependent on each other (Figure 4.3.3). In particular the number of aircraft in the incoming and outgoing banks of particular complexes and their service rates, that is the arrival and departure capacity, respectively, influence these times. The simultaneous presence of a relatively large number of aircraft of a given complex requires use of the relatively large number of gates and ground staff to serve them, both of which cost substantial amounts of money. Under conditions where the planned revenues are not achieved, this might compromise the airline's overall profitability. Therefore, as a remedy, some airlines such as for example, American Airlines (US) have developed the concept of the so-called 'rolling hub' at its Chicago O'Hare and Dallas Forth Worth hubs (Ott, 2003). The idea of this concept is the reduction of the aircraft time spent on the ground, which in turn enables the costs of gates and staff to be reduced and the flying time to be increased, bringing revenues on the one hand and an inherent risk of getting fewer passengers onboard on the other. The reasoning behind this is that the aircraft generate revenues only while they are airborne. Consequently, the number of aircraft as compared with the original complex is reduced, that is a given complex is extended in terms of time t_c (Figure 4.3.2). There are three types of effects: i) savings in the aircraft ground time at both the hub airport and spoke airports – this time can be used for flying and earning revenues; ii) savings of the costs of using gates at the hub airport, including the cost of the aircraft ground servicing staff; and iii) users-passengers may experience an extension of the connecting time between incoming and outgoing flights of a given complex, which in turn can be compensated by the shorter arrival and departure delays. It is assumed that these savings might compensate eventual losses of revenues due to having fewer passengers on particular flights.

From Equation 4.3.7b the savings of time of the aircraft remaining in a given, modified complex and associated revenues due to flying instead of being on the ground at the hub airport can be estimated as:

$$S_h = \sum_{i=1}^{M}[\frac{1}{2}(N_i - n_i)n_i(1/\mu_{ai} + 1/\mu_{di}) + \sum_{j=i+1}^{M}(N_j - n_j)n_j(1/\mu_{aj} + 1/\mu_{dj})]r_i \quad (4.3.13a)$$

where

n_i is the number of flights removed from the complex serving the sub-network (*i*);

r_i is the earning rate of a flight serving the sub-network (*i*) (monetary units per unit of time).

From Equation 4.3.7c, the savings of time of all the aircraft at spoke airports can be determined as:

$$S_{sh} = \sum_{i=1}^{M-1}\sum_{j=1}^{M-1}\frac{1}{2}[(N_i - n_i)n_i(1/\mu_{ai} + 1/\mu_{di})r_i + (N_j - n_j)n_j(1/\mu_{aj} + 1/\mu_{dj})r_j] \quad (4.3.13b)$$

The saving in the cost of using gates by a complex can be estimated as follows:

$$S_g = \sum_{i==1}^{M} n_i \tau_{0i} c_i \quad\quad\quad (4.3.13c)$$

where

c_i is the cost of using the gate at the hub airport by an aircraft serving the sub-network (i).

As can be seen from Equations 4.3.13a–4.3.13b, the savings in time and associated increase in revenues due to flying instead of being on the ground decrease more than proportionally with the increasing number of flights 'extracted' from the original complex. At the same time, the savings of using gates and aircraft ground handling staff increase in proportion with the increasing number of these flights. There is no estimate for reducing the number of passengers on particular flights. Therefore, in order to estimate the full effects during a given period of time (one year), the effects per given complex need to be multiplied by the number of complexes per day and the number of days per year. In the case of American Airlines, the number of flights 'displaced' from the original complexes at its main hubs is one or two per complex, which for about nine complexes per day and 365 days per year gives 6,570 flights per year. Reduction of block time of particular flights through reducing the ground time also diminishes the cost of pilots and cabin crew related to block time. The size of the fleet and associated costs to carry out almost the same transport work (p-km) during a given period of time will also be smaller. In addition to changing the airline tactics, increasing the capacity of the hub airport can contribute to reducing the airline fleet ground time and consequently enabling higher efficiency and effectiveness of operations. In any case, due to obvious advantages and disadvantages of this concept, a balance should be established between the above-mentioned cost savings and potential revenue losses.

4.3.5.2 The 'continuous' hub

The concept of a 'continuous' hub actually implies the point-to-point network with the very high number of incoming and outgoing flights at particular airports served by an airline. These flights are not synchronised and consequently not interconnected. However, their high number enables relatively efficient and effective passenger connections over a given period of time. For example, one of the biggest US low-cost airlines, Southwest Airlines, operates this network concept enabling connections at particular airports within time windows of about four hours. Consequently, in addition to point-to-point flight frequency, one of the most important attributes of this network appears to be the number of connections. Some ideas of how this number can be determined following such a context let t_k be the time window for which the number of possible connections between N_{ik} incoming and N_{kj} outgoing flights at the airport (k) need to be determined. The interval tk is divided into discrete increments Δ_t. The minimum connection time from any incoming to any outgoing flight is y (at Southwest Airlines this time is 45 minutes). Consequently, the number of increments within given time window is $L = |(t_k\text{-}y)/\Delta_t|$. In addition, the number of possible connections after $l\Delta t$ increments from the beginning of the time window t_k can be determined as follows:

$$n_{k,ij} = (\frac{N_{ik}}{t_0}l\Delta t)(\frac{N_{kj}}{t_0})[t_k - (l\Delta t + y)] \quad \text{for } l = 1,2,...,L \tag{4.3.14}$$

where other symbols are as in previous equations.

As can be seen, the number of connections is proportional to the product of the incoming and outgoing flights at the airport (this concept enables Southwest Airline to connect 20–25 per cent between incoming and outgoing flights within its network). Equation 4.3.14 also indicates that at the beginning of the period, passengers from the smaller number of incoming flights have the opportunity to connect to the greater number of outgoing flights. Later, this opportunity becomes opposite, that is the greater number of incoming passengers experiences a diminishing opportunity to connect to the smaller number of outgoing flights. Summing up the number of connections over given increments L gives the total number of connections during the time window t_k. Similarly, the total number of connections between all the incoming and all outgoing flights at an airport over a given period of time can be obtained.

4.3.5 Qualitative Evaluation of Airline Hub-and-Spoke Network

Combining the qualitative values of particular attributes of performances while considering the preferences of the actors involved enables a qualitative evaluation of sustainability, that is the overall social feasibility, of an airline hub-and-spoke network. The importance of particular attributes can be estimated by weighting their positive or negative effects by particular actor(s). In such a context, a higher weight usually means higher importance independent of the type of effect. Due to the lack of quantitative information, only the general direction of influence of a given attribute is provided in Table 4.3.1 below. The sign '+' indicates positive, sign '−' negative, and a blank space no-effect or relevance of particular attribute for particular actors. As can be seen, the attribute 'spatial layout and interconnectivity of nodes' has a positive effect for users-passengers due to the increasing choice of destinations from each, particularly spoke, airport of the network. It may have positive effects for the airline due to connecting the same number of airports by the smaller number of routes-links compared with the equivalent point-to-point network. Some local communities around spoke airports may suffer from the permanent threat of losing air connections and consequently employment at the local scale.

The attribute 'spatial and time concentration of traffic flows/services' appears to be beneficial for users-passengers due to the increased flight frequencies on particular – smaller number of routes. For the airline, it is feasible due to increasing the flight frequencies on the smaller number of routes by the existing fleet. This attribute might appear non-feasible for passengers and employees at the hub airport due to the surface congestion while accessing and leaving the airport.

Table 4.3.1   **Attributes of performances of (air) transport hub-and-spoke network and their expected effects for particular actors**

Attribute of performance	Actor(s)				
	Users – passengers	Infrastructure providers – airport(s), ATC	Transport operators – airline(s)	Local community	Government/ policy makers
1. Spatial layout and interconnectivity of nodes	+		+	–	
2. Spatial and time concentration of demand/services	±	–	+	–	
3. Economics of operations – users and operator costs	+		+		
4. Size and utilisation of airline fleet			–/+		
5. Utilisation of airport infrastructure		–	–		
6. Reliability and vulnerability	–	–	–		
7. Environmental impacts – noise, air pollution, land use				–	–
8. Social effects				+	+

The attribute 'economics of operations - users and operator costs' is beneficial for both airline and users-passengers due to economies of scale, which enables diminishing airfares on the one hand and increasing the airline's fare and frequency competitiveness on the other.

The attribute 'size and utilisation of airline fleet' may have both positive and negative effects. Positive effects include an increased utilisation of the airline fleet thanks to the higher flight frequencies. Negative effects emerge because of carrying out flights on the generally shorter routes and due to the potentially lower utilisation of the smaller aircraft engaged on the shorter routes.

The attribute 'utilisation of airport infrastructure' has a negative effect for the hub airport due to the need for providing the substantial airport infrastructure, landside and airside, on the one hand and its non-balanced utilisation over time on the other.

The attribute 'reliability and vulnerability' has a generally negative effect for users-passengers, airports, and airline due to the additional costs caused by disruptive events.

The attribute 'environmental impacts' has a generally negative effect on the local communities and policy makers. Noise air pollution and land use burden the local community around the hub airport. The growth of these impacts creates problems for local and national policy makers regarding the further development of the air transport sector.

The attribute 'social effects' has positive effects on the local community and policy makers by contributing to the local and global employment.

The variety of the above-mentioned effects of attributes of performances of the hub-and-spoke networks for particular actors reveals the full complexity of evaluating their overall sustainability and reliability, that is the social feasibility. Therefore, each case of these networks should be considered separately using the proposed models for quantifying particular attributes, assessing their relative importance analytically and/or empirically and applying the appropriate evaluation methods.

4.3.6 Concluding Remarks

This section has elaborated the concept for the evaluation of sustainability, that is the overall social feasibility, of an airline hub-and-spoke network, both the conventional model and modified forms such as the 'rolling' and 'continuous' hub-and-spoke network. For evaluation purposes, eight generic attributes reflecting the technical/technological, operational, economic, environmental and social dimension of the network performances have been developed and modelled as follows: i) spatial layout and interconnectivity of nodes; ii) spatial and time concentration of demand and services; iii) economics of operations – users and operator's costs; iv) size and utilisation of airline fleet; v) infrastructure requirements and its utilisation; vi) reliability and vulnerability; vii) environmental impacts – noise, air pollution, land use; and viii) social effects. Qualitative evaluation of particular attributes has indicated prospective advantages and disadvantages of this network configuration for particular actors involved. Nevertheless, a quantitative analysis including an application of the cost-benefit and/or multi criteria decision-making method is needed for each concrete case to enable a final judgement about the overall social feasibility, that is sustainability, of this network.

4.4 Vulnerability of the Airline Hub and Spoke Networks

4.4.1 Background

Large-scale disruptive events such as bad weather, industrial action by aviation staff, catastrophic failures of the system facilities and equipment, air traffic accidents, and/or terrorist treats can cause a significant deterioration of an airline's network and operations, materialised as long delays, rerouting and cancellation of numerous flights. The scale and scope of deterioration of the airline network performance and the related consequences expressed by the cost of the affected actors such as users-passengers, airline, and airports reflect the vulnerability of the system on the one hand and its 'robustness' to save performances under disruptive conditions on the other. In the US, the most frequent disruptive event is severe weather. In Europe, it has been the industrial action of the aviation staff. When major airlines are affected, the entire air transport system in a region is disrupted for a long time. Anyway, the experience up to date has shown that the airline hub-and-spoke networks have been the most vulnerable even to small-scale disruptive events (Schavell, 2000; Allan et al., 2001).

In many considerations of sustainability of the air transport system and its components this issue has been neglected. Therefore, this section presents a model for the assessment of the economic consequences of large-scale disruptions to an airline single hub-and-spoke network, expressed by the airline's costs of delayed and cancelled flights. This model is based on the theory of queuing systems, in which the airline hub airport is considered as a server and the flights bunched as complexes of flights at the hub airport as customers. Most of the material used originates from the author's already published research work (Janić, 2005a).

4.4.2 The System and the Problem

4.4.2.1 Characterisation of airline hub-and-spoke network
The airline hub and spoke network has an exclusive star-shaped layout. In this network, the relatively frequent flights connect spoke airports to the centrally located hub (Kanafani and Ghobrial, 1985; O'Kelly et al., 1997; O'Kelly, 1998). At the hub airport, the flights are clustered into complexes, each consisting of the incoming and outgoing bank of flights. Banks of the same complex are scheduled close together in time (usually one to three hours), enabling an efficient and effective exchange of traffic – passengers and cargo – and resources – aircraft and crew. Usually, the first flight from the outgoing bank cannot be released before the last arriving flight from the incoming bank plus the necessary connecting time (Hall, 2001).

The number of complexes and flights may vary for particular airlines. For example, major European airlines such as KLM (the Netherlands) and Air France (France) schedule six complexes of flights per day at two their main hubs – Amsterdam Schiphol (Amsterdam) and Paris Charles de Gaulle (Paris), respectively. KLM operates 92 (46/46) and Air France 70 (35/35) flights per complex. Lufthansa (Germany) schedules four complexes of flights per day at its main hub – Frankfurt Main airport. Each complex contains 158 flights (79/79), which are realised over a period of 3 hours. Some US majors such as, for example, American and Delta Air Lines schedule 10 and 12 complexes of flights at their main hubs – Dallas/Fort-Worth and Atlanta Hartsfield airport, respectively. The average number of flights per complex is 70 and 90, respectively (AEA, 2001; BTS, 2001; OAG, 2002).

4.4.2.2 Disruptions of an airline hub-and-spoke network and mitigating strategies
Due to the high interconnectivity between flights in the incoming and outgoing bank of particular complexes, disruptive events may affect their regularity and integrity with a varying degree of intensity and scale. Generally, for a given complex, the long delay of one or more incoming flights may cause the long delay of one or more outgoing flights (Hall, 2001). In addition, the delay from an earlier affected complex, if not neutralised by an appropriate time buffer, may propagate and affect some later, otherwise unaffected complexes.

The affected airline uses the so-called 'schedule recovery strategies' to handle disruptions of its network. The Airline Operation Centre (AOC) executes these strategies (Schavell, 2000).

In general, two strategies can be used. The *first strategy* includes 'delaying and cancelling later on' the affected complexes and flights. If a disruptive event overturns the airport capacity or the capacity of the entire system to zero, all foreseen affected complexes are to be temporarily suspended. They will be carried out after the partial or full recovery of capacity. If the disruptive event only diminishes the capacity, all the complexes are still to be realised in the scheduled order but with delays, which, if not neutralised, may spread from the earlier directly affected to the later directly unaffected complexes. Consequently, the very late complexes and flights may come under a risk of cancellation due to the lack of time for their realisation. This strategy enables the airline to save the integrity of the planned schedule, but with the cost of delays and the loss of revenue from the affected – delayed and late cancelled complexes and flights, respectively.

The *second strategy* implies immediate cancellation of the directly affected complexes and flights. This should prevent spreading of the unpredictably long delays of the earlier affected to the later directly unaffected complexes and flights. In most cases, the 'unpredictably long delays' are not particularly specified, which reflects the lack of clear criteria for either delaying or cancellation of the affected complex and/or flight. The airline suffers revenue losses from the cancellations and the costs of repositioning the aircraft after the disruptive event. In addition, the airline may reap some benefits from the avoidable operational costs of the non-realised – cancelled – flights. This strategy compromises the integrity of the airline schedule. Therefore, it is usually used when long disruptive events last for several hours and have a severe impact on the capacity of the hub airport (persistent bad weather, industrial action by aviation staff).

Both the above-mentioned strategies are usually operationalised by rerouting and ground holding the affected flights. When the disruptive event closes the hub airport, the already airborne flights from the incoming bank of a given complex are either delayed for a while or rerouted to other alternative airports. In the case of perceivably long delays, the affected flights are rerouted immediately and their arrival at the affected hub airport is cancelled. The revenues from these flights are mostly saved. However, there might be additional costs of repositioning the aircraft and crew from the temporary airports to the hub afterwards, and the cost of compensation to the affected passengers to reach their intended destinations. The other still non-airborne flights-aircraft belonging to the same incoming bank are ground-held at their origin-departure airports at least for the time of the disruptive event (Terrab and Odoni, 1993). The passengers from these flights may be patient and stay around the airline. In such a case, the airline prospectively counts on their revenue. Otherwise, the airline suffers revenue losses. At the hub airport, the outgoing bank of the same complex carried out by the same aircraft as the cancelled incoming bank has to be either delayed or cancelled. The 'feeding' aircraft from the incoming bank has not arrived and the aircraft from the outgoing bank cannot depart due to the hub's closure. The affected passengers may stay around the airline or give up.

If particular spokes are severely affected and long delays perceived, the corresponding individual flights of a given complex and the succeeding flights of the later complexes carried out by the same aircraft will be cancelled. This will prevent spreading of the long delays from an individual affected flight to other flights of the same and later complexes. Such serial cancellations temporarily reduce the number of flights in the later complexes and 'isolate' the affected spokes (Schavell, 2000).

4.4.2.3 Scale of consequences

In general, a flight is considered delayed if it departs or arrives 15 or more minutes behind schedule (ITA, 2000). In the US, bad weather causes about 70–75 per cent and the airport and airspace congestion about 20–30 per cent flight delays. In Western Europe, weather causes 1–4 per cent and the airport and airspace congestion 30–40 per cent flight delays (AEA, 2001; BTS, 2001; EEC, 2001b).

During the past decade and a half, in the US, the proportion of cancelled and rerouted flights compared with the proportion of delayed flights has been relatively modest. The proportion of cancelled flights has been three to 10 times greater than the proportion of the rerouted flights. The number of cancellations has generally increased while the number of rerouted flights has remained relatively stable (BTS, 2001). Some figures reported by the US majors for the period 1996–1999 indicated that the total direct annual cost of irregular operations was about USD1.9bn. The delay cost accounted for approximately USD900, cancellations USD800–900, and redirections USD50–60mn. These estimates include the total delay, cancellation and rerouting costs as the primary and the secondary-downstream events along the aircraft itineraries (Lettovsky et al., 1999; RC, 1999; Schavell, 2000).

4.4.3 Model of Large-Scale Disruption of Airline Network

4.4.3.1 State of the art

The large-scale disruption of an airline hub-and-spoke network usually happens when a disruptive event significantly deteriorates the practical capacity of a hub airport. Due to this and other operational and planning reasons of the airport, airline and air traffic control, this capacity and the associated aircraft delays have been extensively investigated. Determination of the airport capacity and delays has been a very complex task because of the difficulties in handling many influencing factors such as weather, run away configuration, the arrival/departure traffic mix, the aircraft fleet mix and the air traffic controller's workload appropriately. The investigation has resulted in developing many analytical, simulation and optimisation models of the airside capacity and delay for the individual airport and the airport networks.

Usually, the analytical models have obtained a two-value parameter on the airport practical capacity – one for the arrival and another for the departure capacity. Both have been expressed by the maximum number of aircraft handled during a given period of time (one every 15 minutes) under given conditions (Tosic and Horonjeff, 1976; Andreata and Romanin-Jacur, 1987; Bianco and Bielli, 1992; Richetta and Odoni, 1993). Some important inputs have been constant demand, the average delay per operation, the relationship between arrivals and departures, fleet mix, and the runway configuration (Blumstein, 1960; Harris, 1972; Hockaday and Kanafani, 1974; Newell, 1979; Swedish, 1981; Janić and Tosic, 1982; Gilbo, 1993, 1997). In addition, some recently known models for planning purposes have been the Quasi-Analytical Models of Airport Capacity and Delay (FAA Airport Capacity Model, LMI Runway Capacity Model, and DELAYS model) (Odoni and Bowman, 1997).

Simulation computer-supported models have been developed to handle more realistically the airport airside operations, capacity and delays. They have provided more detailed simulation of the aircraft movements at an airport and in its vicinity.

Some important models have been those simulating Airport Operations at a High (Hermes and The Airport Machine) and Intermediate Level of Detail (NASPAC, FLOWSIM and TMAC), and those simulating Airport and Airspace Operations at a High Level of Detail (Taam and SIMMOD) (Odoni and Bowman, 1997).

The optimisation models for airport airside capacity and delays have been developed in the scope of the Air Traffic Flow Management (ATFM) models. In these models, the airport capacity has been considered as the changing (stochastic) variable, which reflects an uncertainty in predicting the flight delays and their costs. Under such circumstances, these delays and costs have been minimised with respect to the possibility of spreading – propagation – between the flight origin and destination airports. For such purposes, the ATFM models for a single airport, the airport network and the airport network including the airspace between airports have been developed (Andreata and Romanin-Jacur, 1987; Richetta and Odoni, 1993, 1994; Terrab and Odoni, 1993; Vranas et al., 1994, 1994b; Richetta, 1995).

Recently, some models for handling the influence of significant deterioration of airport capacity on the airline schedule have been elaborated (Beatty et al., 1998; Mayer et al., 1999; Schavell, 2000; Schaefer and Millner, 2001; Welch and Lloyd, 2001). In particular, quantification of the costs of large-scale disruptions to an airline network has been pointed out as a challenging task for the following reasons: i) each large-scale disruptive event is unique in terms of the time it happens, duration and intensity of impact; ii) the cost of disruptions is mostly known afterwards and is highly variable for different airlines; iii) data to estimate some of the above costs might be missing; iv) some existing models for estimating these costs mostly relate to the very specific – concrete – cases of airport and airline disruptions (Beatty et al., 1998; Schavell, 2000; Allan et al., 2001).

This section develops a generic model to assess the economic consequences – costs – of large-scale disruption to an airline single hub-and-spoke network. The model can contribute to the planning purposes and enable sensitivity analysis of the disruption costs regarding the changes of the main influencing factors.

4.4.3.2 Assumptions
The following assumptions are introduced in developing the model:

- An airline schedules a series of complexes of flights at its hub airport during a given period of time as shown in Figure 4.4.1;
- The airline tends to achieve the objectives such as 'saving integrity', 'avoiding mixing' and 'maintaining the scheduled order' at 'all costs' (Mayer et al., 1999). 'Saving integrity' implies keeping the scheduled number of flights in each complex regardless of the intensity of the disruption. 'Avoiding mixing' implies the strict allocation of aircraft fleet to complexes of a given type in order to keep the impact of disruptive events relatively 'isolated'. 'Maintaining the scheduled order' implies carrying out the affected complexes according to the planned order regardless of the intensity of the impact of the disruptive event;
- The disruptive event in terms of time, duration and intensity of impact occurs according to a given scenario and is predictable in advance;

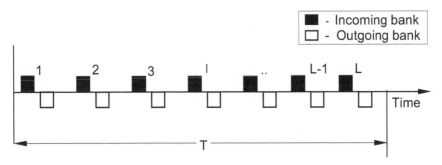

Figure 4.4.1 Scheme of the complexes of flights scheduled by an airline at its hub airport

- For particular scenarios, the disruptive event may affect the practical capacity of the airline hub airport and/or the capacity of the surrounding airspace by different intensities – from modest or no reduction to the overturning to zero. For a given scenario, this impact is constant for the duration of the disruptive event. The former intensity of impact still enables the airport and airspace operation although with reduced capacity. The latter one prevents any airport and airspace operation; and
- The affected airline uses one of the two above-mentioned mitigating strategies to handle the large-scale disruption of its network.

4.4.3.3 The model structure
The model is based on the queuing theory. The complexes of flights scheduled at the hub airport during a given period of time are regarded as customers and the hub itself as a server. The times of the first incoming flights of each complex at the hub airport form the customer arrival process. The times of the first outgoing flights of each complex from the hub airport form the customer departure process.

The planned service time of each complex at the hub airport is defined as the difference between its arrival and departure time. This time depends on many factors, but mostly on the practical capacity of the hub airport achieved under regular – the most common – operating conditions (Newell, 1979; Gilbo, 1993).

A disruptive event usually affects (diminishes) the airport's practical capacity and prolongs the scheduled service time of the particular complexes-customers. The difference between the actual-affected and the planned-scheduled service time defines the delay of a given complex. If this delay spreads from the earlier affected to the later unaffected complexes of the same or different type, the virtual queue of complexes waiting for service builds up. Consequently, this model structure seems to deal with the delay propagation throughout an airline network analogously as the models of delay propagation through an airport network (Beatty et al., 1998; Yu, 1998; Campbell et al., 2000; Johnston, 2000; Rosenberger et al., 2000; Schaefer and Millner, 2001).

The model structure includes modelling a) the service time of the scheduled – regular – and affected – irregular – complex; b) delays of the particular complexes; and c) the costs of the affected (delayed and cancelled) complexes.

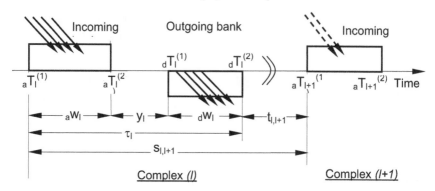

**Figure 4.4.2 The time characteristics of the two successive complexes (*l*) and
          (*l*+1)**

Service time of scheduled and affected complexes Scheduled complexes Modelling
of the scheduled service time of a given complex is based on the scheme in Figure
4.4.2 where particular symbols have the following meaning (Hall and Chong,
1993).

l is the 'position' of a given complex in the series of L complexes scheduled
          during period T;

$_aT_l^{(1)}$, $_aT_l^{(2)}$ is the time of the beginning and the end of the incoming bank of
          complex (*l*), respectively, measured from the beginning of period T;

$_dT_l^{(1)}$, $_dT_l^{(2)}$ is the time of the beginning and the end of the outgoing bank of
          complex (*l*), respectively, measured from the beginning of period T;

$_aw_l$, $_dw_l$ is the time window for realising the incoming and outgoing bank of
          complex (*l*), respectively;

y_l is the time window between the latest (last) incoming and the earliest (first)
          outgoing flight of complex (*l*) ($y_l \geq 0$);

τ_l is the scheduled service time of complex (*l*), defined as the time between its
          earliest (first) incoming and the latest (last) outgoing flight;

$t_{l,l+1}$ is the time between two successive complexes (*l*) and (*l*+1) as the gap
          between the last outgoing flight of complex (l) and the first incoming flight
          of complex (*l*+1), respectively; and

$s_{l,l+1}$ is the time between the beginning of two successive complexes (*l*) and (*l*
          +1), respectively.

As shown in Figure 4.4.2, the scheduled service time of complex (l), τ_l is equal to the
sum of the time windows $_aw_l$, y_l, and $_dw_l$ as follows:

$$\tau_l = {_aw_l} + y_l + {_dw_l} \qquad (4.4.1)$$

The time windows $_aw_l$ and $_dw_l$ are determined as follows:

$${_aw_l} = ({_aF_l} + {_af_l})/{_a\mu_l} \text{ where } {_a\mu_l} = \min\left[1/{_a\delta_l}; {_aM_l}\right] \qquad (4.4.2a)$$

and

$$_d w_l = (_d F_l + _d f_l) / _d \mu_l \text{ where } _d \mu_l = \min \left[1 / _d \delta_l ; _d M_l \right] \tag{4.4.2b}$$

where

$_a F_l, _d F_l$ is the number of flights in the incoming and outgoing bank of complex (l), respectively;

$_a f_l, _d f_l$ is the number of incoming and outgoing flights of other airlines, respectively, mixed with the incoming and outgoing flights of complex (l);

$_a \mu_l, _d \mu_l$ is the service rate of the incoming flights $_a F_l$ and outgoing flights $_d F_l$, respectively, of complex (l);

$_a M_l, _d M_l$ is the practical arrival and departure capacity, respectively, of the hub airport or the 'critical' surrounding airspace;

$_a \delta_l, _d \delta_l$ is the average time separation between the aircraft in the resulting incoming and outgoing traffic stream consisting of the flights of complex (l) and the flights of other airlines, respectively.

In Equation 4.4.2, the number of flights $_a F_l$ and $_d F_l$ corresponds to the number of spokes connected to the hub. The service rates $_a \mu_l$ and $_d \mu_l$ depend on the necessary time separation between the aircraft in the incoming and outgoing bank, respectively, or on the airline specification. The practical capacities $_a M_l$ and $_d M_l$ depend on the main factors influencing the airport's practical capacity. In general, these capacities may depend on each other (Newell, 1982; Gilbo, 1993). The 'critical' airspace can either be some low altitude sector, terminal airspace or only the busiest inbound and outbound routes around the hub airport. The practical capacity of this airspace mostly depends on the aircraft separation rules and the air traffic control capacity (workload) (Janić and Tosic, 1982; Hall and Chong, 1993; Mayer et al., 1999; Welch and Lloyd, 2001).

Affected complexes The service time of an affected complex is determined by the 14 detailed analytical expressions (developed by the author) to embrace the possible influence of a disruptive event on the scheduled service time of a given complex. Figure 4.4.3 illustrates an example: the disruptive event starts before the starting of the incoming bank and ends during the outgoing bank of a given complex.

The service time of the affected complex (l) is determined as follows:

$$\tau' = \begin{bmatrix} [(_a F_l + _a f_l) / _a v_l + y_l + (T_0 + d - _d T_l^{(l)}) + [(_d F_l + _d f_l) - (T_0 + d - _d T_l^{(l)})_d v_l] / _d \mu_l \text{ for } _a v_l, _d v_l > 0 \\ (T_0 + d - _a T_l^{(l)}) + \varepsilon B_l + (_a F_l + _a f_l + _a A_l) / _a \mu_l + y_l + (_d F_l + _d f_l + _d A_l) / _d \mu_l \text{ for } _a v_l, _d v_l = 0 \end{bmatrix} \tag{4.4.3}$$
$$\text{for } T_0 \le _a T_l^{(l)} \text{ and } _d T_l^{(l)} \le (T_0 + d) \le _d T_l^{(2)}$$

where

$_a A_l, _d A_l$ is an additional incoming and outgoing traffic, respectively;

B_l is the time of starting either the first incoming or outgoing flight of complex (l) after reopening the hub airport;

ε is a binary variable taking the value *1* if the first affected complex is (l) and otherwise *0*;

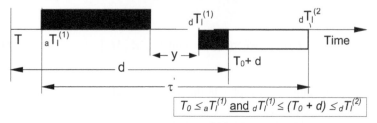

Figure 4.4.3 An example of the possible relationships between the disruptive event and the service time of an affected complex

$_a v_l,\ _d v_l$ is the service rate of the incoming and outgoing bank of complex (l), respectively, during the disruptive event;

T_0 is the time of the beginning of the disruptive event;

D is the duration of the disruptive event.

In Equation 4.4.3, the additional traffic $_a A_l$ and $_d A_l$ reflects the intentions of many airlines to operate as soon as possible to/from the just 'recovered' airport. The time B_l may vary from a few minutes to an hour or so. In the former case, the first airborne incoming flight in the vicinity of the airport lands quickly or the first outgoing flight is almost immediately ready to depart. In the latter case, the first flight arrives from the closest spoke where it has been ground-held or the first outgoing flight needs some preparation time. The affected airport capacity, $_a v_l$ and $_d v_l$ can be determined as in Equation 4.4.2 by replacing the capacities $_a M_l$ and $_d M_l$ with their 'affected' constant counterparts $_a V_l$ and $_d V_l$, respectively $(0 \le {_a V_l} \le {_a M_l};\ 0 \le {_d V_l} \le {_d M_l})$ (Newell, 1982; Gilbo, 1993).

Delays of the affected complexes   Let (l) and $(l+1)$ be the two successive complexes as shown in the above-mentioned Figure 4.3.4. According to that figure, the arrival delay of a complex $(1+1)$, $_a d_{l+1}$ can be determined as follows:

$$_a d_{l+1} = \max\left[0;\ _a d_l + (t'_{l,l+1} - t_{l,l+1})\right] = \max\left[0;\ _a d_l + (t'_{l,l+1} - t_{l,l+1}) + (\tau'_l - \tau_l - b_l)\right] \quad (4.4.4a)$$
for $l = 1, 2, \ldots, L - 1$

Analogously, the departure delay of complex $(l+1)$, $_d d_{l+1}$ can be determined as:

$$_d d_{l+1} = \max\left[0;\ _a d_{l+1} + (\tau'_{l+1} - \tau_{l+1} - b_{l+1})\right] \quad (4.4.4b)$$
for $l = 1, 2, \ldots, L - 1$

where

$\tau_l,\ \tau'_l$ is the scheduled and the affected service time of complex (l), respectively;

$b_l,\ b_{l+1}$ is the time 'buffer' for neutralising the short foreseeable delay of complexes (l) and $(l+1)$, respectively;

$t'_{l,\,l+1}$ is the revised time between the complexes (l) and $(l+1)$ after complex (l) is affected by the disruptive event;

$_a d_l,\ _d d_l$ is the arrival and the departure delay of complex (l), respectively.

In Equation 4.4.4, each flight of the affected complex 'carries' the same delay between the hub and a given spoke. There, the previously outgoing delay is turned into the incoming delay. After the aircraft turnaround service, the previously incoming delay, if not neutralised, turns again into the outgoing delay, and so on.

Airline costs of the affected complexes   Delayed complexes   The total airline costs of all the affected complexes during period T is determined as follows:

$$C_d = \sum_{l=1}^{L} \begin{bmatrix} {}_a d_l \sum_{n=1}^{{}_a F_l} [{}_a c_{\ln} ({}_a d_{\ln})] + \sum_{n=1}^{{}_a F_l} {}_a P_{\ln} ({}_a d_l) {}_a \lambda_{\ln} {}_a N_{\ln} {}_a r_{\ln} + \\ {}_d d_l \sum_{n=1}^{{}_d F_l} [{}_d c_{\ln} ({}_d d_l)] + \sum_{n=1}^{{}_d F_l} {}_d P_{\ln} ({}_d d_l) {}_d \lambda_{\ln} {}_d N_{\ln} {}_d r_{\ln}] \end{bmatrix} \qquad (4.4.5)$$

where

   ${}_a c_{\ln}({}_a d_l)$, ${}_d c_{\ln}({}_d d_l)$ is the average unit cost of delay of an affected incoming and outgoing flight (n) of complex (l), respectively;

   ${}_d P_{\ln}({}_a d_l)$, ${}_d P_{\ln}({}_d d_l)$ is the proportion of passengers who have given up from an affected incoming and outgoing flight (n) of complex (l), respectively ($0 \leq {}_d P_{\ln}({}_a d_l)$, ${}_d P_{\ln}({}_d d_l) \leq 1$);

   ${}_a \lambda_{\ln}$, ${}_d \lambda_{\ln}$ is the load factor of an affected incoming and outgoing flight (n) of complex (l), respectively ($0 \leq a \lambda_{\ln}$, ${}_d \lambda_{\ln} \leq 1$);

   ${}_a N_{\ln}$, ${}_d N_{\ln}$ is the aircraft capacity (seats) of an affected incoming and outgoing flight (n) of complex (l), respectively;

   ${}_a r_{\ln}$, ${}_d r_{\ln}$ is the average revenue per passenger of an affected incoming and outgoing flight (n) of complex (l), respectively.

Other symbols are analogous to those in Equations 4.4.2–4.4.4.

   In Equation 4.4.5, if complex l is not affected, its arrival and departure delays and corresponding costs will be zero. In general, the delay costs ${}_a c_{\ln}({}_a d_l)$ and ${}_d c_{\ln}({}_d d_l)$ may be the increasing or constant functions of the magnitude of delay (ITA, 2000). They may be dependent on the airline operating costs (aircraft, crew), type of delay (ground, airborne), the aircraft seat capacity, type of flight (short, medium, long haul; domestic, international), and the airline (major, regional, low-cost) (ITA, 2000; Schaefer and Millner, 2001). The proportion of the passengers who give up, ${}_d P_{\ln}({}_a d_l)$ and ${}_d P_{\ln}({}_d d_l)$, may also be increasing functions of the magnitude of delay.

Cancelled complexes   Let (k) be the first and Q the last affected complex to be cancelled immediately. The total cost of the cancelled complexes is as follows:

$$C_c = \max \left\{ 0; \sum_{l=k}^{Q} \sum_{n=1}^{{}_a F_l} [{}_a \theta_{\ln} {}_a C_{\ln} + (1 - {}_a \theta_{\ln})({}_a q_{\ln} {}_a \lambda_{\ln} {}_a N_{\ln} {}_a r_{\ln} + {}_a C_{\ln} - {}_a C_{\ln}^*)] + \sum_{n=1}^{{}_d F_l} \begin{pmatrix} {}_d q_{\ln} {}_d \lambda_{\ln} {}_d N_{\ln} {}_d r_{\ln} \\ + {}_d \gamma_{\ln} C_{\ln} - {}_d C_{\ln}^* \end{pmatrix} \right\} \qquad (4.4.6)$$

for $l \in Q$ and $k \leq Q \leq L$

where

$_a q_{ln}$, $_d q_{ln}$ is the proportion of passengers giving up from the cancelled incoming and outgoing flight (n) of complex (l), respectively ($0 \leq _a q_{ln}$, $d\,q_{ln} \leq 1$);

$_a C_{ln}$, $_d C_{ln}$ is the aircraft and crew repositioning cost from the cancelled incoming and outgoing flight (n) of complex (l), respectively;

$_a C_{ln}^{*}$, $_d C_{ln}^{*}$ is the avoidable cost of the cancelled incoming and outgoing flight (n) of complex (l), respectively;

$_a \theta_{ln}$ is a binary variable, which takes the value 1 if the flight (n) of complex (l) is redirected and the value 0 otherwise.

$_d \gamma_{ln}$ is a binary variable, which takes the value 1 if there is the cost to provide the departure flight (n) of complex (l) with the necessary resources, and 0 otherwise.

The other symbols are analogous to those in Equation 4.4.5.

In Equation 4.4.6, the costs of cancelled flights include the losses of revenues from the passengers who have given up, the cost of repositioning the temporarily rerouted aircraft, and the avoidable costs from the non-realised flights.

4.4.4 Application of the Model

4.4.4.1 Inputs

The model is applied to a large European airline – Lufthansa (Germany) – operating from its main hub at Frankfurt Airport (Germany). The large-scale disruptive events such as the winter snowstorms and/or icy-fogs in the late autumn/early winter and industrial action of the aviation staff may significantly affect the airport's practical capacity and consequently the airline schedule. The airline operates four complexes of flights over the period $T = 24$ hours/day: the first starts at 07.00, the second at 11.30, the third at 15.30, and the last at 19.30 The series starts with the incoming bank of the first and ends with the outgoing bank of the last (fourth) complex.

The time between the beginning of complexes 1–2 and complexes 2–3–4 is $s^{*} = 270$ and 240 min, respectively. Since each complex lasts for $\tau^{*} = 180$ min, the time between complexes 1–2 is $t^{*} = s^{*} - \tau^{*} = 270 - 180 = 90$ min, and between the complexes 2–3–4, $t^{*} = s^{*} - \tau^{*} = 240 - 180 = 60$ min. These times are always kept constant regardless of the intensity of the impact of a disruptive event. The time buffers to neutralise the short foreseeable delays are $b = 0$ for all complexes.

Each complex consists of 158 flights (79/79). It is carried out on two independent parallel runways during the time windows of $w_a = w_d = 75$ min and $y = 30$ min. The flights are clustered into the short- and medium-haul and the long-haul flights (Lufthansa, 2003; OAG, 2002). The long haul flights are handled during the first 45 minutes of each incoming and during the last 45 minutes of each outgoing bank. The short- and medium-haul (European) flights are handled during the remaining 30 minutes of each bank's time window. Consequently, the service rate of each incoming and outgoing bank of flights is $\mu_a = \mu_d = 79$ flights/75 min $\cdot$ 60 min ≈ 63 flights/h. In addition to the clustering into two broad categories, the flights of a typical incoming/outgoing bank are characterised by their number in the particular markets, the route length, the aircraft seat capacity, load factor, the cost of delay, revenue, and the avoidable and repositioning costs (Table 4.4.1) (OAG, 2002; Janić, 2003; Lufthansa, 2003).

Table 4.4.1 Characteristics of typical incoming/outgoing bank of complexes of flights in a given example

Flight category	Fights per category	Market(s)	Aircraft type	Average seat capacity (seats) N	Average load factor (%) λ	Average route length (km) ρ	Average delay cost ($ US/min) c_a, c_d^1	Revenue ($ US/flight) R_a, R_d^2	Repositioning cost ($ US/flight) C_a/C_d^3	Avoidable cost ($ US/flight) C_a^*/C_d^{*4}
1	16	Germany, The closest EU[5]	CRJ 100/200/700, Avro Rj85	77	61.0	330	40	3,197	3,178/0	2,384/2,384
2	12	Central EU, Central Europe	B737, A319/320/321	135	61.0	675	87	10,819	10,688/0	8,016/8,016
3	12	Scandinavia	B737, A320/321	135	61.0	968	91	14,770	14,514/0	10,886/10,886
4	12	Mediterranean and Eastern Europe	A319/320/321	154	61.0	1635	102	25,444	21,655/0	12,993/12,993
5	8	Middle East and Russia	A300/A310, B757	246	71.3	4,290	128	79,787	10,848/0	0/55,993
6	8	North America – East Coast, India	A310/340	246	80.1	6,275	101	93,930	12,187/0	0/62,923
7	3	North America – Central, Africa	B747, A340	318	76.0	7,518	99	112,115	14,948/0	0/73,378
8	8	South America, North America – West Coast, Japan, South-East Asia	B747, A340	318	79.2	9,623	71	104,909	15,577/0	0/66,168

[1] For both arrival and departure flight: $c_{a/d} = (1/60) \, [c_u(\rho)N\lambda\rho]/(0.62 + 0.002,1\rho)$ where $c_u(\rho) = 0.218e^{-0.000186\rho}$, $300 \leq \rho \leq 10,000$ km (Janić, 2003; Lufthansa, 2003)

[2] For both arrival and departure flight: $R_{a/d} = r(\rho)N\lambda\rho$, where $r(\rho) = 0.218e^{-0.000168\rho}$, $300 \leq \rho \leq 10,000$ km (Janić, 2003; Lufthansa, 2003)

[3] The operational cost between the hub and other airports for flight categories (1)–(4) and the operational cost between the secondary and primary hub ($\rho = 300$ km) for the flight categories (5)–(8). The flight operational cost is: $C(\rho) = c_u(\rho)N\lambda\rho$ (Janić, 2003; Lufthansa, 2003)

[4] The avoidable cost C_a^*, C_d^* of all flight categories (1)–(4) and only of the outgoing flight categories (5)–(8) amount about 75 per cent of the operational costs (fuel and oil, crew, maintenance, navigational charges and handling cost are included). For the incoming flights of the categories (5)–(8), $C_a^* = 0$

[5] EU – European Union

The airline data on the proportion of passengers who have given up from affected flights are not available. Nevertheless, it is assumed that all passengers give up from long-delayed flights of the categories 1–2, that is $p_a = p_d = 1$, and no passengers give up from the affected flights of the categories 3–8, that is $p_a = p_d = 0$.

All the passengers from cancelled incoming and outgoing flights of the categories 1–4 give up, as well as from all outgoing flights of categories 5–8, that is $q_a = q_d = 1$. No passengers give up from incoming flights of the categories 5–8. These flights are assumed to be cancelled at the primary hub (Frankfurt) but actually they are rerouted to the airline's secondary hub – Munich airport, that is $q_a = q_d = 1$. The assumption that the passengers from flight categories 1–2 in Table 4.4.1 give up is based on the availability of the convenient surface – high-speed rail – alternative at Frankfurt Airport.

The avoidable and repositioning cost of the cancelled flights, C_a^*, C_d^* and C_a, C_d, respectively, are based on the flight operational costs, similarly to the delay cost c_a, c_d (Table 4.4.1).

The scenarios of occurrence of the disruptive event, A), B), and C), are selected to anticipate the event's inherently stochastic and unpredictable nature. According to these scenarios, the disruptive event starts at about 07.00 a.m, midday, and 05.00 p.m., respectively. In each scenario, the event lasts for $d = 2, 4, 6$, and 8 hours. The disruptive event is supposed to deteriorate the airport capacity of 63 flights/h by an intensity of 0 *per cent*, 40 *per cent* and 70 *per cent*. This gives the deteriorated capacities of $v_a = v_d = 0, 25$, and 44 flights/h, respectively.

Consolidation of the particular banks of flights starts $B = 60$ minutes after the end of the disruptive event. This time is needed to move the flights from the closest airports to the Frankfurt hub or the time to prepare the aircraft already at the hub airport for departure. The complexes are not mixed with the flights of other airlines and there is no additional traffic, that is $f_a = f_d = 0$ and $\Delta_a = \Delta_d = 0$, respectively (OAG, 2002).

4.4.4.2 Analysis of results
The results from the experiments with the model are shown in Figures 4.4.4, 4.4.5, and 4.4.6.

Figure 4.4.4 shows the impact of the characteristics of the disruptive event and the airline mitigating strategies on the costs of the affected complexes. The disruptive event overturns the capacity of the hub airport to zero. As can be seen, the costs of disruption are of the order of millions USD. With both mitigating strategies, these costs generally increase with the increasing duration of the disruptive event and with the earlier time of its occurrence during the day, and vice versa. The strategy 'immediate cancellations' always produces lower costs than the strategy 'delays and cancellations later on', which generally confirms the airline's preference for the former strategy.

The absolute differences in the costs from the two mitigating strategies are significant, from a few to several million USD. These differences increase with the increasing duration of the disruptive event and diminish as the time of its occurrence gets later. For both strategies the influence of the costs of the cancelled complexes and flights becomes similar and dominant. For example, in scenario C) when the disruptive event lasts for eight hours, the exclusive use of the strategy 'immediate

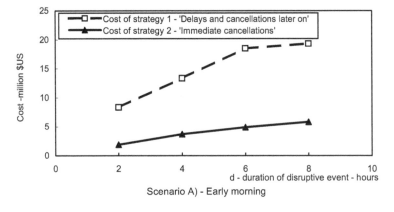

Scenario A) - Early morning

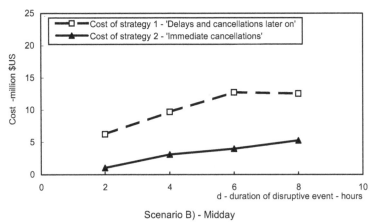

Scenario B) - Midday

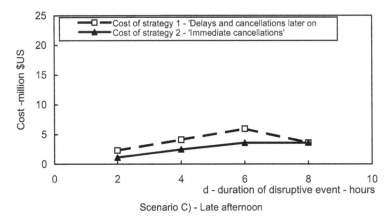

Scenario C) - Late afternoon

Figure 4.4.4 Influence of the characteristics of the disruptive event and airline mitigating strategies on the total disruption costs – the hub airport is closed in a given example: a) Scenario – Early morning; b) Scenario – Midday; c) Scenario – Late afternoon

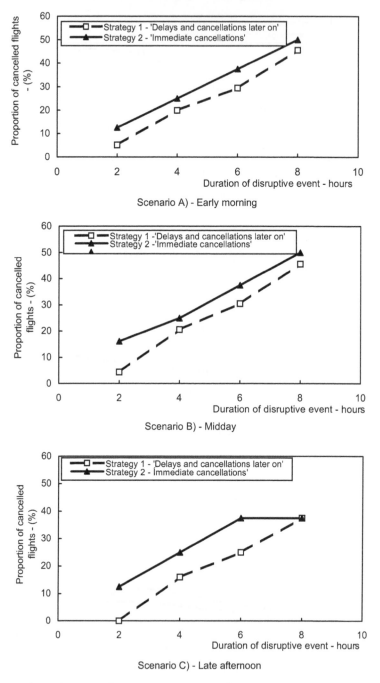

Scenario A) - Early morning

Scenario B) - Midday

Scenario C) - Late afternoon

Figure 4.4.5 Influence of the characteristics of the disruptive event and airline mitigating strategies on the proportion of cancelled flights – the hub airport is closed in a given example: a) Scenario – Early morning; b) Scenario – Midday; c) Scenario – Late afternoon

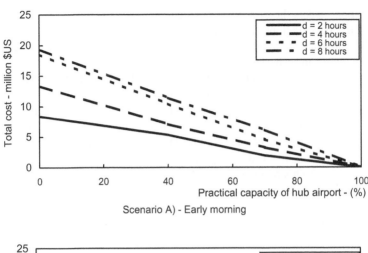

Scenario A) - Early morning

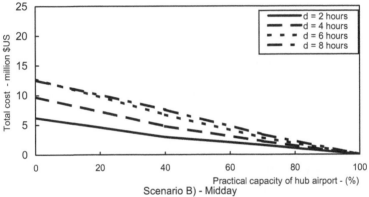

Scenario B) - Midday

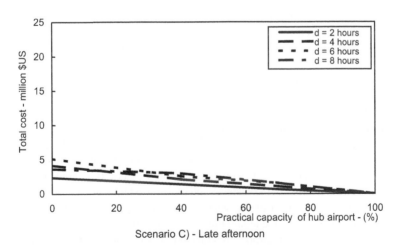

Scenario C) - Late afternoon

Figure 4.4.6 Influence of the characteristics of the disruptive event on the total disruption cost in a given example: a) Scenario – Early morning; b) Scenario – Midday; c) Scenario – Late afternoon

cancellation' makes the disruption costs equal. Generally, the costs of the strategy 'immediate cancellations' are lower than the costs of the strategy 'delays and cancellations later on' by about 3–4 times in scenario A), 2–6 times in scenario B) and 1.5–2 times in scenario C).

Figure 4.4.5 shows the impact of the characteristics of the disruptive event and the airline mitigating strategies on the proportion of cancelled flights. The disruptive event overturns the capacity of the hub airport to zero. The number of cancelled flights is always lower with the strategy 'delays and cancellations later on'. The difference in the number of cancelled flights for both strategies decreases with the increasing duration of the disruptive event and its later start during the day, and vice versa. For both strategies, the proportion of cancelled flights is the greatest in scenario A), when the disruptive event starts in the morning. It is the smallest in scenario C) when the disruptive event starts in the late afternoon. For the strategy 'delays and cancellations later on', the proportion of cancelled flights in the total number of 632 flights (four complexes) varies between 4 and 46 per cent, that is from 28 to 288 flights (~ 1.8 complexes) given the duration of the disruptive event. Under similar conditions, for the strategy 'only cancellations', this proportion varies between 12 and 50 per cent, that is from 79 to 316 flights (1–2 complexes).

Figure 4.4.6 shows the influence of the deterioration of the airport's capacity and the duration of the disruptive event on the total cost of the delayed and cancelled complexes and flights. The mitigating strategy 'delays and cancellations later on' is applied. As can be seen, the disruption costs decrease in line with the diminishing intensity of the impact of the disruptive event. These costs are highest when the impact is so strong as to overturn the airport capacity to zero and are practically nonexistent if the impact is negligible.

In scenarios A) and B), for the given intensity of impact, the costs of disruption are generally higher for the longer disruptive events. In scenario C), the costs of disruption can be lower even if the disruptive event is longer. This is due to the very strong – diminishing – influence of the cancellations of the very late complexes and flights on the disruption costs. In addition, the longer disruptive event occurring earlier in the day affects more complexes (flights) and imposes higher costs on the airline.

4.4.5 Concluding Remarks

This section has presented an analytical model for the quantification of the economic consequences of large-scale disruptions of an airline single hub and spoke network, measured by the total airline costs of delayed and cancelled complexes of flights. This issue has often been neglected while considering the sustainability of operations of the air transport system and its components and their ability to resist (that is their robustness) different types of the system's internal and external disruptions. The former have been under the control of the system operators while the latter have not. The model has been based on the queuing theory, in which particular complexes of flights scheduled at the airline hub airport are considered as customers and the airport itself as a server. The disruptive event affects the airport capacity with a different intensity, prolonged the scheduled service time of particular complexes and caused their delays and cancellations. The size and the number of complexes, characteristics of the disruptive event in terms of duration, intensity of impact and the time of occurrence, the cost of flight delays and the cost of flight repositioning, the flight revenue, and global airline mitigating strategies have constituted the input for the model.

The results, related to a large European airline, have embraced the costs of delayed and cancelled complexes and flights caused by the disruptive event, which occurred according to different scenarios. The disruption costs range from a few to several million USD. They have generally increased with the increasing duration and intensity of the impact of the disruptive event on the airport capacity as well as with the earlier happening of this event. When the disruptive event overturns the airport capacity to zero, the mitigating strategy 'immediate cancellation' produces a higher proportion of cancelled flights and consequently significantly lower costs of disruption than the strategy 'delays and cancellations later on'. This indicates the general preference for this (former) strategy.

In all scenarios, the larger and more expensive complexes and flights with the higher proportion of passengers who give up have higher disruption costs. These costs could be diminished by some avoidable costs of the cancelled flights and by convenient rerouting of flights.

4.5 Flying at the Fuel Non-Optimal Altitudes

4.5.1 Background

In many cases, the actual aircraft four-dimensional (4D) flight paths between given origin and destination airports deviate from the planned ones, mostly in terms of extra time and due to flying at the fuel non-optimal altitudes – flight levels. Prolongation of the flight times is caused by the various reasons such as weather conditions (strong head instead of tail wind for example), en route and terminal (arrival) delays due to the airspace congestion, and eventual bypassing of the flying restricted (military or other restricted) zones. Flying at the fuel non-optimal flight levels is mainly caused by traffic congestion in the en route airspace. In any case, any deviation results in extra fuel consumption, which in turn causes extra emissions of air pollutants. In particular, increased fuel consumption due to flying at the fuel non-optimal flight levels has specific impact because of depositing additional (extra) air pollutants in troposphere.

This section aims at modelling the aircraft additional fuel consumption while flying at the fuel non-optimal flight levels in en route airspace, that is during cruising phase of flight. The main part of the section refers to the author's earlier published research work (Janić, 1994, 2001).

4.5.2 Modelling Aircraft Extra Fuel Consumption at Fuel Non-Optimal Flight Levels

4.5.2.1 The system
Each flight consists of five phases: taking-off from the origin airport, climbing to the cruising altitude-flight level, cruising at the fuel-optimal flight level, descending towards destination airport, and approach and landing. The actors involved in carrying out flights such as airlines, Air Traffic Control, airports, and users, are interested in

the optimality of flights in terms of their duration influencing fuel consumption, cost, and the associated emissions of air pollutants. In particular, airlines intend their flights to always follow the fuel optimal trajectories in order to minimise fuel consumption and consequently the operational costs. The flights should therefore be as short as possible, preferably along the shortest (direct) routes, without delays in particular phases, and at the fuel optimal flight levels-altitudes. Air Traffic Control (ATC) tries to enable such required optimality of flights by providing the shortest possible – preferably great circle – routes and enabling flying at the fuel optimal flight level as much as possible. Airports are particularly interested in delays as an important time component of flights. They usually intend to make them as short as possible by providing sufficient capacity. Users prefer a flight that is as short as possible, thus saving their valuable time (EEC, 2004). In general, if all flights were optimal, the system costs would be minimal, and the associated fuel consumption and related emissions of air pollutants only at the necessary level.

In the scope of endeavours to minimise fuel consumption, cruising at fuel-optimal flight trajectories appears to be particularly important. In the conventional airspace organisation, this phase takes place along the airways defined by radials of the ground navigational facilities such as NDB, VOR, VOR/DME (ICAO, 1998). Thanks to onboard avionics and also by means of inertial and satellite navigation the aircraft use information from these facilities and perform flights either facility-to-facility or directly between origin and destination airports defined by aerial 2D, 3D and 4D RNAV. While on the airways the aircraft are vertically separated by 1,000 ft (1 ft – 0.305 m). Flights with headings 0–179° use odd flight levels (FL 130, 150...., 330, 350) while flights with headings 180–359° use even flight levels (FL 140, 160,). In some controlled airspaces such as those in Europe and the US all flight levels above FL290 are now usable according to the same scheme. In general, the fuel optimal cruising flight levels are lower for shorter than for longer flights, although this also depends on the aircraft type. It implies that the cruising level depends on the flight length. Some evidence in Europe shows that flights lasting between 80 and 100 minutes usually optimally cruise from FL 280 to FL320. The other group of flights lasting from 80 to 130 minutes use flight levels FL320-FL380 (EEC, 2004).

The ATC separates aircraft on the same flight level by the time- or distance-based horizontal separation rules. Distance based separation rules are applicable when radar coverage of the airspace is provided. At the same time, aircraft on different flight levels not changing their altitudes are vertically separated by the separation minima between particular flight levels. Between aircraft changing flight levels and those remaining in the horizontal flights the longitudinal distance- or time-based separation rules are always provided (ICAO, 1998). Figure 4.5.1a shows a possible traffic scenario along an airway.

As can be seen, some aircraft change altitudes while some others stay at the previously assigned ones. The aircraft changing altitudes are those that have previously been assigned fuel non-optimal flight levels. Under such circumstances two flows of traffic can be formed along an airway: i) those aircraft keeping a constant flight level; and ii) those intending to change flight levels while flying along the airway. The ATC/ATM uses different tactics to facilitate the flight level change. One, which can be used when radar coverage is provided, includes vectoring the

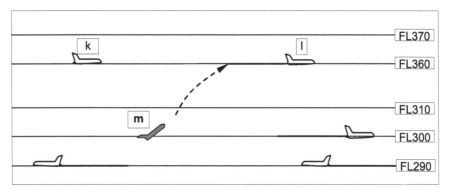

a) Scenario of changing flight levels along given airway

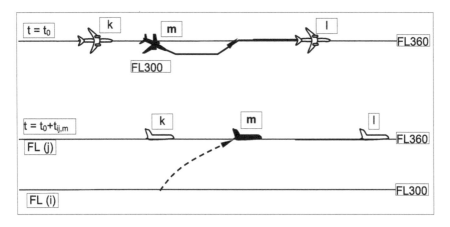

b) Aircraft separation during changing flight levels

Figure 4.5.1 Scenario of changing the flight level along a given airway: a) Horizontal projection; b) Vertical projection

aircraft intending to change altitude to the offset airway parallel to an airway while maintaining the appropriate (required) lateral separation during the time of altitude change. Figure 4.5.1 (a, b) shows a simplified scheme. The aircraft (m) moves from flight level (i) to flight level (j). It carries the flight level change in time $t_{ij/m}$ and includes itself between the aircraft (k) and (l) already cruising at level (j). This tactic, if applied to several aircraft simultaneously, spatially separates the flows of aircraft intending to change altitude and the flows of aircraft remaining at the same flight levels. In addition, aircraft moving in different directions are 'extracted' to the offset airways on different sides of the airway.

Assuming that the flows already at the fuel optimal flight levels should not be interrupted, a sufficiently large gap at the desired level must exist in order for an aircraft to move from its current to its desired fuel-optimal flight level without creating conflicts in the existing traffic flow. This implies that the aircraft, which has just reached the desired flight level, is longitudinally separated from the aircraft in front and behind according to the ATC/ATM minimum longitudinal separation rules. The aircraft changing altitude moves along a three-segment trajectory considered in the horizontal plane as follows: i) the segment created by the radar vector which gets the aircraft off the airway to the parallel off-set airway; ii) the segment along the off-set airway enabling the aircraft to climb or descend to the desired-fuel optimal flight level-altitude; and iii) the segment created again by the radar vector, which includes the aircraft in the traffic flow at the desired flight level of the airway without conflicts. This scenario of changing altitude is adopted in developing the model for estimating the fuel consumption due to flying at the fuel non-optimal flight levels. It should be mentioned that several aircraft from the same or different levels may need to change flight level for the same or different fuel optimal flight levels at the same time, which might create a significant workload for the ATC (Janić, 1994).

4.5.2.2 State of the art and assumptions
The academic research on the quantification of an aircraft's extra fuel consumption due to flying at the fuel non-optimal flight levels has been relatively scarce. The exception is the research from which most of this material is extracted (Janić, 1994). Within the system, extensive research has been undertaken aiming to assess the quality of services in terms of the efficiency and effectiveness of the planned 4D aircraft trajectories both in Europe and the US. The results have shown deviations of the actual from the planned flight trajectories in nearly all four dimensions, which have resulted in increased fuel consumption and the associated emissions of air pollutants. The main causes for these deviations have been identified as the lack of sufficient system capacity at some airports, the restrictions of using the airspace in particular areas which prevents flying along the shortest-direct trajectories, and an inherent imperfectness of the ATC/ATM function (FAA, 2000; EEC, 2004).

In developing the model, the following assumptions are adopted:

- The time of the aircraft entering and leaving the given airway represents two stochastic processes, which might have a different realisation at particular time intervals (hour, day, year). They are approximated by the equivalent deterministic processes;
- The aircraft positions at each flight level of an airway can be represented by the spatial Poisson process;
- Several aircraft at the same flight level might need to change altitude at the same time. They form groups of requests for the altitude change;
- Successive groups of requests for altitude change at each flight level form a Poisson process of requests for the altitude change. They belong to the same class of requests and are considered to be mutually independent, that is the moment, origin and size of one group does not depend on the others;
- The occurrences of groups belonging to different classes taking place at different flight levels are also independent of each other;
- The aircraft changing flight level appear at particular locations at the time ATC expects them; and

- ATC has sufficient capacity to handle the increased workload while flight levels change. This implies that possible over load at ATC is not a cause of possible failures to change altitude.

4.5.2.3 The model structure

Using scenarios of changing flight levels along given airways and the above-mentioned assumptions, the model for quantifying the aircraft fuel consumption due to flying at the fuel non-optimal flight levels along a given airway based on the dynamic queuing system can be developed. The model consists of two parts. In the first part the customers are the groups of requests for a flight level change. In each group, the individual aircraft are appropriately separated and thus able to realise the transfer other almost simultaneously without interfering with each. However, particular groups may interfere, which might cause a delay at the beginning of the flight level change. They are waiting at their current flight levels, representing the waiting room of the queuing system. The service channel is the space between the current and the desired flight level the aircraft uses for the altitude change. The time from the moment when the first aircraft from the group starts the flight level change to the moment when the last aircraft from the group completes the altitude change represents the group's service time. In the second part, the customers are the aircraft staying in the horizontal flight at particular flight levels while on the airway. The service channels are the flight levels with the service time equivalent to the minimum inter-arrival time between successive aircraft at the given level(s). There might be simultaneous entry to particular flight levels as well as staying there for the time needed to pass along the airway.

The groups of requests for a flight level change along a given airway, which arise during given period T can be classified into mutually dependent and mutually independent. This is influenced by the relation between particular aircraft in each group. Let (i, J) and (k, L) be two classes of requests for the flight level change moving in the same direction along a given airway. These groups are mutually dependent as long as there is the common airspace that particular aircraft from both groups can simultaneously use for flight level changes, that is the following condition specifies this:

$$i \leq L \text{ and } J \geq k \text{ for } (i, k, J, L) \in U \tag{4.5.1}$$

i, k is the flight level at which the group for the flight level change (i, J) and (k, L) occur, respectively;

J, L is the desired fuel optimal flight level for the class of requests (i, J) and (k, L), respectively;

U is the number of flight levels along the airway.

Otherwise, the requests (i, J) and (k, L) are independent. If the aircraft from both classes move in opposite directions, they are independent. Mutually independent classes of requests can be served simultaneously since there is always the vertical separation. The order of service is FCFS ('First-Come – First-Served'). The mutually dependent classes or requests are served either according to FCFS rule or priority. In the case of prioritising particular classes the priority rank is assigned to each of them before starting services. The priority is non-pre-emptive for a given set of classes of requests. The classes with the same priority rank are served according to the FCFS rule.

Serving independent classes of requests   Let $(m-1)$ and (m) be two successive groups for the altitude change arising at flight level (i) at time $T_{i,m-1} = (T - t_{i/m,m-1})$ and $T_{i,m} = (T)$, respectively ($t_{i/m,m-1}$ is the time between raising the successive requests $(m-1)$ and (m) at the flight level (i)). The waiting time for class (m), due to servicing class $(m-1)$, can be estimated by a known recursive formula as follows (Newell, 1982):

$$d_{i,m} = \max[0; d_{i,m-1} + t_{i,m-1} - (T_{i,m} - T_{i,m-1})] \tag{4.5.2}$$

where
$\quad d_{i,m-1}$ is the waiting time of request $(m-1)$ from flight level (i); and
$\quad t_{i,m-1}$ is the service time of the $(m-1)$ group of requests arising at flight level (i);

There might be M_i requests for the flight level change arising at the level (i) during the observed period.
    In Equation 4.5.2, the service time $t_{i,}$ $m-1$ can be determined as follows:

$$t_{i,m-1} = \max(\overline{t}_{ij,m-1}) \tag{4.5.3}$$

$$(i, j) \in N; i \neq j$$

where
$\quad \overline{t}_{ij,m-1}$ is the average time an aircraft needs to pass from flight level (i) to flight
$\qquad$ level (j).

Group (m) consists of subgroups represented by the aircraft requesting the change from the current flight level (i) to other levels. The sub group consists of the aircraft demanding to change the flight level (i) for another (the same) flight level. Its size is $q_{ij,m}$. Consequently, the size of the group of requests from level (i) to all other flight levels can be determined as:

$$q_{i,m} = \sum_{j=1}^{U} q_{ij,m} \tag{4.5.4}$$

In Equation 4.5.4 the size of the group of requests for the altitude change cannot be greater than the maximum number of aircraft simultaneously flying at flight level (i). On the other hand the number of the aircraft simultaneously at that flight level is constrained by the length of the airway and ATC minimum longitudinal separation rules. One should also understand that the number of aircraft at particular flight levels will change after the flight level changes have taken place, with probability p_{ijm} for sub-group (ij) within group (m) (Janić, 1994). Consequently, the expected number of requests (aircraft) unable to change flight level can be determined as:

$$\overline{q}_{ij,m} = (1 - p_{ij,m})q_{ij,m} \qquad (4.5.5)$$

The service rate of the request $q_{ij,m}$ is determined as:

$$\mu_{ij,m} = p_{ij,m}q_{ij,m} / \bar{t}_{ij/m} \qquad (4.5.6)$$

The waiting time due to the inability to make the flight level change can be determined as follows (Janić, 1994):

$$d_{ij,m} = [1/p_{ij,m} - 1]\bar{t}_{ij,m} \qquad (4.5.7)$$

Combining Equations 4.5.2 and 4.5.7 provides the total waiting time for aircraft from the sub-group of requests $q_{ij,m}$ as follows:

$$\bar{d}_{ij,m} = d_{i,m} + d_{ij,m} \qquad (4.5.8)$$

In Equation 4.5.5 the probability of the successful realisation of the flight level change $p_{ij,m}$ can be determined according to the scenario of realising the flight level change shown in Figure 4.5.1b. That implies that two conditions must be fulfilled in order for an aircraft to change altitudes: i) existence of an appropriate (safe) space or time gap at the required flight level at the moment of its arrival; and b) being in the correct position at the current flight level at the moment of starting the altitude change in order to be able to pick up the convenient gap at the desired flight level.

The probability of realisation of the first event can be determined as (Janić, 1994):

$$p_{ij,m}^1 = P(S_j > 2\delta) = \exp(-2N_j\delta / L) \qquad (4.5.8a)$$

where
  S_j is the random variable representing the spacing (that is distance) between aircraft at flight level (j);
  δ is the minimum longitudinal separation between aircraft in the horizontal flight at each flight level of the airway;
  N_j is the number of aircraft at flight level (j) at the moment of the arrival of a new aircraft; and
  L is the length of the airway.

The probability of realisation of the second event, that is that an aircraft is in the suitable position to successfully realise the altitude change, can be determined as:

$$p_{ij,m}^2 = 1 - \exp[q_{ij,m}S_j / X] \qquad (4.5.8b)$$

where all symbols are analogous to those in the previous expressions (Janić, 1994).

Consequently, the probability of realising the flight level change (ij) is equal to the product of the probabilities 4.5.8a and 4.5.8b.

The average service time an aircraft in group $q_{ji,m}$ spends passing from flight level (i) to flight level (j) along the off-set airway shown in Figure 4.5.1b can be determined as the average product of the proportions of particular aircraft categories in the group, the climbing speeds of each category, and the difference in the current and required flight levels (Janić, 1994).

Serving dependent classes of requests   Dependent groups (classes) of requests for the altitude change are served according to some priorities. This implies that each group from the set of groups simultaneously requiring the altitude change is assigned a priority rank. The assignment of priority ranks and consequently the service order is carried out according to the pre-defined characteristics of each class of requests, which enables the minimisation of waiting for the altitude change and the associated extra fuel consumption as well. For example, the rank of class of requests (m) from the flight level (i) is equal to (Janić, 1994):

$$r_{i,m} = (1/T_{i,m}) \sum_{j=1}^{N} q_{ij,m} P_{ij,m} f_{ij,m} \tag{4.5.9a}$$

where
$\quad f_{ij,m}$ is the average extra fuel consumption (units of fuel per unit of time) by an average aircraft from the group of requests arising at flight level (i).
$\quad N$ is the number of flight levels the aircraft from group (m) at flight level (i) require to pass.

After determining the ranks for all classes and groups, the order of service is determined according to the decreasing values of the parameter $r_{i,m}$, that is the lower the value of the parameter, the lower the service priority rank. For example, the highest priority rank is assigned according to the following rule:

$$r_{i,m} = \max[r_{i,k}]_{k \in M} \tag{4.5.9b}$$

For example, the group with rank (r) will have to wait until all groups with the higher ranks ($r - 1$) are served. Its waiting time will be approximately equivalent to the sum of the service times of the higher priority groups. The fuel consumption of this group will be equal to the product of the waiting time and the average unit fuel consumption. The total extra fuel consumption can be obtained by summing up the extra fuel consumption of the groups requiring the altitude change over a given period of time (Janić, 1994).

The other sub-groups will be assigned the priority ranks in decreasing order of $r_{k/z}$, for example the sub-groups with lower priority parameter values will be assigned lower priority ranks. If sub-group (z) is assigned priority rank (r) it will have to wait for all the preceding groups and sub-groups requesting altitude change to be serviced (Janić, 1994). The total extra fuel consumed for each group due to its flying at fuel non-optimal altitude can be computed on the basis of estimated delays and extra fuel consumption (Janić, 1994).

4.5.3 Application of the Model

4.5.3.1 Inputs

The model is applied to three traffic scenarios on an airway with a length of 450 nm (nm − nautical miles) where the aircraft use eight flight levels, from the lowest FL 260 up. The number of aircraft in the group of requests for the level change is equal to the number of aircraft currently on that level. In the first scenario, four classes of requests occur at FL260 ($i = 1$), FL270 ($i = 2$), FL280 ($i = 3$), and FL290 ($i = 4$). In each class the following group of requests appear: group (46) (passage from FL290 to FL330), group (15) (passage from FL260 to FL310), group (26) (passage from FL270 to FL330), and group (37) (passage from FL280 to FL350). The classes are all considered as independent.

According to the second scenario, two classes of requests are ($i = 1$) and ($i = 3$) with groups (15) and (37) occurring simultaneously and being treated as dependent on each other. Group (37) always has a higher priority over group (15) according to the criteria for ranking the groups.

In the third scenario it is assumed that both dependent groups from the second scenario (37) and (15) are served according to the FCFS priority rule.

The average interval between the successive groups of requests for the altitude change at the same flight level is $T = 15$ and 5 minutes for four classes in the first,

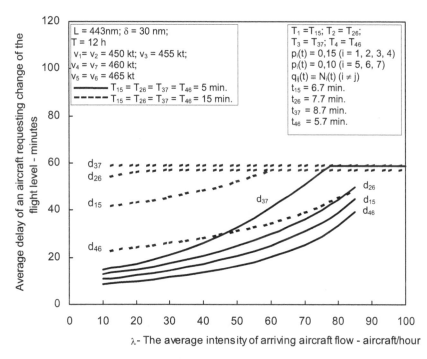

Figure 4.5.2 Dependence of the average delay of an aircraft requesting altitude change on the intensity of traffic flow on a given airway

Compiled from Janić (2001)

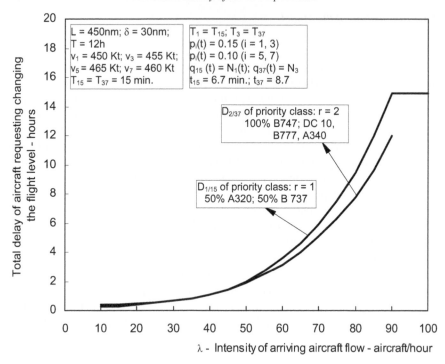

Figure 4.5.3 Dependence of the delay of aircraft assigned a priority rank on the average intensity of aircraft flow on the airway

Compiled from Janić (2001)

and $T = 15$ minutes for two classes in the second and third scenario. The rate of climb for each aircraft requiring the altitude change is adopted to be 600 ft/min. After taking into account the climb rate and length of trajectories needed for the altitude change the average service time of particular groups of request is estimated as follows: $t_{46} = 5.7$ min, $t_{15} = 6.7$ min, $t_{26} = 7.7$ min, and $t_{37} = 8.7$ min.

The extra fuel consumption is calculated for only two groups (15) consisting of medium jets of A320 and B737 types (each with proportion of 50 per cent) – 2.2. kg/min, and for the group (37) consisting exclusively of heavy aircraft (B747, DC-10, B777, and A340 type) – 12.8 kg/min. The ATC applies the distance based separation minima of $\delta = 30$ nm. The time interval of recording the process of serving particular requests for the flight level change is 12 hours (Janić, 1994).

4.5.3.2 Results

Parts of the results from the model are presented in Figures 4.5.2, 4.5.3 and 4.5.4. Two groups of requests regarding the aircraft types are considered: i) the group of requests consisting only of large wide-body aircraft of the type DC-10, B747, B777, A340, and so on. ; and ii) the group of requests consisting only of the medium-sized aircraft of type A320 and B737 types. Figure 4.5.2 shows the dependence of the delay of the aircraft requesting the flight level change (vertical axis) on the average traffic flow on the airway (horizontal axis).

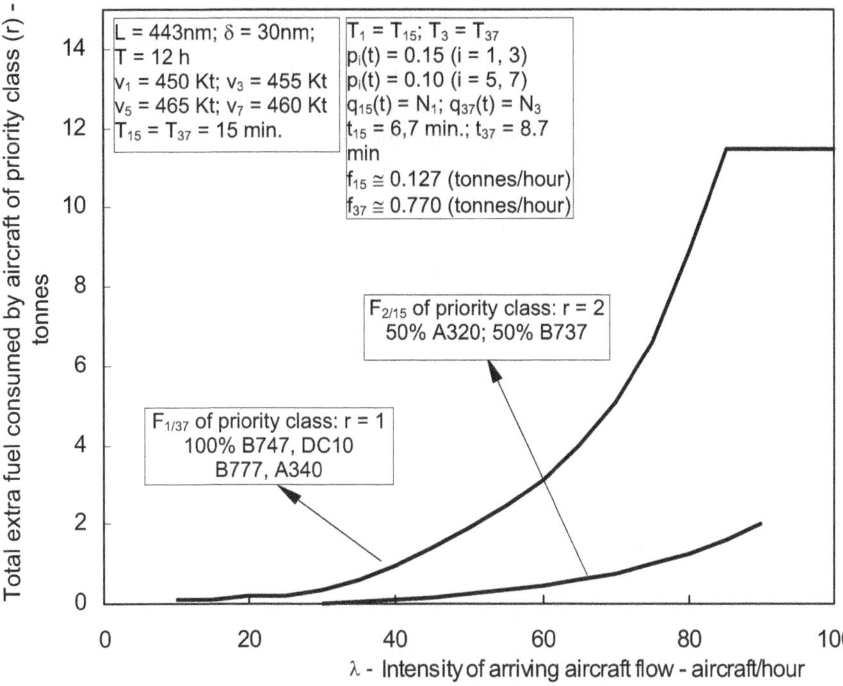

L = 443nm; δ = 30nm;
T = 12 h
v1 = 450 Kt; v3 = 455 Kt
v5 = 465 Kt; v7 = 460 Kt
T15 = T37 = 15 min.
T1 = T15; T3 = T37
pi(t) = 0.15 (i = 1, 3)
pi(t) = 0.10 (i = 5, 7)
q15(t) = N1; q37(t) = N3
t15 = 6,7 min.; t37 = 8.7 min
f15 ≅ 0.127 (tonnes/hour)
f37 ≅ 0.770 (tonnes/hour)
F2/15 of priority class: r = 2
50% A320; 50% B737
F1/37 of priority class: r = 1
100% B747, DC10
B777, A340
Total extra fuel consumed by aircraft of priority class (r) - tonnes
λ - Intensity of arriving aircraft flow - aircraft/hour
0
2
4
6
8
10
12
14
0
20
40
60
80
100

Figure 4.5.4 Dependence of the total extra fuel consumed on the priority class and the intensity of aircraft flow

Compiled from Janić (2001)

It is assumed that the intensity of requests for the flight level change increases together with the increasing intensity of traffic flow on the airway. This implies that the same proportion of aircraft independent of their number always requires a flight level change. The average delay per aircraft requesting altitude change increases more then proportionally with the increase in the traffic density on the airway. This delay is also dependent on the intensity of demand for the flight level changes, duration of service time per group of requests as well as on the traffic density on the required flight levels influencing the probability of successive inclusion into the flows there. In addition, the given example also shows that some aircraft remain at the current flight level all the time while on the airway.

Figure 4.5.3 shows the relationships between the average delay of the aircraft requesting altitude change and the intensity of traffic flow moving along the airway. As can be seen, the aircraft requesting the flight level change from flight level (3) to flight level (7) (all aircraft are of type DC-10, B747, B767, B777) who are assigned higher priority ranking might have longer delays imposed on them than the aircraft with the lower priority rank. According to the structure of both classes of requests for the flight level change the aircraft with the higher priority rank will be penalised more by the extra fuel consumption than the aircraft from another group. The differences in the total fuel consumption between these two groups of requests are shown in Figure 4.5.4 (Janić, 1994). As can be seen the absolute amount during a flight of one hour at the fuel non-optimal flight levels could be substantive and have two impacts: i) the increased emission of greenhouse gases such as CO_2 and H_2O, which are deposited in the troposphere; and ii) increased airline operation costs due to the higher fuel consumption.

The former is increasingly important due to the overall awareness of the specific impact of the air transport system on the environment and the constant rates of emission of CO_2 (3.16 kg per 1 kg of fuel) and H_2O (1.24 kg per 1 kg of fuel). The latter becomes particularly important for airlines under conditions of rising fuel prices taking place nowadays. In general, the airline cost per flight has increased with the increasing time of flying at fuel non-optimal flight levels. In the given example, if a given flight spends either 15 minutes or one hour at the fuel non-optimal flight level, this cost will increase by about 0.5 and 3.5 per cent, respectively, depending on the difference between the current and desired flight level (Janić, 1994). These results are similar to those obtained in the most recent studies (EEC, 2004).

4.5.4 Concluding Remarks

Aircraft cruising at fuel non-optimal flight levels cause extra fuel consumption and the associated emissions of greenhouse gases, and increase the airline operating costs. To quantify these quantities of extra fuel, the model based on the dynamic queuing system was developed. In this model, an airway is considered as a system. Flying in a given airway is possible at several flight levels. At each of them a request to change flight level can emerge. These requests are considered as customers and the airspace they use to realise the flight level change as the service channel. In such a context, simultaneous requests for changing flight levels at each flight level are considered as either dependent or independent. In case of serving dependent requests, priority rules aiming to minimise the total extra fuel consumption from all the affected request-aircraft have been introduced.

The results from several experiments with the model applied to three traffic scenarios on a given airway have shown that the extra fuel consumption increases with the increasing number and size of groups of aircraft requiring the flight level change, differences between the current and required flight levels (that is length of the groups' service time), and traffic intensity on the airway (airspace). Consequently, the associated emissions of air pollutants also increase, including the airline operating costs. Since the quantity of extra consumed fuel per flight has been shown to be relatively substantive, it appears to be all the more important to enable the congested airspace to be used flexibly through using more direct routes on which the aircraft-flights would be able to use the fuel-optimal cruising altitudes for as long as possible.

4.6 Innovative Technologies and Land Use at Airports

4.6.1 Background

One of the most important factors influencing the capacity of airports is the number and configuration of the runways. Others are the ATC separation rules, technologies for aircraft primary navigation, air traffic surveillance, communications and information, the mix of the aircraft fleet regarding the wake-vortex categories and the aircraft approach/departure speeds, proportion of the arrival and departure demand, the ATC tactics of sequencing particular operations, and other economic and social-environmental constraints (EEC, 1997; ICAO, 2004).

The number of runways depends on the airport size, that is on the volume of airport traffic demand, and indirectly on the availability of land for expansion of the airport infrastructure, and vice versa. The configuration of the runways is influenced by the meteorological conditions (prevailing winds and visibility) given the required airport annual utilisation rate of nearly 100 per cent. The runway configuration can include a single, two or more parallel, intersecting and converging/diverging runways and/or their combinations.

In particular, dual parallel runways can be closely, intermediate and far-spaced. In the former case, the separation between the runway centre-lines varies from 700 to 2,499 ft (1 ft = 0.305 m); in the second case it varies from 2,500 to 4,299 ft; in the last case it is 4,300 ft or more (ICAO, 2004). In Europe, the main four continental hubs also operate dual parallel runways: Frankfurt-Main (Germany) a pair of closely, and London Heathrow (UK), Paris Charles de Gaulle (France), and Amsterdam Schiphol Airport (the Netherlands) a pair, two pairs, and three pairs of far-spaced (independent) parallel runways, respectively (NASA, 1998). Currently, at the busiest hub airports in the US 28 pairs of closely-, 10 pairs of intermediate- and 28 pairs of far-spaced parallel runways are used (EEC, 1997).

The separation between centrelines, technologies influencing the ATC separation rules and weather conditions (IMC (IFR) and VMC (VFR)) influence the degree of dependence of operations on dual parallel runways. Table 4.6.1 shows an example for the US airports (ICAO, 1998).

This section presents a model of the ultimate capacity of dual dependent parallel runways at an airport. In the context of dealing with the sustainability of the air transport system and its components, in this case the airport, the main intention

Table 4.6.1 Dependency of operations on dual parallel runways at US airports (IFR conditions)

Degree of (inter)dependency of operations				
Separation between runway centrelines (ft)	Arr-Arr	Dep-Dep	Arr-Dep	Dep-Arr
700–2,499	Like single runway	Departure clears the runway(s)	Arrival clears the runway(s)	Departure clears the runway(s)
2,500–3,399	Dependent: Lateral diagonal separation	Independent	Independent	Independent
3,400–4,299	Dependent: Lateral diagonal separation without PRM Independent with PRM	Independent	Independent	Independent
≥ 4,300	Independent	Independent	Independent	Independent

Arr – Arrival; Dep – Departure; PRM – Precision Runway Monitor

is to show the contribution made by new technologies and operational procedures to the increasing of the existing runway system, and consequently the temporary mitigation of pressure for new land for building new runways. There is always a concern that in addition to taking land, a new runway can cause increasing and/ or redistribution of noise, thus affecting more people. Therefore, findings in this section could be particularly important for airports closely surrounded by potentially or actually affected populations, such as many in Europe and US. The material is mainly taken from the author's previous research work (Janić, 2005).

4.6.2 The Runway System and Ultimate Capacity

Various navigation, surveillance, communications and information technologies are already in place, or planned to be put in place, in order to diminish the degree of dependency of the dual-dependent parallel runways. They are broadly classified as i) the air traffic flow management tools; ii) the ATC/ATM traffic surveillance equipment-devices; iii) the improved navigation precision equipment-devices; and iv) the 'mixed' traffic surveillance and conflict alert equipment-devices. Table 4.6.2 gives classification of some of these technologies.

The airport runway's ultimate capacity is usually expressed by the maximum number of aircraft accommodated under conditions of constant demand for service. This implies that the aircraft continuously use the runway system over a given period of time (usually one hour or 15 minutes) separated by the ATC minimum (safety) separation rules (ICAO, 1998). At US airports this capacity is essentially based on VMC and VFR and only exceptionally on IMC and IFR (Instrumental Flight Rules). At European airports, the capacity is exclusively based on IMC and IFR independent of the actual weather conditions (Newell, 1982; FAA, 1993; NASA, 1998; Hammer, 2000; Janić, 2001; Ignaccolo, 2003).

4.6.3 Modelling the Ultimate Capacity of Dual Parallel Runways

4.6.3.1 State of the art

Modelling an airport's ultimate (runway) capacity has occupied the airport, ATC and airline operators, planners, analysts and academics for a long time. These efforts have resulted in developing numerous analytical and simulation models of two broad classes for: i) calculating the (runway) capacity of individual airports and the capacity of airport networks (10); and ii) optimisation of the utilization of the airport (runway) capacity under changing influencing factors (Andreatta, 1987; Bianco and Bielli, 1992; Richetta and Odoni, 1993, 1994; Terrab and Odoni, 1993; Vranas, Bertsimas and Odoni, 1994; Richetta, 1995).

Specifically, the analytical models for calculation of the airport runway's ultimate capacity have provided the two-value parameter – one for the arrival and another for the departure capacity (Blumstein, 1960; Harris, 1972; Hockaday and Kanafani, 1974; Newell, 1979; Swedish, 1981; Janić and Tosic, 1982; Gilbo, 1993; Donohue, 1999). Some other models such as the FAA Airport Capacity Model, LMI Runway Capacity Model, and DELAYS as 'Quasi-Analytical Models of Airport Capacity and Delay', developed mainly for the airport (runway) planning purposes and

Table 4.6.2   Technological systems to decrease degree of dependency of dependent runways

Air traffic flows management tools – by ATC/ATM on the ground	• CTAS (Centre/TRACON Automation System) assists in optimising the arrival flow and runway assignment; • Integrated Arrival and Departure manager enables replacing FCFS (First-Come–First-Served) rule with the successive sequencing rule of the aircraft of the similar wake vortex and approach/departure speed characteristics.
Air traffic surveillance equipment – ATC/ ATM on the ground	• RADAR of improved precision enables reduction of the minimal separation between aircraft from 3 to 2.5 nm; • PRM-Precision Runway Monitor consisting of a beacon radar and computer predictive displays enables the independent use of dual- and triple-dependent parallel runways spaced less than 4,300 ft.
Improved aircraft avionics – onboard	• FMS 4D – Flight Management System enables more precise following of the time schedule according to the flight plan, which reduces the position error of arrivals at the final approach gate; • Waas – Wide Area Augmentation System improves basic GPS accuracy both horizontally and vertically; • AILS – Airborne Information for Lateral Spacing improves the navigation precision while approaching to the closely spaced parallel runways; • TCAS – Traffic Alert and Collision Avoidance System shows the spatial relation of two aircraft and provides instructions to avoid potential conflicts; • LVLASO – Low Visibility Landing and Surface Operating Program reduces, controls and predicts the runway occupancy time; • ADS-B – Automatic Dependent Surveillance Broadcasting improves situation awareness both onboard and on the ground and is used independently, but in addition to TCAS and enhanced CDTI (Cockpit Display of Traffic Information); • CDTI – Cockpit Display of Traffic Information provides an integrated traffic data onboard the aircraft, which reduces the separation rules between aircraft.
'Mixed' traffic surveillance and conflict alert equipment – devices – by ATC/ATM on the ground and onboard	• Distributed Air Ground solution combines *ADS/B*, TCAS, and Free Flight devices enabling simultaneous aircraft-*ATC/ ATM* traffic surveillance, alerting and resolution of potential conflicts.

Source: FAA (1993); NASA (1998); Ignaccolo (2003)

based on the analytical single-runway capacity model, have calculated the so-called 'capacity coverage curve' including the associated aircraft delays (Newell, 1979; Odoni and Bowman, 1997). In parallel, separate models of the ultimate capacity of the apron/gate complex and the system of taxiways have been developed. Only recently, the efforts have been made to integrate these analytical models into the 'airport integrated-strategic planning tool' (Stamatopoulos, Zografos and Odoni,

2004). Such integration has however been achieved by developing the computer-supported simulation models for calculating the airport capacity and delay at i) Low (Hermes and The Airport Machine), ii) Intermediate (NASPAC, FLOWSIM and TMAC), and iii) High Level of Detail (Taam and SIMMOD) (Odoni and Bowman, 1997; Janić, 2001; Wu and Caves, 2002; Ignaccolo, 2003). In comparison to the analytical models, these models have studied the airport airside operations in much greater details. In some cases, they have seemed to require a relatively long time for familiarisation, time-consuming preparation of input, consequently relatively high costs, and produced too detailed output, which paradoxically made the strategic planning choices more complex and time consuming than otherwise (Odoni and Bowman, 1997; Stamatopoulos, Zografos and Odoni, 2004).

4.6.3.2 Objectives and assumptions

The above-mentioned overview of the modelling of the airport runway system capacity has demonstrated the scarcity of less complex convenient analytical (dedicated) models for calculating the ultimate capacity of dual dependent (parallel) runways that still exists, with the exception of the FAA Airfield Capacity Model dating from the 19 seventies (Swedish, 1981; NASA, 1998). This also coincides with recent efforts for a more comprehensive integration of the analytical models of different parts of the airport airside area into a convenient 'tool-kit,' which could be used for preliminary estimates in strategic airport planning, before the use of the simulation models (software packages) such as for example SIMMOD, Taam and Airport Machine (Swedish, 1981; Stamatopoulos, Zografos and Odoni, 2004). Therefore, the objectives of this section are:

- Developing an analytical model for calculating the ultimate capacity of a dual dependent parallel runway system;
- Carrying out a sensitivity analysis of this capacity respecting the most important influencing factors and particularly new technologies and innovative operational procedures in use: and
- Synthesising the 'capacity coverage curve'.

In developing the model the following assumptions are adopted (Tosic and Horonjeff, 1976; Janić and Tosic, 1982).

- The geometry of dual parallel runways – the staggering distance, spacing of runway centrelines, length of the final approach path(s), and location of the runway landing exits – is given;
- The runways operate according to a given degree of inter-dependency (Table 4.6.1);
- The arriving aircraft use the ILS (Instrumental Landing System), enabling them to follow the standard descent angle of 2.5–3 ° along a straight-line path between the final approach gate and the landing threshold;
- Depending on use of the technologies and related operational procedures, the arriving aircraft may be separated by the ATC longitudinal, lateral-diagonal

$E_{I/J}$, E_k — Final approach gate of aircraft I and J, and k, respectively
$T_{I/J}$, T_k — Landing threshold of aircraft I and J, and k, respectively
$\gamma_{I/J}$, γ_k — Length of common approach path of aircraft I and J, and k, respectively
d — Spacing between RWY 1 and RWY 2
$l_{IJ/k}$ — Initial longitudinal ATC separation rules between aircraft I and J
$S_{Ik}{}^0$, $S_{kJ}{}^0$ — Initial longitudinal "spacing" between aircraft Ik and kJ, respectively
$z_{I/k}$ — Staggering distance between the threshold of aircraft I and k

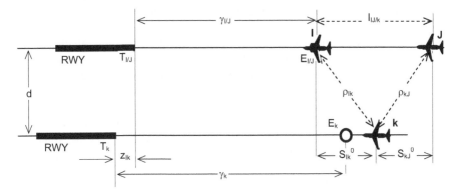

Figure 4.6.1 Geometry of landings on dual dependent parallel runways

radar-based and/or vertical separation rules; the departing aircraft are usually separated by the time-based ATC separation rules;

- Successive arrivals (and successive departures) are alternated on each parallel runway; and
- The aircraft arrive at the specified locations of their prescribed paths almost precisely when the ATC (controller) expects them to arrive; this may become more realistic by using innovative (advanced) technologies.

4.6.3.3 The model for arrivals

Basic structure Computation of the ultimate capacity of the airport runway system is usually based on the computation of the minimum inter-arrival times between successive aircraft operations at the selected 'location' or 'point' in space they all pass through, under conditions of constant demand service and non-violation of the ATC minimum separation rules. This 'location' or 'point' for counting the operations and calculating the capacity is usually either the landing threshold or the 'final approach gate'.

Specifically, on dual dependent parallel runways, the two parallel arriving and/or departing traffic flows are simultaneously handled while maintaining generally the ATC longitudinal (that is in-trail) separation rules between the aircraft in the same and the lateral-diagonal or vertical separation rules between the aircraft in different (parallel) flows. Figure 4.6.1 shows the basic geometry of the dual parallel runway system and a 'string' of three landing (generally different) aircraft types in the horizontal plane. The aircraft *I* and *J* land on runway RWY1 and the aircraft *k* on runway RWY2 in the order *I-k-J*. Aircraft *I* and *J* follow the final approach path from the entry gate $E_{I/J}$ to the landing threshold $T_{I/J}$ of RWY1. Aircraft *k* follows the parallel final approach path from the entry gate E_k, to the landing threshold T_k of RWY2. While airborne the aircraft 'string' *I-k-J* must be separated as follows:

- Sequence *IJ* (*I* is leading and *J* is trailing aircraft) by the ATC longitudinal separation rules; and
- Sequence *Ik* (*I* is 'ultimately' leading and *k* is 'ultimately' the trailing aircraft) and the sequence *kJ* (*k* is 'ultimately' leading and *J* is 'ultimately' the trailing aircraft) by the ATC lateral-diagonal or vertical separation rules.

Since one of the separation rules must always exist between these aircraft, the lateral-diagonal (or vertical) separation rules for the sequences *Ik* and *kJ* might depend on the minimum longitudinal (in-trail) separation rules for the sequence *IJ*, and vice versa. In addition, these rules might be influenced by the runway centreline spacing (d) and staggering distance (z). Consequently, the inter-arrival time of the sequence *IJ* at the threshold T_{IJ} might depend on the inter-arrival time of the sequences *Ik* at the thresholds T_{IJ}/T_k and the sequence *kJ* at the threshold T_k/T_{IJ}, as the 'reference location(s)' for calculating the arrival capacity (Vranas, Bertsimas and Odoni, 1994).

When the ATC/ATM minimum separation rules are applied, the time intervals between operations at the threshold T_{IJ} of RWY1 and threshold T_k of RWY2 will be minimal and consequently the accommodated aircraft flow will be at the level of ultimate capacity. As in the analytical models of a single runway, let $_a t_{ij/k}$ be the inter-arrival time of aircraft types I and *J* at the threshold T_{IJ} of RWY1, dependent on aircraft *k* landing at the threshold T_k of RWY2 (Hockaday and Kanafani, 1974; Wu and Caves, 2002). The time $t_{ij/k}$ is equal to: $_a t_{ij/k} = _a t_{ik} + _a t_{kj}$, where t_{ik} is the inter-arrival time between aircraft *i* and *k* at their landing thresholds T_{IJ} and T_k, respectively; $_a t_{kj}$ is the inter-arrival time between aircraft *k* and *J* at their landing thresholds T_k and T_{IJ}, respectively. The times $_a t_{ik}$, $_a t_{ij/k}$ and $_a t_{kj}$ should guarantee that: i) the ATC airborne minimum separation rules must exist over the period $t \in (0; t_1 = \gamma_1/v_1)$ – while aircraft *I* is between the approach gate E_{IJ} and threshold T_{IJ}, and over the period $t_2 = t_1 + _a t_{Ik}$ – until aircraft *k* passes threshold T_k; and ii) any two aircraft do not occupy the runway system at the same time, that is aircraft *I* must clear RWY1 before aircraft *k* passes over the threshold of RWY2 and aircraft *k* must clear RWY2 before aircraft *J* passes over the threshold of RWY1. Consequently:

$$_a t_{Ik} = min \left(_a t_{ik/min}; _r t_{Ik} \right)$$
$$_a t_{IJ/k} = min \left(_a t_{IJ/k/min}; _a t_{Ik} + _r t_{kJ} \right) \text{ and} \qquad\qquad (4.6.1a)$$
$$_a t_{kJ} = min \left(_a t_{kJ/min}; _r t_{kJ} \right)$$

where

$_a t_{Ikmin}$, $_a t_{kJ/min}$ is the minimum inter-arrival time of the aircraft sequences *Ik* and *kJ* at the corresponding thresholds respectively, dictated by the minimum ATC airborne lateral-diagonal separation rules;

$_a t_{IJ/k/min}$ is the minimum inter-arrival time of the aircraft sequence *IJ* at the corresponding threshold dictated by the ATC minimum airborne longitudinal (in-trail) separation rules;

$_r t_{Ik}$, $_r t_{kJ}$, $r\, t_{IJ}$ is the inter-arrival time at the thresholds for the aircraft sequences *Ik*, *kJ*, and *IJ*, respectively, based on the ATC minimum runway occupancy rule.

The runway occupancy rules $_r t_{Ik}$, $_r t_{kJ}$, $r\, t_{IJ}$ must fulfil the conditions:

$$_r t_{Ik} \geq t_{aI}; \quad _r t_{IJ/k} \geq _r t_{Ik} + t_{ak}; \quad _r t_{kJ} \geq t_{aJ} \tag{4.6.1b}$$

where

t_{aI}, t_{ak}, t_{aJ} is the runway occupancy time of the arriving aircraft *I*, *k* and *J*, respectively.

Existing analytical models of the runway capacity assume the independency of aircraft types in particular sequences. Analogously, in the 'string' *I-k-J*, the type of leading aircraft *I* does not depend on the type of trailing aircraft *J* and the type of the 'intermediate' aircraft *k* does not depend on the aircraft types *I* and *J*. Consequently, the probability of the occurrence of a 'string' of aircraft of types *I-k-J* can be determined as follows:

$$p_{IJ/k} = p_I p_J p_k \tag{4.6.2}$$

where

p_I, p_k, p_J is the proportion of aircraft types *I*, *k* and *J* in the mix, respectively.

When the minimum times $_a t_{IJk}$ and probabilities p_{IJk} are calculated for all combinations of aircraft types *I*, *j* and *k*, the expected inter-arrival time at the threshold T_{IJ} of RWY1 as the capacity-calculating 'reference location' can be computed as:

$$\bar{t}_a = \sum_{IJk} {_a t_{IJ/k}} p_{IJ/k} \tag{4.6.3}$$

Since the time $\bar{t}_a$ simultaneously materialises (that is is replicated) at the threshold T_k of RWY2, the arrival capacity of dual dependent parallel runways can be calculated as:

$$\lambda_a = 2 / \bar{t}_a \tag{4.6.4}$$

The inter-arrival times at the 'capacity counting' location The inter-arrival time $_a t_{IJk}$ for the aircraft sequence *IJ* at the threshold T_{IJ} of RWY1 influenced by the aircraft *k* approaching to RWY2 (Figure 4.6.3) can be defined as follows (Tosic and Horonjeff, 1976; Janić and Tosic, 1982):

$$_a t_{IJ/k} = \frac{l_{IJ/k} + \gamma_{IJ}}{v_J} - \frac{\gamma_{IJ}}{v_I} \tag{4.6.5a}$$

where

$l_{IJ/k}$ is the 'initial' longitudinal (in-trail) separation between the aircraft types I and J measured along the path of aircraft J at the moment $t = 0$ when the aircraft type I is at the final approach gate $E_{I,J}$ of RWY1;

γ_{IJ} is the length of the final approach path of aircraft I and J; and

v_I, v_J is the average approach speed of aircraft I and J, respectively.

In Equation 4.5.6a, the 'separation' $l_{IJ/k}$ must meet the following conditions (Figure 4.6.2):

$$l_{IJ/k} = \max \ (\delta_{IJ}; S_{Ik}^0 + S_{kJ}^0) \tag{4.6.5b}$$

where

δ_{IJ} is the ATC minimum longitudinal (in-trail) separation rule applied to the aircraft sequence $I\,J$;

S_{Ik}^0, S_{kJ}^0 is the 'initial spacing' between the aircraft sequence Ik and kJ, respectively, at the moment $t = 0$ when aircraft I is on the gate E_{IJ} of RWY1.

Similarly as in Equation 4.5.6a, the inter-arrival times $_a t_{Ik}$ and $_a t_{kJ}$ at the thresholds T_I/T_k and T_k/T_J, respectively, can be determined as:

$$_a t_{kJ} = \frac{S_{Ik}^0 + S_{kJ}^0 + \gamma_{IJ}}{v_J} - \frac{S_{Ik}^0 + \gamma_k}{v_k} \quad \text{and} \quad _a t_{Ik} = \frac{S_{Ik}^0 + \gamma_k}{v_k} - \frac{\gamma_{IJ}}{v_I} \tag{4.6.5c}$$

where the symbols are analogous to those in Equations 4.6.5a, b.

In addition to the ATC separation rules applied to aircraft of different wake-vortex categories, the 'spacing' S_{Ik}^0 and S_{kJ}^0 in Equations 4.6.5b, c is the subject of combinations of different aircraft types I, k and J in terms of the final approach speed. In general, the aircraft can relate to each other as either 'fast' – F or 'slow' – S. This gives eight possible combinations of three aircraft in the 'string' IkJ. In the first four combinations, the aircraft I and J are considered as either 'slow' or 'fast' and the aircraft k as 'slow', that is S-s-S, S-s-F, F-s-S and F-s-F. In the next four combinations, the aircraft k is considered as 'fast,', that is S-f-S; S-f-F; F-f-S; and F-f-F.

Consequently, after setting $_a t_{Ik} = S_{Ik/\min}^0 / v_k$ and $_a t_{kJ} = S_{kJ/\min}^0 / v_J$ for the sequences $v_I < v_k$ and $v_k < v_J$, respectively, and $S_{Ik}^0 = S_{Ik/\min}^0$ and $S_{kJ}^0 = S_{kJ/\min}^0$ for the sequences $v_I \geq v_k$ and $v_k \geq v_J$, respectively, the 'initial' separations S_{Ik}^0 and S_{kJ}^0 are determined from Equation 4.6.5c as follows:

a) $S_{Ik}^0 = S_{Ik/\min}^0 + \gamma_{IJ}(v_J/v_k) - \gamma_k$ and $S_{kJ}^0 = S_{kJ/\min}^0$ for $v_I < v_k < v_J$

b) $S_{Ik}^0 = S_{Ik/\min}^0$ and $S_{kJ}^0 = S_{kJ/\min}^0 + S_{Ik/\min}^0(v_J/v_k - 1) + \gamma_k(v_J/v_k) - \gamma_{IJ}$ for $v_I \geq v_k < v_J$ $\quad$ (4.6.5d)

c) $S_{Ik}^0 = S_{Ik/\min}^0$ and $S_{kJ}^0 = S_{kJ/\min}^0$ for $v_I \geq v_k \geq v_J$

where

$S_{Ik/\min}^0, S_{kJ/\min}^0$ is the minimum 'spacing' between aircraft Ik and kJ, respectively.

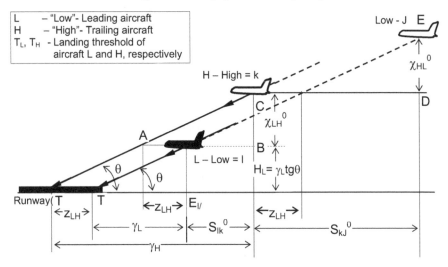

Figure 4.6.2 Application of the vertical separation rules to the successive landings on dual close parallel runways

Both 'spacing' $S_{lk/\min}^0$ and $S_{kJ/\min}^0$ can be based on the ATC longitudinal, lateral-diagonal and vertical minimum separation rules.

Specifically, if the vertical separation rules are applied as shown in Figure 4.6.2, the leading aircraft l is considered as 'low' – L – and the trailing aircraft k is considered as 'high' – H.

Each aircraft follows the final approach part at the ILS standard descent angle $\theta = 2.5°- 3°$. The ATC establishes the minimum vertical separation rules between these aircraft either at time $t = 0$, when aircraft L, aiming to land on the threshold $T_L \equiv T_{lJ}$ of RWY1, is at the entry gate E_{lL} (Figure 4.6.2), or at the time $t_l = \gamma_l/v_l$ when it is just over the threshold $T_L \equiv T_k$. In any case, the aircraft $H \equiv k$, to land at the landing threshold $T_H \equiv T_k$ of RWY2 displaced by z_{LH} is vertically separated from the aircraft $L \equiv l$ by the vertical separation rule χ_{LH}^0. A similar principle, applied to the sequence kJ, considers the leading aircraft k as 'low' – L – and the trailing aircraft J as 'high' – H. In the case of the staggered (displaced) threshold for H aircraft, sufficient (safe) landing distance on RWY2 needs to be ensured. From the triangles ABC and CDE in Figure 4.6.2 the 'spacing' $S_{lk/\min}^0$ and $S_{kJ/\min}^0$ are determined as follows (Tosic and Horronjeff, 1976):

$$S_{lk/\min}^0 = \chi_{LH}^0 / tg\theta - z_{LH} \quad \text{for} \quad L \equiv l; H \equiv k$$
$$S_{kJ/\min}^0 = \chi_{HL}^0 / tg\theta + z_{HL} \quad \text{for} \quad H \equiv k; L \equiv J \tag{4.6.5e}$$

where $0 \le z_{(\bullet)} \le \chi_{(\bullet)}^0 / tg\theta$.

Each $S_{lk/\min}^0$ and $S_{kJ/\min}^0$ in Equations 4.6.5a–e, based on any of the ATC separation rules, can be determined as:

$$S^0_{(\bullet)/min} = \begin{cases} \delta_{(\bullet)}, & \text{for the longitudinal (in-trail) separation} \\ \sqrt{\left[\rho^0_{(\bullet)}\right]^2 - d^2}, & \text{for the lateral-diagonal separation} \\ \chi_{(\bullet)}/tg\theta \pm z_{(\bullet)}, & \text{for the vertical separation} \end{cases} \qquad (4.6.5f)$$

where

> $\delta_{(\bullet)}$ is the ATC minimum longitudinal (in-trail) separation rule applied to successive aircraft using any of two parallel runways;
>
> $\rho^0_{(\bullet)}$ is the ATC minimum lateral-diagonal separation rule applied to successive aircraft arriving using different parallel runways;
>
> d is the spacing between dual dependent parallel runways.

4.6.3.4 The model for departures

Generally, the successive departures on the dual dependent parallel runways can be carried out according to the ATC minimum separation rules, respecting their wake-vortex categories and securing that no more than one aircraft occupies one of the runways during any time (Table 4.6.2). If the runway threshold of either runway is the 'location' for calculating the capacity, the minimum inter-departure time between the aircraft I and J departing from any runway can be determined as follows:

$$_d t_{IJ} = \max[\,_d t_{IJ/min}; t_{dI}; \Delta t_{dI} + t_{dJ/taxi}] \qquad (4.6.6)$$

where

> $_d t_{IJ/min}$ is the ATC minimum time-based separation rule between successive departing aircraft types I and J;
>
> t_{dI} is the runway occupancy time by departing aircraft I;
>
> $t_{dj/taxi}$ is the taxiing time needed for aircraft J to reach its departure threshold;
>
> Δ_{tdI} is the time aircraft I 'affects' aircraft J taxiing towards its departure threshold.

Condition 4.6.6 satisfies the ATC/ATM minimum separation rule including possible 'interference' between departing aircraft while taxiing towards their departure positions. For example, if aircraft J needs to cross the departure runway of aircraft I in order to reach its departure position, it should wait at last until aircraft I clears the crossing location. For the intermediate and far spaced parallel runways (Table 4.6.1), the successive departures separated according to Equation 4.6.6 can be realised independently (and simultaneously) on each runway.

If the proportion of aircraft types I and J in the mix is p_I and p_J, respectively, then similarly as in Equation 4.6.2, $p_{IJ} = p_I p_J$. Consequently, the average inter-departure time $\bar{t}_d$ can be calculated as:

$$\bar{t}_d = \sum_{IJ} p_{IJ} \,_d t_{IJ} \qquad (4.6.7)$$

Similarly to Equations 4.6.3 and 4.6.4, by using Equation 4.6.7, the departure capacity can be calculated as:

$$\lambda_d = \begin{cases} 1/\overline{t_d} \text{ for the closely spaced parallel runways} \\ 2/\overline{t_d} \text{ for the intermediate and far spaced parallel runways} \end{cases} \quad (4.6.8)$$

4.6.3.5 The model for mixed operations
On dual closely spaced dependent parallel runways, a single departure might be inserted between successive arrivals *Ik* and/or *kJ*. For example, to realise such a departure between the arrivals *Ik*, arrival *I* must clear the runway RWY1 after landing and arrival *k* should be no closer than the prescribed minimum ATC separation rules from threshold T_k of RWY2 (Figure 4.6.1) at the moment when the departure from either runway is cleared. The following condition, enabling realisation of *m* departures within the arrival sequence *Ik*, must be fulfilled:

$$t_{Ik} \geq {}_{ad}t_{Ik} = \varphi \, t_{aI} + \varepsilon \, t_{d/taxi} + (m-1)\frac{{}_d\delta_k}{v_k} \quad (4.6.9)$$

where

$_{ad}t_{Ik/min}$ is the ATC minimum separation rule enabling realisation of *m* departures between the arrival sequence *Ik*;

t_{aI} is the runway occupancy time by the arrival *I*;

$t_{d/taxi}$ is the time needed by the departing aircraft to reach its departure threshold;

φ is a binary variable, which takes the value '0' if the departure is carried out independently of arrival *I* and otherwise takes the value '1';

ε is a binary variable, which takes the value '0' if the departing aircraft does not need extra taxiing time to reach its departure threshold and the value '1,' otherwise;

$_d\delta_k$ is the minimum longitudinal separation between the given departure and the arrival *k*;

v_k is the approach speed of the arrival *k*.

At closely spaced dual parallel runways it is often $\varphi = 1$; at intermediate spaced runways $\varphi = 0$. In both cases ε can be either 1 or 0. Condition 4.6.9 is analogous for arrivals *kJ*. From Equation 4.6.9, the inter-arrival time $_a t^*_{IJ/k}$ between the arrival sequence *IJ* at the threshold of RWY1 influenced by arrival *k*, which enables realisation of *m* departures between the sequence *Ik* and *n* departures between the sequence *kJ* can be determined as follows:

$$_a t^*_{IJ/k} = {}_{ad}t_{Ik}(m) + {}_{ad}t_{kJ}(n) \quad (4.6.10)$$

Consequently, the average inter-arrival time for all possible combinations *IkJ*, $\overline{t^*_a}$ and corresponding landing capacity λ^*_a can be computed from Equations 4.6.3 and 4.6.4, respectively.

Let p_{dm} and p_{dn} be the probabilities of realising m and n departures between the arriving sequence Ik and kJ, respectively, that is between arrivals IJ. For close parallel runways the departure capacity λ_d^* dependent on the arrival capacity λ_a^* will be as follows:

$$\lambda_d^* = \lambda_a^* \left(\sum_{m=1}^{M} m p_{dm} + \sum_{n=1}^{N} n p_{dn} \right) \tag{4.6.11}$$

where M and N are the number of types of time gaps in which a given number of departures can be realised.

By combining Equations 4.6.4 and 4.6.11, similarly to in the single runway capacity model (Blumstein, 1960), the capacity of dual dependent parallel runways for the mixed operations can be determined as:

$$\lambda^* = \lambda_a^* + \lambda_d^* = \lambda_a^* \left(1 + \sum_{m=1}^{M} m p_{dm} + \sum_{n=1}^{N} n p_{dn} \right) \tag{4.6.12}$$

4.6.4 Application of the Model

4.6.4.1 Inputs
The proposed model is applied to calculating capacity of Frankfurt Main airport (Germany), as one of the busiest European hubs. Figure 4.6.3 shows a simplified scheme of the airport layout.

Currently, the airport operates three runways: two parallel with a length of 12,000 ft (4,000 m) − RWY07L/R and RWY25L/R − used for both landings and take-offs, and the third one of the length of 14,800 ft (4,500 m) − RWY18 − used exclusively for take-offs. In addition, there is the 'hidden' fourth RWY26L of the length of 8,200 ft (2,500 m) used exclusively for landings. Physically, this is part of RWY25L with a displaced landing threshold of 5,000 ft (1,500 m). The parallel runways RWY07L/R and RWY25L/R are closely spaced for $d = 1,700$ *ft (518 m)* and staggered (offset) for $z = 500$ *ft (150 m)*. When the runways RWY26L/25R are used, the staggering (offsetting) distance is $z = 4,900$ *ft (1,500 m) (30)*. In addition, a new runway with a length of *9,200 ft (2,800 m)* is planned at the North-West of the present airport area (the dotted line in Figure 4.6.3). It is parallel to the current RWY07L/R and RWY25L/R and spaced at $d = 4,600$ *(1,400 m)*, which will enable independent operations of both runway systems. The new runway will be used exclusively for landings.

For both existing landing runways, the length of the common final approach path is $\gamma_{(\cdot)} = 6.6$ *nm* (nm − nautical mile). 'Fine-tuning' of the separation between landing aircraft starts at around *40−50 nm* from the airport.

The aircraft fleet at Frankfurt airport consists of three aircraft wake-vortex categories with the characteristics given in Table 4.6.3.

The ATC applies longitudinal horizontal (δ), lateral-diagonal (ρ) radar-based and the vertical (χ) separation rules to the arrivals on two parallel runways. The longitudinal separation rules dependent on the wake-vortex categories are given in Table 4.6.4.

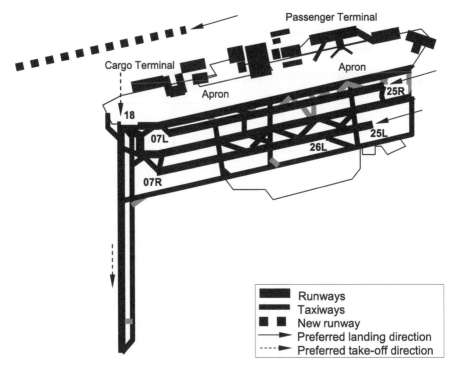

Figure 4.6.3 Simplified layout of Frankfurt Main Airport (Germany)
Compiled from EUROCONTROL (1998); Fraport (2004)

Table 4.6.3   Aircraft Classification in given example: Frankfurt Main Airport (Germany)

Wake-vortex category	Type	Proportion in the mix (%)	Average approach speed (kt)[1]	Runway (landing) occupancy time (s)
Heavy	A300-600; A330, A340; B767, 747, 777	20	140	60
Large	B737; A319, 320, 321	60	130	55
Small	ATR 42,72; AVRO jet, Dash 8	20	110	45

[1] *kt – knot; Compiled from Fraport (2004)*

When RWY25R and 26L are used, the ATC applies the vertical separation to the arrival sequences Heavy/Large, Heavy/Small and Large/Small at the final approach gate of $\chi_{(*)} = 1,000\,ft$ (see Figure 4.6.2). Such separation provides the minimum horizontal spacing measured along the path of the trailing aircraft of about three instead of four, five or six nm, respectively (that is $S_{(*)/min}{}^0 = \chi/tg\theta = 1,000/tg3^0 = 3\ nm$).

When the minimum lateral-diagonal separation rules between a pair of aircraft on two parallel approach paths is $\rho = 2\ nm$ and technologies ADS-B and CDTI were available, the so-called 'paired approach procedure,' enabling application of vertical,

Table 4.6.4   The ATC minimum separation rules between successive arrivals of different wake-vortex categories: Frankfurt Main Airport (Germany)

| Aircraft sequence *I/J* | | Category | |
	Heavy	Large	Small
Heavy	4	5	6
Large	3	4	4
Small	3	3	3

All values in nm-nautical miles
Compiled from Fraport (2004)

diagonal and longitudinal separation rules, could be applied (NASA, 1998; Hammer, 2000). In this case the paired aircraft would be aircraft *I* approaching RWY25R (leading) and aircraft *k* approaching RWY26L (trailing) (Figure 4.6.2). The ATC minimum 'spacing' between such paired aircraft *Ik* would be $S^0_{Ik/min}$ = *1 nm* (that is ρ_{Ik} ~ *1.01 nm*) and the vertical separation $\chi_{I\text{-}L/k\text{-}H}$ = *1,000 ft* ($\forall$ *Ik*). The minimum longitudinal spacing between aircraft type *k* of the pair *Ik* using RWY26L and aircraft type *J* as the leading aircraft of the successive pair using RWY25R, that is between two successive pairs, would be $S^0_{kJ/min}$ = 3 nm ($\forall$ *kJ*). This implies in each of two flows approaching two parallel runways that the successive aircraft would be longitudinally separated by about δ_{IJ} = *4 nm*. Such pairings imply only the application of sequencing tactics, when all the aircraft of the same category are successively cleared to land. In addition, the ATC can apply FIFO (First-In-First-Out) tactic, implying the same order of clearing for landings as the order of arriving at the final approach gate (Figure 4.6.1) (Newell, 1982; Janić, 2001).

The ATC time-based separation rules between successive take-offs either from two parallel runways RWY25L/R or from the departures-only RWY18 are given in Table 4.6.5.

Also in this case, the ATC can use FIFO and sequencing tactics to clearing departures.

Thanks to the very well-developed system of taxiways, any two departing aircraft do not interfere with each other while arriving at their departure thresholds (that is Δt_{dl} = 0); under constant demand a departure always waits for the preceding arrival

Table 4.6.5   The ATC minimum separation rules between successive departures of different wake-vortex categories: Frankfurt Main Airport (Germany)

| Aircraft sequence *I/J* | | Category | |
	Heavy	Large	Small
Heavy	90	120	120
Large	90	90	120
Small	45	45	45

All values in 'seconds'
Compiled from Fraport (2004)

Table 4.6.6   Scenarios of using the runway system in a given example: Frankfurt Main Airport (Germany)

Scenario	Characteristics[1]
A	• RWY25R/L–26L are used exclusively for landings; • RWY18 is used exclusively for take-offs; • The ATC applies longitudinal, lateral-diagonal and vertical separation rules.
B	• RWY25R/L–26L are used exclusively for paired landings supported by ADS-B and CDTI; • RWY18 is used exclusively for take-offs; • The ATC applies longitudinal, lateral-diagonal and vertical separation rules.
C	• RWY25R/L–26L or RWY18 are used as a single runway for both landings and taking-offs; • The ATC applies only longitudinal separation rules.
D	• RWY25R/L–26L and new North-West runway are used exclusively for landings and RWY18 exclusively for take-offs; • The ATC applies longitudinal, lateral-diagonal and vertical separation rules.

[1] Based on the preferred runway use due to the prevailing weather conditions

to clear the runway system (that is $\varphi = 1$); and the departure is always ready when the convenient time 'gap' occurs (that is $\varepsilon = 0$).

4.6.4.2 Results

Based on the above-mentioned inputs, the capacity can be calculated for different scenarios of using the runway system at Frankfurt Main airport, as given in Table 4.6.6. Each scenario includes application of FIFO and sequencing tactics of both arrivals and departures.

The results are shown in Figure 4.6.4. In particular, Figure 4.6.4a shows the capacity of existing runway system under conditions of using the existing (Scenario A) and advanced technology (Scenario B). In Scenario A, when the FIFO tactic is applied, the landing capacity is 39, take-off capacity 40, and the total capacity 79 ops/h. However, the incumbent airline Lufthansa operates 'waves' (that is 'banks') of incoming and outgoing flights. Each incoming bank is composed of three groups of successive arrivals scheduled in order: i) heavy − from the long (intercontinental and continental) routes; ii) large − from the intermediate (continental) routes; and iii) small − from short (domestic and continental) routes (Fraport, 2004). The last group of arriving aircraft then departs first and the first as the last. Such a pattern requires the ATC to apply the sequencing tactic, which produces an arrival capacity of 40, departure capacity of 44, and the total capacity of 84 ops/h. The airport's declared capacity is 80 ops/h, which is based on the premise that during the longer period of several successive hours, the departure capacity should be equal to the arrival capacity (NASA, 1998; Ignaccolo, 2003; Fraport, 2004). Consequently, the sequencing tactic produces a higher arrival, departure and total capacity by about 3 per cent, 10 per cent and 6 per cent as compared with the FIFO tactic. In Scenario B an application of the paired approach could increase the arrival capacity to 64 ops/h (60 per cent greater than otherwise), departure capacity would remain 44 ops/h while the total capacity would increase to 108 ops/h (that is by about 29 per cent than otherwise) (Scenario B).

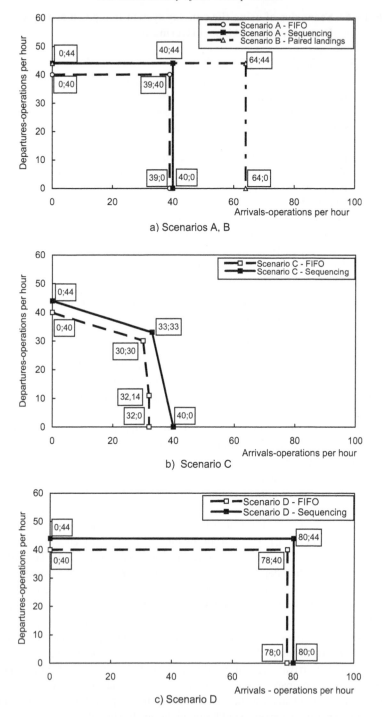

Figure 4.6.4 Capacity coverage curves for given scenarios of using the runway system in a given example: Frankfurt Main Airport (Germany)

Figure 4.6.4b shows that applying only longitudinal separation rules and FIFO tactics to the arrivals diminishes the corresponding capacity to 32 ops/h (Scenario C). However, it is possible to realise 11 departures between successive arrivals, which, combined with the departure capacity of RWY 18 of 40 ops/h gives the total departure capacity of 51 ops/h, and the total capacity of 83 ops/h. For an equal proportion of the arrival and departure demand, the capacity for arrivals and departures is 30 ops/h, and the total capacity 60 ops/h. Sequencing tactics produce an arrival capacity of 40 and a departure capacity of 44 ops/h. For the equal proportion of the arrival and departure demand, the corresponding capacities are 33, and the total capacity 66 ops/h (this is for about 10 per cent higher than in FIFO tactic).

Figure 4.6.4c illustrates the contribution of the new runway to the capacity (Scenario D). As can be seen, when FIFO tactic are applied the arrival capacity could be increased to 78, departure capacity would remain at 40, and the total capacity would increase to 118 ops/h. Sequencing tactics would raise the arrival capacity to 80, departure capacity to 44, and the total capacity to 124 ops/h. In both cases, this would be an increase of about 100 per cent and 50 per cent in the arrival and total capacity, respectively.

4.6.5 Concluding Remarks

This section has developed an analytical model for calculating the ultimate capacity of dual dependent (parallel) runway system. The model has particularly considered the influence of new technologies, enabling the application of the ATC different minimum separation rules (longitudinal, lateral-diagonal and vertical), procedures, and the ATC tactics of aircraft sequencing on the capacity. Increase in the capacity thanks to new technologies and operational procedures have been shown, indicating their potential for relieving pressure in the short-term on building new runways and consequently taking new land, as well as spreading the noise to new areas around an airport. Conclusions based on the model applied to Frankfurt Main airport (Germany) are as follows:

- The model has provided very similar (identical) results to those in practice. Therefore, it is reasonable to expect that it will also be able to perform similarly when applied to other cases. Such capabilities qualify the model for airport strategic planning while looking for relatively effective and efficient (cheap) decision making options as compared with more complex, comprehensive and expensive models (software packages) such as SIMMOD, Taam and/or Airport Machine.
- The capacity of two dependent parallel runways in the given example has always been slightly greater when the ATC sequencing tactic is applied instead of FIFO to either arrivals (about 3 per cent) or departures (10 per cent). However, the selection of tactic is usually dependent on the pattern and structure of demand in terms of the aircraft fleet mix.
- Introduction of the vertical (in addition to the longitudinal and lateral-diagonal) separation rules combined with using the displaced landing threshold has increased the arrival capacity by about 21 per cent for FIFO and by about 25 per cent for the sequencing tactic compared with using only the longitudinal

separation rules. The paired approach supported by ADS-B and CDTI could increase the landing capacity significantly, by about 60 per cent.

- Addition of a new (independent) parallel runway to the existing runway system can significantly increase the landing and total capacity, by about 100 per cent and 50 per cent, respectively.
- The 'capacity coverage curves' obtained from the model can be used as direct input to estimate arrival and departure delays, optimal allocation of capacity to the arrival and departure demand, and for calculating the taxiway and apron/gate complex capacity.

4.7 Optimisation of Utilisation of the Airport Capacity

4.7.1 Background

Congestion and delays in the air transport system may cause significant costs to air transport operators – airports, air traffic control, airlines, and users – air travellers and freight shippers. Generally, the most important factors causing congestion and delays are growing demand, scarcity of the air transport system capacity, an increased concentration of airline operations such as hub-and-spoke networks, and the various social/environmental constraints affecting the efficient and effective use and very often the necessary physical expansion of the current system capacity (Caves and Gosling, 1999).

Congestion and delays may have impacts on almost all the actors involved in the process: users consider them as loss of time, airlines consume additional fuel if delays are carried out airborne; the additional fuel consumption causes emissions of air pollutants and consequently a greater burden on the environment; the airports consider expansion of the airside and landside infrastructure which requires additional land, and would eventually spread impacts such as noise to a wider population around them.

In general, the airports are the most congested parts and they employ various options to mitigate congestion and delays on both the demand and the capacity side. On the demand side these might be the so-called demand management measures such as 'slot-control', 'charging peaks', 'auction of slots', and 'charging congestion'. On the capacity side, these are innovative technologies and advanced ATFM procedures enabling more efficient utilisation of existing capacity, and physical expansion of the airport airside and landside infrastructure (Janić and Stough, 2003).

Mitigation of congestion and delays at airports under existing conditions generally implies balancing the available capacity with the expected demand. According to the time horizon, this takes place at three levels: i) the strategic level over a period of about six months or longer (Helme and Lindsay, 1992); ii) the tactical level over several hours or one day – a Ground-Holding Program is one of the examples (Terrab and Odoni, 1993); and the operational level over a period of several minutes, which mainly includes 'on-line' control of aircraft separation (Janić, 2004).

This section presents a heuristic algorithm for allocating the existing airport runway capacity to the expected demand at the tactical level. The algorithm aims to minimise the total cost of aircraft/flight delays over a given period of time under specified circumstances (Gilbo, 1993, 1997). In addition to showing how the existing capacity of an airport can be used more efficiently and consequently the pressure for new land for expansion diminished, at least temporarily (in the near term), developing such an algorithm has some other reasons: firstly, seemingly, there is an inherent complexity to the previously developed operation research models based on linear and integer programming, which might appear relatively difficult to understand and consequently increase the hesitance of the air traffic control operators who mostly prefer to operate according to their rule of thumb (Gilbo, 2003). Secondly, in order to provide an acceptable solution for reducing the use of the rule of thumb, the intention is to eventually offer a sufficiently feasible alternative, which closely reflects the rule of thumb of the ATFM operators and produces 'sufficiently satisfactory' and 'comparable' results to those obtained from the more complex optimisation models. Finally, due to the fact that the above-mentioned optimisation models have recently been tested at some US congested airports (Gilbo, 2003), the proposed algorithm could eventually be used as an optional alternative. The material is mainly taken from the author's already published work (Janić, 2004).

4.7.2 The Airport Capacity

At an airport's airside area, congestion and aircraft/flight delays happen whenever demand exceeds the available capacity of particular components – the runway system, taxiways and/or the apron (EEC, 2001b; Janić, 2001; FAA, 2002). In this situation the airport runway system's capacity can be defined as the maximum number of aircraft/flights that can be accommodated during a given period of time – one, half or quarter of an hour – under the prevailing operational, economic and environmental conditions (Blumstein, 1960; Janić and Tosic, 1982, Caves and Gosling, 1999; FAA, 2002).

The most important operational conditions include the minimum ATFM separation rules applied to the constant flows of the arrival and departure aircraft/ flights, which pass through the 'reference location' for measurement of the capacity. This 'location' is usually the runway threshold. If the minimum separation rules are applied to all aircraft/flights in the flows, the maximum number of aircraft/ flights will be accommodated, each usually with the specified average delay. This number represents the airport runway system's practical capacity (Blumstein, 1960; Newell, 1979). In addition, other factors may influence this capacity. These are the number and configuration of runways, the composition of the aircraft fleet in terms of approach/departure speeds, type of radio-navigational (ground and airborne) facilities, weather, and particularly the mixture of the arrival and departure flights which influences the order of sequencing of particular types of operations (Gilbo, 1993). Regarding the above-mentioned factors, the so-called airport runway capacity envelopes shown in Figure 4.7.1 can be synthesised (Newell, 1979; FAA, 2002).

In general, for other constant conditions, the ATFM can use an alternating and consecutive tactic to serve a given mixture of the arrival and departure flows (Newell, 1979). The alternating tactic handles each arrival at the soonest possible time before or after handling a departure, or vice versa, that is a 'trade-off' is made between the

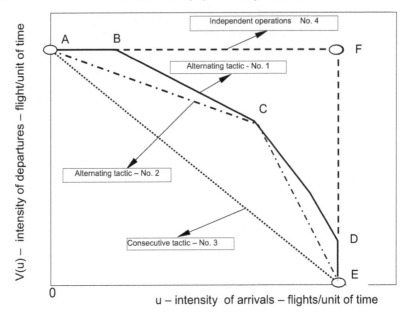

Figure 4.7.1 Some typical shapes of the airport runway capacity envelope

arrival and departure flows. The capacity envelope 1 (the heavy line ABCDE) and 2 (the dotted line ACE) in Figure 4.7.1 represents such cases, each for a given fleet mix. The consecutive tactic serves all departures before or after all arrivals of the given mixture. Envelope 3 (the slant dotted line AE) represents this case, which may often be characterised by significant arrival and/or departure delays. In addition, the arrivals and departures can also be handled independently of each other at separate runways operating, for example, in a segregated mode. Capacity envelope 4 (the horizontal and the vertical dotted line AF and EF, respectively) represents this case.

4.7.3 A Heuristic Algorithm

4.7.3.1 Problem and state of the art
The capacity envelopes in Figure 4.7.1, defined for given conditions and time period (usually one or a quarter of an hour), provide information about the capacity limits of the airport runway system. They are to be allocated to the expected demand with respect to given criteria in order to achieve the specified objectives. The capacity allocation implies determining and using a pair of the inter-dependent values from a given capacity envelope – one for the arrival and another for the departure capacity. Usually, the objectives include balancing and/or diminishing (minimising) the arrival and departure queues, delays and associated costs, and improvement of the utilisation of the existing runway system capacity, both over a given period of time. In such a context, delays can be ground-held and/or airborne, each with a specific, different unit cost.

Over the past decade, allocation of the airport's runway system capacity by using the operations research models (integer and linear programming) and optimal control theory has become one of the challenging issues for both researchers and the ATFM practitioners in the US (Gilbo, 2003). In Europe, for most airports, this is still a matter of individual interest for some academics and the rule of thumb for the ATFM practitioners.

Up to date relatively substantive research has been carried out. For example, at an early stage Helme and Lindsay (1992) specified that modelling congestion and delays in the national airspace system (US) should reflect realistic situations. Consequently, numerous analytical and simulation models were developed but very few explicitly treated the runway arrival and departure capacities as interdependent parameters (Odoni and Bowman, 1997). Nevertheless, one of the earliest academic efforts in this direction was Newell's (1979) concept of the inter-dependent airport arrival and departure runway system's capacity. These capacities were mainly influenced by the ATFM separation rules, sequencing of a mixture of the arrival and departure flows, the aircraft fleet structure, and the runway configuration. Gilbo (1993) carried out the statistical analysis confirming existence of the capacity envelopes at some US airports. Currently, the Federal Aviation Administration (FAA, 2002) regularly updates these envelopes for 29 of the busiest US airports. In addition, the potential use of these envelopes was also investigated. Gilbo (1993) developed the linear programming model for optimal allocation of the interdependent airport arrival/departure capacity to the expected demand, which minimised the total number of arrival and departure queues and delays over a given time period. Gilbo (1997) used previous ideas to develop an integer programming model for the optimal allocation of airport capacity, which included the restrictive capacity of the arrival and departure fixes in the vicinity of given airport (that is terminal area). Gilbo and Howard (2000) extended the two previous models by including prioritisation of the flights of different airlines. Most recently, Gilbo (2003) reported on the pilot of optimisation models at one congested US airport.

Dell'Olmo and Lulli (2001) developed a dynamic programming model to handle the optimal allocation of the airport runway capacity and an integer-programming model for optimal allocation of this capacity in an airport network. Velazco (1995) developed an analytical queuing model to estimate delays under conditions of prioritising arrivals and departures at a single runway airport. According to an optimal strategy different operations were equally prioritised during the peaks while the arrivals were given non pre-emptive priority during the off-peaks.

Despite being very sophisticated, the above optimisation models, based on linear, integer and dynamic programming for the optimal allocation of the airport arrival and departure capacity also seemed to be relatively complex in terms of understanding and implementation. In particular, it appeared to be the case for the relatively non-familiar users – the ATFM operators – also preferring to use their rules of thumb (Gilbo, 2003). In addition, the steady-state queuing models were adequate only for cases not requiring allocation of the airport capacity since the demand was always lower than the capacity.

4.7.3.2 Objectives and assumptions
The objectives are to:

• Develop a heuristic algorithm for allocation of the airport runway system capacity based on the heuristic ('greedy') criteria (rules) closely reflecting the rule of thumb of the ATFM operators. The algorithm should minimise the total costs of aircraft/flight delays during a given period of time under the specified conditions. A heuristic 'greedy' approach implies making a choice of what looks the 'best' at the moment, that is making the 'best' choice locally hoping that it will lead to an optimal solution globally. The current 'best' choice is ultimately made without considering the outcomes of other choices, which makes this algorithm inherently different from the linear and dynamic programming technique where in each step a choice depends on the choices made in other steps (Johnsonbaugh and Schaefer, 2003), and

- Compare the results from the proposed heuristic algorithm with those from the earlier developed linear and integer-programming optimisation models in order to investigate the algorithm's performances for the purpose.

The proposed algorithm is based on the following assumptions:

- The airport runway system capacity envelope can be synthesised using the minimum ATFM time separation rules applied to particular combinations of sequences of the arrival and departure aircraft/flights at the 'reference location' selected for the capacity calculation;
- Allocation of the airport runway system capacity to the expected demand over a given period of time (for example, several hours during the day) is carried out sequentially in the shorter sub-periods (for example, a quarter of an hour), which enables proper handling of fluctuations in both demand and capacity;
- The service priority rule for each aircraft/flight in either arrival or departure flow is 'first-come – first-served;' and
- The cost of a unit of delay of an arrival and of a departure aircraft/flight is determined as a constant average within a given sub-period of time.

4.7.3.3 The structure of the algorithm

The proposed heuristic algorithm consists of four components, each specified for a given period of time T as follows: i) the objective function as the sum of the total cost of the arrival and departure delays, which should be minimised; ii) the arrival and departure demand – aircraft/flight flows – to be served; iii) the airport runway system capacity envelope; and iv) the heuristic ('greedy') criteria (rules) for allocation of the airport capacity aiming at minimising the objective function under given circumstances.

The objective function   The time period T in which allocation of the runway system capacity needs to be carried out is divided into N equal sub-periods Δ_{tk}, that is $N\Delta t_k = T$. Consequently, the objective function can be formulated as follows:

$$C_T = \sum_{k=1}^{N} [c_{ak}Q_{ak} + c_{dk}Q_{kd}]\Delta t_k \qquad (4.7.1)$$

where

$\quad C_T$ is the total cost of delays of the arrival and departure aircraft/flights during time T (\$);

c_{ak} is the average cost of a unit of delay of the arrivals in the sub-period Δ_{tk} ($/time unit);

c_{dk} is the average cost of a unit of delay of the departures in the sub-period Δ_{tk} ($/time unit);

Q_{ak} is the arrival queue at the end of sub-period Δ_{tk} (aircraft/flights);

Q_{dk} is the departure queue at the end of sub-period Δ_{tk} (aircraft/flights).

In Equation 4.7.1, the arrival and departure queue Q_{ak} and Q_{dk}, respectively, can be determined as follows:

$$Q_{ak} = \max\left[0; Q_{a/k-1} + (a_k - u_k)\Delta t_k\right] \tag{4.7.2a}$$

and

$$Q_{dk} = \max\left[0; Q_{d/k-1} + (d_k - v_k)\Delta t_k\right] \tag{4.7.2b}$$

where

$Q_{a\,k-1}$ is the arrival queue at the end of sub-period Δ_{tk-1} (aircraft/flights);

$Q_{d\,k-1}$ is the departure queue at the end of sub-period Δ_{tk-1} (aircraft/flights).

a_k is the arrival demand in the sub-period Δ_{tk} (aircraft/flights per unit of time);

d_k is the departure demand in the sub-period Δ_{tk} (aircraft/flights per unit of time);

u_k is the arrival capacity allocated to the arrival demand in sub-period Δ_{tk} (aircraft/flights per unit of time);

v_k is the departure capacity allocated to the departure demand in the sub-period Δ_{tk} (aircraft/flights per unit of time).

In Equation 4.7.2, the departure queue Q_{dk} physically takes place at the given airport. The arrival queue Q_{ak} may take place in the vicinity of the given airport, in the relatively remote airspace from the given airport or at the aircraft/flight origin airports if a Ground Holding Programme (GHP) is applied (Terrab and Odoni, 1993). In any case the cost of a unit of the ground-held delay is usually considered lower than the cost of a unit of airborne delay, that is $c_{dk} \leq c_{ak}$ (ITA, 2000).

The runway system capacity envelope   The arrival and departure capacity in Equation 4.7.2, u_k and v_k, generally depend on each other. They can be determined from the capacity envelope under conditions of alternating an uneven mix of the arrival and departure aircraft/flight flows. In Figure 4.7.1, the capacity envelope 1 appears typical for such a case. It consists of:

- The horizontal (first) segment AB indicating the insertion of arrivals between the minimally separated successive departures without displacing these departures;
- The slant (second) segment BC referring to alternating (that is trading-off) the greater number of departures and the lower number of arrivals;

- The slant (third) segment CD referring to alternating (that is trading-off) the greater number of arrivals and the lower number of departures; and
- The vertical (last) segment DE indicating the insertion of departures between the minimally separated successive arrivals without displacing these arrivals.

The segments BC and CD define the so-called capacity 'trading-off' area. When they exist, the proposed algorithm and the previously mentioned optimisation models make sense. Otherwise, allocation of the airport runway system capacity is straightforward.

The analytical representation of the segments (i)–(iv) of the capacity envelope 1 in Figure 1 comes from the modified approach of Newell (1979) in which the departure capacity v in Equation 4.7.2 can be expressed as a function of the arrival capacity u as follows:

$$v(u) = \begin{cases} 1/t_{dd}, & \text{for } 0 \le u \le u_0 \\ (1/t_{dd})[1-(u-u_0)((t_{ad}+t_{da}-t_{dd})/[1-u_0(t_{ad}+t_{da})])] & \text{for } u_0 \le u \le 1/(t_{ad}+t_{da}) \\ [1/(t_{ad}+t_{da}-t_{aa})][(1-v_0t_{aa})-[1-v_0(t_{ad}+t_{da})t_{aa}u]] & \text{for } 1/(t_{ad}+t_{da}) \le u \le 1/t_{aa} \\ v_0, & \text{for } u = 1/t_{aa} \end{cases} \quad (4.7.3)$$

where

t_{dd} is the minimum time separation between any two successive departures (time units);

t_{ad} is the minimum time separation between an arrival and a succeeding departure (time units);

t_{da} is the average minimum time separation between a departure and a succeeding arrival (time units);

t_{aa} is the minimum time separation between any two successive arrivals (time units);

u_0 is the number of arrivals inserted between the minimally separated successive departures without displacing these departures (aircraft/flights per time unit);

v_0 is the number of departures inserted between the minimally separated successive arrivals without displacing these arrivals (aircraft/flights per time unit);

u is the airport arrival capacity (aircraft/flights per unit of time);

$v(u)$ is the airport departure capacity as a function of the arrival capacity (aircraft/flights per unit of time);

The minimum time intervals between successive operations are measured at the 'reference location' for calculation of the airport runway system capacity (Blumstein, 1960; Janić and Tosic, 1982; Venkatakrishnan, Barnett and Odoni, 1993, FAA, 2002).

In Equation 4.7.3, $u_0 = (1/t_{dd})\sum_{l=1}^{L} lp_{a/l}$ and $v_0 = (1/t_{aa})\sum_{m=1}^{M} mp_{d/m}$ where $p_{a/l}$ and $p_{d/m}$ are the probabilities of inserting l and m arrivals and departures, respectively, between any two successive minimally separated departures and arrivals, respectively ($l \in L$; $m \in M$). If $u_0 = v_0 = 0$, only the 'trading-off' segments exist, that is capacity envelope 1 transforms into capacity envelope 2 in Figure 4.7.1.

The heuristic ('greedy') criteria and algorithmic steps The heuristic ('greedy') criterion for allocation of the airport runway system's capacity as a 'core' of the proposed heuristic algorithm is based on the analogy from priority queuing systems (Winston, 1994). Essentially, in the present case this rule calls for minimising the differences between the rate of inflow and the rate of outflow of the aircraft/flight delay costs in each subsequent sub-period Δ_{tk}, which in turn should lead to minimisation of the objective function 4.7.1 over the entire period T.

For example, the rate of inflow of delay costs in the sub-period Δ_{tk}, ultimately being 'out of control' of the ATFM operators, can be expressed as follows:

$$R_{\inf/k} = c_{ak}[Q_{a/k-1}/\Delta t_k + a_k] + c_{dk}[Q_{d/k-1}/\Delta t_k + d_k] \quad [\$ac/\min^2] \quad (4.7.4a)$$

Similarly, the rate of outflow, that is 'expulsion' of delay costs during the sub-period Δ_{tk} can be expressed as follows:

$$R_{out/k} = c_{ak}u_k + c_{dk}[v_k(u_k)] \quad [\$ac/\min^2] \quad (4.7.4b)$$

Consequently, minimisation of the difference between 4.7.4a and 4.7.4b can be achieved by maximising 4.7.4b, which is 'under control' of the AFTM operators. Maximisation of 4.7.4b implies selection of a pair of the interrelated capacities u_k and $v_k(u_k)$ from the capacity envelope 3. In order to minimise the objective function 4.7.1, such a selection of the capacities should be carried out in each sub-period Δ_{tk} of period T. Consequently, the proposed heuristic algorithm consists of the following steps:

Step 1   Initialisation of the sub-period Δ_{tk} ($\Delta_{tk} \in T$).
Step 2   Specification of the airport runway system's capacity envelope 3 for the sub-period Δ_{tk}.
Step 3   Composing the 'capacity vector' S_k whose elements are the arrival and departure capacities selected from the capacity envelope 4.7.3 as follows:

$$S_k = [u_{kj}; v_{kj}(u_{kj})] \quad \text{for } j \in J_k \quad (4.7.5a)$$

where J_k is the number of the selected pairs of the capacity values in the sub-period Δ_{tk}.
Step 4   Calculating the 'greedy vector' $\theta_k\{\theta_{kj}\}$ for the sub-period Δ_{tk} using information on the current arrival and departure queues Q_{ak} and Q_{dk} from Equation 4.7.2 and the 'capacity vector' S_k from Equation 4.7.5a as follows:

$$\theta_{kj} = | c_{ak}(Q_{ak}/\Delta_k - u_{kj}) | + | c_{dk}[Q_{dk}/\Delta_k - v_{kj}(u_{kj})] | \quad \text{for } j \in J_k \quad (4.7.5b)$$

Step 5   Extracting the minimum element of the 'greedy vector' θ_k as $\theta_k \equiv \min\{\theta_{kj}\}$ and the corresponding associated capacity values $[u_{kj}^*; v_{kj}^*(u_{kj}^*)]$.

Step 6   Calculating the arrival and departure queues (and delays) from 2) using the outcome from Step 5, that is the selected capacities $[u_{kj}^{*}; v_{kj}^{*}(u_{kj}^{*})]$.

Step 7   Repeating Steps 1–6 until the end of time period T.

Step 8   Calculating the total (minimum) cost of the arrival and departure delay over time period T from Equation 4.7.1.

Determination of the airport capacity in Step 4 and Step 5 enables simultaneous minimisation of the queues and delays and maximisation of utilisation of the available airport capacity. Namely, if the current demand is greater than the airport's capacity envelope, the greater capacity (the values from the envelope) will be allocated to minimise queues and delays. If demand is below the capacity envelope, only the necessary (minimum) capacity will be allocated to serve it. This characteristic makes the proposed 'greedy' criterion comparable to the logic of the above-mentioned optimisation linear and integer programming model(s).

4.7.4 Application of the Proposed Algorithm

The proposed heuristic algorithm is applied to traffic scenarios at three US airports: i) Chicago O'Hare International Airport (ORD), one of the world's busiest airports (FAA, 2002); ii) Newark International Airport (EWR), one of the three biggest airports serving the New York area with significant and frequent aircraft/flight delays (Allan et al., 2001); and iii) the Lambert St. Louis International Airport (STL), where the pilot demonstration of the optimisation models was first carried out (Gilbo, 2003). The three airports were selected intentionally to enable comparison of the results from the proposed heuristic algorithm with the results from the earlier developed optimisation models (Gilbo, 1997; Gilbo and Howard, 2000; Gilbo, 2003). Consequently, the same input as in cases of using the above-mentioned optimisation models is used. The corresponding outputs from the latter models are used as the benchmarking cases for validation of the proposed heuristic algorithm.

4.7.4.1 Inputs
The input for each of the three airport cases contains four groups of information: i) airport layout indicating a typical pattern of use for the runway system; ii) the capacity envelopes synthesised by Equation 4.7.1 for each 15-minute sub-period of the given time period; and iii) the arrival and departure demand in terms of the number of aircraft/flights in each 15-minute sub-period of the given time period; and iv) the cost of a unit of the arrival and departure delays.

Figure 4.7.2(a, b, c) contains the information on (i)–(ii). Figure 4.7.3 (a, b, c) provides the information on (iii).

Airport layout and the capacity envelope   Figure 4.7.2 shows the airport layouts and the capacity envelopes in the given cases (Gilbo, 1993; Gilbo and Howard, 2000; FAA, 2000; Gilbo, 2003). Figure 4.7.2a-i shows that Chicago O'Hare Airport (ORD) operates six runways. Two pairs of runways are busy most of the time. Runways 4R/9R are typically used for arrivals and runways 4L/9L for departures. Runways 32R/32L are used less frequently (FAA, 2002). The capacity envelopes for

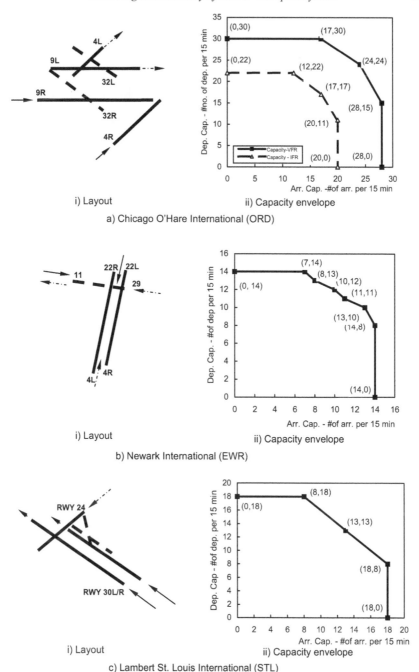

i) Layout   ii) Capacity envelope

a) Chicago O'Hare International (ORD)

i) Layout   ii) Capacity envelope

b) Newark International (EWR)

i) Layout   ii) Capacity envelope

c) Lambert St. Louis International (STL)

Figure 4.7.2 Schemes of the layouts and capacity envelopes at three airport cases: a) Chicago O'Hare International (ORD); b) Newark International (EWR); c) Lambert St. Louis International (STL)

Compiled from FAA (2002); Gilbo (1997); Gilbo and Howard (2000); Gilbo (2003)

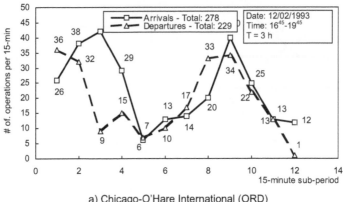

a) Chicago-O'Hare International (ORD)

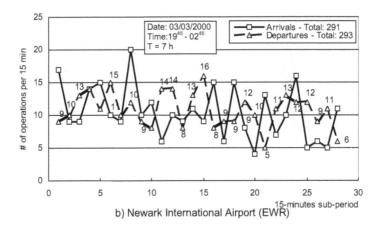

b) Newark International Airport (EWR)

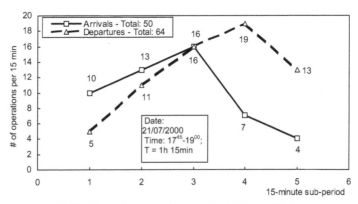

c) The St. Louis Lambert International Airport (STL):

Figure 4.7.3 The initial demand at three airport cases: a) Chicago O'Hare International (ORD); b) Newark International (EWR); c) Lambert St. Louis International (STL)

Compiled from Gilbo (1997); Gilbo and Howard (2000); Gilbo (2003) such an operational pattern under VFR and IFR conditions are synthesised in Figure 4.7.2a-ii.

Figure 4.7.2b-i shows that Newark Airport (EWR) operates three runways. Under VFR conditions, runways 22L and 11 are used for arrivals and runways 22R and 29 for departures. Under IFR conditions, runways 4R and 4L are used for arrivals and departures, respectively. Figure 4.7.2b-ii shows the capacity envelope for this latest (IFR) case.

Figure 4.7.2c-i shows that Lambert St. Louis Airport (STL) operates four runways: two parallel (30L/12R or 30R/12L) and the third one crossing them (24/06) are active, and the fourth one (31/13) is not currently in use. The capacity envelope for typical use of the runway system under VFR conditions is shown in Figure 4.7.2c-ii.

Arrival and departure demand   Figure 4.7.3(a–c) shows the arrival and departure aircraft/flight demand at Chicago O'Hare, Newark and Lambert St. Louis Airport, respectively, for the selected day and period of time. As can be seen, at each airport, the number of either the arrival or departure aircraft/flights varies during the selected time period of the specified day.

Cost of a unit of aircraft/flight delay   The average cost of a unit of the departure and/ or arrival delay is generally taken as the relative values as follows: i) $c_a = c_d = 1.0$; and ii) $c_a = 2c_d$ $(c_d = 1.0)$. The former case implies the application of GHP when the expected arrival delays are realised on the ground at the aircraft/flight origin airports with the engines switched-off. The latter case implies that the expected arrival delays are airborne. After introducing GHP this strategy is rarely applied due to savings in the fuel consumption and associated costs.

4.7.4.2 Results
By using the above input, five numerical experiments are carried out using the proposed heuristic algorithm. The aggregated outputs are provided for validating the proposed algorithm. These outputs are given in Table 4.7.1 and Figure 4.7.4 together with the results from the linear and integer-programming optimisation model as the benchmarking cases (Gilbo, 1997; Gilbo and Howard, 2000; Gilbo, 2003).

Figure 4.7.4 shows the total sums of the arrival and departure queues and delays at the end of the given period of time (7 h in the former and 1 h and 15 min in the later case) obtained from the heuristic algorithm and benchmarking model (for the comparative purposes, the results for Chicago O'Hare Airport are also shown).

As can be seen, there is the very high similarity these outputs. For example, at Newark Airport, the heuristic algorithm produces a total of 151 and the benchmarking model a total of 146 units of 15-minute delays. The difference is about 3.4 per cent in favour of the benchmarking model. At St. Louis Lambert Airport, both methods produce the same aggregate number of units of 15-minute delays, 12. The results from the five experiments for the three airports are additionally summarised in Table 4.7.1.

As can be seen, the absolute and relative differences in the total queues (arrival + departure) obtained from the two methods are relatively insignificant. In all experiments, the heuristic algorithm performs weaker than the benchmarking

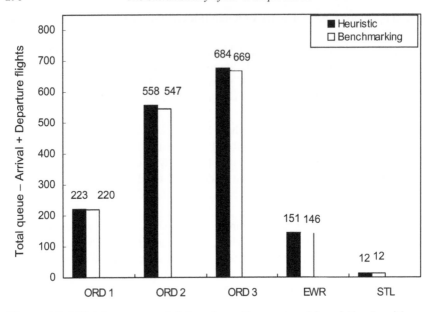

Figure 4.7.4 Total queues and delays from the proposed heuristic algorithm and benchmarking model in the given examples

Table 4.7.1 Summary of performances of the proposed heuristic algorithm and benchmarking optimisation model

Experiment/ case	The total queue (aircraft/flights)		Relative difference in the total queue (%)	Absolute difference in the total delay (min)	Absolute difference in the average delay (min/op)
	H^1	B^1			
ORD 1	223	220	+1.4	+45	+0.090
ORD 2	558	547	+2.0	+165	+0.326
ORD 3	684	669	+2.2	+225	+0.444
EWR	151	146	+3.4	+175	+0.128
STL	12	12	0	0	0

[1]H – Heuristic results; B – Benchmarking results

models as was intuitively expected. However, the differences in the total output vary by about 1.4–2.4 per cent in all cases, which seems acceptable. Setting up the acceptable margins of acceptability of the difference in results is something for a separate discussion since, based on such small number of experiments, it appears impossible to specify here the generic conditions for the required performance of the proposed algorithm.

The differences in the total delays seem also to be sufficiently significant to disfavour the proposed heuristic algorithm (0–225). However, considered per individual operation – arrival or departure, these differences appear in seconds, which looks acceptable.

4.7.5 Concluding Remarks

This section has elaborated the heuristic ('greedy') algorithm for allocation of the airport runway system's capacity to the expected demand during congestion periods when demand exceeds the capacity. The objectives were to minimise the cost of aircraft/flight delays and maximise utilisation of the available capacity under given (specified) conditions. The main reasons for developing the algorithm were to look for a relatively simple but suitable alternative to the already established optimisation methods based on linear and integer programming. Since using the 'greedy' criteria relates more closely to the rule of thumb of the ATFM operators, the proposed algorithm was also expected to be possibly acceptable for practical purposes. In addition, more efficient and effective utilisation of airport capacity should reduce the airport environmental burden in terms of acquiring new land for expansion, and thus contribute to the overall airport social feasibility, that is sustainability.

The heuristic algorithm was applied to three airports in the US due to the availability of relevant results obtained from the optimisation models used as the benchmarking case for validation of the proposed algorithm.

Comparison of the results from five experiments showed a very high similarity in the total results although distribution of delays and associated costs between particular types of operations was relatively significant. In these experiments the heuristic algorithm produced weaker results by about 1.4–2.4 per cent than the benchmarking models as was intuitively expected. Such results could indicate that under some conditions the proposed heuristic algorithm can perform 'relatively satisfactory'. However, such a judgement is based on relatively limited validation of the algorithm. Therefore, in order to reach more final conclusions about the overall feasibility of the proposed heuristic algorithm more experimentation is needed. This should also include setting up the acceptable thresholds on differences between the results from the proposed heuristic and other methods.

4.8 Demand Management at Airports by Charging for Congestion

4.8.1 Background

Congestion occurs due to the imbalance between people's needs to travel by a given transport mode at specific times more than at others during the day, week, and/or year and the available capacity of the transport system capacity to handle these needs. Congestion occurs when the volume of needs exceeds the available capacity. One of the main reasons for such a travel pattern is that people, while maximising the benefits of their choices, often neglect the interests and choices of others. Consequently, among other factors, they bunch around the service facilities because they do not pay the full social cost of their choice.

Over the last decade, both in Europe and the US, air transport congestion has increased, mainly due to growing air transport driven by the preferences of the travelling public and constraints in the capacity of air transport infrastructure to appropriately accommodate them. Since the demand for air transport is continuing to grow, the efficient measures for matching the capacity to growing demand need to be implemented. The physical expansion of the infrastructure is one of the long-term measures. Two short-term measures, particularly useful for airports as the 'critical' air transport infrastructure, are enhanced utilisation of the existing airport capacity due to technological and operational innovations and demand management.

At many airports, physical expansion of the infrastructure may be difficult or even impossible in the short-term due to political and environmental constraints such as excess noise, air pollution, land use, patterns and conventions. More efficient utilisation of the existing airport capacity based on new technologies and procedures produces limited effects. Demand management has recently been considered as a potentially useful short-term measure (Adler, 2002; DeCota, 2001; FAA, 2000).

Generally, demand management at airports implies utilising administrative and economic instruments to ensure that congestion is contained within the prescribed limits. The administrative measures imply negotiations among airlines, airport, and air traffic control about the volume of demand, its distribution during the day, and an acceptable level of congestion and delays. Such an airport is considered as 'fully co-ordinated' with access opened exclusively for the slot holders. Recently, this administrative system has been criticised as anti-competitive and in contradiction to the widely spreading air transport liberalisation policies. As an alternative, economic instruments, which embrace measures such as the slot auction and charging for congestion, are considered.

The auction of slots can enable efficient usage of existing slots and provide more efficient distribution of the contingents of new slots. Charging for congestion can be applied to the price-sensitive demand when other demand-management measures are inefficient. At present, scheduling flights at congested airports to handle passenger preferences in terms of the arrival times only takes into account the flight's private benefits. However, there are some external, that is social, costs. These are the costs of marginal delays, which each flight imposes on other flights during the congestion period. In order to deter or 'penalise' such flights the corrective charge equivalent to the total cost of marginal delays is considered. Together with the private costs, this charge constitutes the total social costs of each flight taking part in the congestion. In any case, these costs include the cost of delay time for both the airline (aircraft) and passengers. From the airline's point of view, if the total social costs of a given flight are higher than its perceived benefits, the flight will not take place, indicating efficiency of the congestion charge, and vice versa.

This section presents a model of congestion charging at an airport, which actually means internalising the congestion costs as externality. The congestion cost embraces the cost of marginal delays, which each aircraft/flight imposes on other flights at the airport during the congested period. The majority of the material is extracted from work already published by the author (Janić, 2005).

4.8.2 Airport Demand, Capacity and Congestion

An airport's demand is represented by the flights scheduled during a given period of time to handle people's travel needs. Usually, at most large European and US airports most flights are scheduled by one or a few airlines, their subsidiaries and alliances.

The available arrival and departure slots, that is the airport's declared capacity, enables the handling of these flights. In Europe, the number of arrival and departure flights accommodated at an airport during a given period of time (usually one hour) under specified conditions determines the airport's declared capacity. This capacity is based on IMC and IFR. Usually, this capacity is an agreed value between the airlines, airports and air traffic control (EEC, 2002). In the US, the agreed airport capacity usually has two values: the 'optimal' one determined for VMC and VFR, and the 'reduced' one determined for IMC and IFR (FAA, 2000). The demand and the airport capacity may have different relationships during a given period of time, which may cause congestion and consequently delays to incoming and outgoing flights. Demand is usually balanced with the airport declared capacity in an effort to prevent unacceptable congestion. In both Europe and the U. S this is carried out as a multi-stage process in order to prevent the escalation of congestion and delays above prescribed limits (Liang, Marnane and Bradford, 2000; FAA, 2001, 2002a; ATA, 2002; EEC, 2002; Janić, 2003).

Congestion occurs when demand exceeds the airport capacity. The periods of congestion are called peaks. At most airports, the peaks happen in the morning, possibly at midday and in the late afternoon/early evening as a reflection of people's preference to travel at some times more than others during the day. At most airports the peaks are recognisable, but at some they are not since they may be so long that they spread over the whole day (Janić, 2005).

Congestion causes time deviations to the flight's time for passing through the reference point (location) where possible delays are registered, usually the airport arrival/departure gates. These deviations are called delays. At most airports, the agreed delay threshold of either arrival or departure flights is a period of 15 minutes. In general, delay times increase with increasing congestion (AEA, 2001; BTS, 2001; EEC, 2001b; FAA, 2003a).

Generally, delays are expressed as the average delay per flight as well as the average delay per delayed flight. The former is the total delay divided by the number of all flights during a given period of time. The latter is the number of delayed flights divided by the total number of flights per period of time (FAA, 2003a; EEC/ECAC, 2002). In addition, delays can be classified into arrival and departure delays as well as according to the specific causes (AEA, 2001; EEC/ECAC, 2001; Janić, 2005).

In addition to the size, the nature of the congestion may also be relevant when considering the potential introduction of economic measures of demand management. For example, consider a hub airport with only one dominant airline, where congestion and delays still occur due to the airline's scheduling practice. In such case, the congestion costs are already internalised and any further charge may either destroy the present schedule or constrain its further development. In the case of two dominant airlines operating at the same hub airport, caution is needed about the real interference of their flights. Despite operating simultaneously at the airport and competing amongst each other, their flights might not interfere if they use separate parts of both the airport's airside and landside area. In such a case, the situation is similar to the case of the single dominant airline and therefore there is no sense in additionally managing demand and congestion by economic measures. Consequently, it appears that the economic measures of demand management, including congestion charging, may only be effective at airports with relatively free access of a large number of airlines operating point-to-point networks with a large number of flights.

4.8.3 Charging for Congestion at Airports

People prefer to travel by air at some times more than at others. The airlines satisfy such needs by scheduling flights with a given seat capacity. When the number of flights exceeds the airport's capacity congestion occurs, indicating a constraint on the airport's capacity. The obvious consequence of congestion is delays to flights – aircraft and passengers. Congestion happens during the period of time called a peak when many aircraft/flights may interfere with each other. A measure of such interference is the demand/capacity ratio, that is the capacity utilisation ratio, defined as the quotient between demand and capacity, which may take values lower, equal to or greater than one. Specifically, if the number of flights is equal to the airport capacity, this ratio is equal to 1.0 or 100 per cent (Newell, 1982). In addition, during some peaks this ratio may reach or even exceed this value, indicating the potential for long delays (FAA, 2001, 2002a). Such cases need additional management of demand, capacity and/or of both. Charging for congestion could be one of the short-term measures (Vickery, 1969). However, although it is a theoretically mature economic concept, congestion charging is still not implemented at any of the congested airports. The main reasons seem to be an ultimate collision with the airport's overall objectives, the complexity of measuring the charging-relevant conditions, ambiguity of the concept and barriers within the industry.

4.8.3.1 Ultimate collision with the airport objectives
The prime aim of most airports is to grow, due to the internal – economic – as well as wider external – economic and political – regional and national interests. Such a policy combined with the regional policies encourages attracting the greatest possible traffic volumes. If such volumes occur according to expectations then congestion is likely to become a problem and the physical expansion of the airport's capacity is the usual long-term solution. At present two factors favour such a policy at many airports: first, the pricing of services, based on recovering the airport average or marginal operational costs, neither controls nor deters excessive demand, that is the pricing mechanism does not force users-flights to cover the full social costs of their choice; second, the benefits of airport growth often over-shadow the real dimension of the associated environmental and social costs – damages. Consequently, constraint of demand by charging for congestion might be considered as a threat to prospective growth.

Nevertheless, this attitude is likely to change particularly at congested airports with no congestion-free alternative and no land for further physical expansion, and also with the current short-term environmental and social barriers, which may become permanent. Under such circumstances, a reasonable non-threatening measure for managing growth appears to be stimulating changes to the fleet structure by introducing larger aircraft carrying out a smaller number of flights but serving almost the same or higher numbers of passengers. Charging for congestion, combined with other measures of demand management may alleviate such a fleet change, including redistribution of demand during the day, both of which should mitigate congestion and reduce delays.

4.8.3.2 Difficulty of measuring the charging-relevant conditions
Peaks in which the demand/capacity ratio may be very high are shown to differ at particular airports in terms of frequency and duration, type of operation, and the airlines and aircraft involved.

Short, sharp and frequent peaks are mostly artificially created by the airline hub-and-spoke operations. Infrequent and sometimes long but less sharp peaks are created by an airline's point-to-point operations. In the former case, a single airline or a few airlines operating hub-and-spoke networks internalise the costs of marginal delays of their flights. Thus, it is questionable if such peaks should be 'modified' by imposing a congestion charge on the existing and new-additional flights, since it might compromise the efficiency and development of these networks. Some airlines that are not willing to accept such a compromise may simply consider leaving the airport. In the case of infrequent but longer peaks, the flights of several airlines may interfere and impose marginal delays on each other during congestion. Since the costs of these delays are not internalised, the situation seems to be convenient for charging. If charging is to occur, however, a criterion for determining the level and causes of the congestion needs to be specified. Currently, congestion causing delays of up to 15 minutes is not relevant. At most airports, such delays occur when the demand/capacity ratio reaches around 85–90 per cent. Much greater congestion causing longer delays happens when demand/capacity ratio has reached or exceeds 100 per cent. Consequently, this level could be considered as a congestion-charging threshold. However, such an increase is very often caused by a combination of factors, some being out of the control of air transport operators such as, for example, bad weather. It may particularly affect the airport capacity and consequently increase the demand/capacity ratio above the prescribed threshold. The partial contribution of this as well as other factors is sometimes difficult to precise extract and the responsibilities allocated and consequently charged (ACI Europe, 2002; Odoni and Fan, 2002, Janić, 2003).

According to the type of operations, airlines and aircraft involved, congestion and delays caused by both regular – planned – and irregular – disturbed – operations of one or several airlines may be transferred between arrival and departure flights, spill out from one peak to another, and be transferred between airports along the aircraft daily itinerary. Under such circumstances, the allocation of responsibilities for the contribution of each single flight to the marginal delays of other flights throughout the network and setting up local congestion fees may become extremely complex.

4.8.3.3 Ambiguity of the concept
Charging of airport congestion seems to be an ambiguous concept for several reasons. Firstly, a charge should be equivalent to the cost of the marginal delays, which a flight imposes on other flights during the congestion. The aim is to make this flight unprofitable and thus prevent its access to the airport during the congestion period. This seems to be in conflict with the currently guaranteed freedom of ultimately unlimited access to airports under ICAO agreements (Corbett, 2002). Second, the congestion charge is expected to be effective, which in the case of market imperfections may not be the case, that is the charge may simply either be too low to be effective or too high to unwillingly damper the extra elastic demand. Third,

the relations seem not to be clear and transparent between the congestion charge and the other internalised externalities at the airport, such as noise and air pollution including the possible relations of these to existing schemes of charging for airport services based, for example, on the aircraft take-off weight. Fourthly, a real benefit from the congestion charge may be questionable since, for example, under current conditions passengers, airline and airport benefit from a given flight while the flights and passengers on whom the marginal delays are imposed lose out. If a charge is introduced, a flight might not be realised and the other flights and passengers would benefit while the passengers from that particular flight, airline and airport would lose out. In any case, if a flight is not realised the congestion fee would not be collected. Thus, it appears to be a virtual fee, which raises the question of a fair distribution of the charge's benefits and losses. Other questions may relate to how the collected congestion fees are spent. On the one hand, the airport could use this fund for capacity expansion. If that happens then one of the sources of airport revenues – the congestion charge – would probably vanish, at least for a while. On the other hand, under current policies, the airport is not very likely to consider allocation of the fee outside the industry – for developing non-air alternatives. Finally, at this moment, it seems rather sensitive to impose the additional charges on the economically and financially vulnerable airline industry even though, from a social perspective, this might contribute to the more efficient utilisation of the scarce airport infrastructure capacity and thus improve the system's overall performance.

4.8.3.4 Barriers within the industry

The above-mentioned reasons have already contributed to building up strong barriers to charging for congestion within the air transport industry. These are (Adler, 2002):

- The institutional, organisational, political and legal barriers intended to protect monopolistic powerful hub airports in Europe, and the hubbing airlines both in Europe and the US. These barriers also include the lack of harmonisation of charging conditions across European countries and across airports of different size both in Europe and the US;
- Unacceptability of the concept for large airlines, their alliances and lobbying groups due to the lack of similar concepts deployed by other transport modes, including roads, both in Europe and the US; and
- Technological barriers to the collection of relevant data on the actual causes of airport congestion and delays including precise data on the capacity of European airports under different circumstances. However, relatively useful databases for this purpose exist in the US (FAA, 2003a).

4.8.4 Modelling Congestion Charging at an Airport

4.8.4.1 The state-of the art and assumptions

The concept of charging for congestion at airports has been substantially investigated. Economic theory supports a thesis that the optimal use of a congested transport facility – in this case an airport – cannot be achieved unless each user-flight pays social costs equivalent to the marginal delay costs it imposes on all other users-flights during the congestion period. In the 1990s, the cost of marginal delays was considered as an externality to be internalised together with other externalities such as air pollution, noise and air traffic accidents (Vickery, 1969; Daniel, 1995; EC, 1997; ECMT, 1998; Daniel and Pahoa, 2000; Daniel, 2001; Odoni and Fan, 2002; Adler, 2002; Brueckner, 2002). In addition, the problem of congestion charging at airports where airlines might have different levels of dominance in terms of the market share and under the assumption that they have already internalised their congestion cost has been elaborated (Brueckener, 2002).

The modelling of congestion charging at an airport, as presented in this section, builds upon previous research and is based on several assumptions, which are seemingly able to handle the realistic congestion situation, as follows:

- At the airport, both the demand and the capacity change over time and both are known for a typical or representative congestion period. Their noticeable variations happen even during short periods of time such as one hour or less;
- The demand profile during the congestion period can be obtained either from the published airport and airline schedule or by recording the process of handling flights at the airport in question;
- Each flight taking part in the congestion is specific regarding the average operational cost and revenue, both of which depend on the airline, aircraft type-capacity, and the number of passengers on board. This implies differences in users-flights, which has not always been sufficiently emphasised in previous research, particularly not in the models based on the road congestion analogy;
- The airport runway system is a critical element of flight congestion and delays. Its capacity may change during the congestion period. The runway system capacity can be determined for given IMC or VMC conditions using the past, current or forecasted – expected – data obtained from the airport and/ or air traffic control operator. This capacity reflects the service time of specific arriving and/or departing flights;
- The congestion that occurs as the demand/capacity ratio approaches or exceeds 1.0 (100 per cent), with the potential for serious delays, is considered relevant for charging. This congestion can be of any type in terms of form and duration, thus including airports serving both point-to-point and hub-and-spoke networks. These characteristics of congestion can be used as criteria for selecting the candidate airports for charging congestion, which has not been particularly discussed in previous research (Adler, 2002); and
- The number of flights is large – at least several dozen, which makes congestion predictably dependent on the variations and positive difference between demand and capacity. For example, for the non-stationary Poisson arrival/ departure processes, if the numbers of users-customers during a given period are greater, the random variations of such greater numbers will be smaller. This enables application of a queuing model based on diffusion approximation, considered in this paper as the most convenient analytical tool to estimate the congestion and delays relevant for charging (Newell, 1982; Hall, 1991).

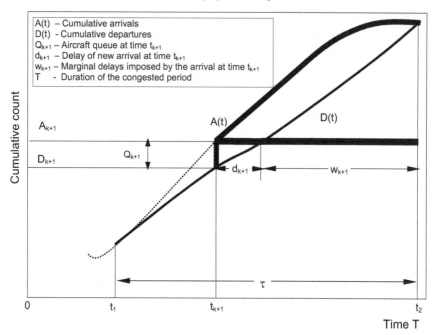

Figure 4.8.1 Typical congestion process at an airport

4.8.4.2 The modelling components
Modelling congestion charging at an airport implies estimating the system's total social costs of delays during a congested period, which embrace i) the private cost of delay of each flight, and ii) the cost of marginal (additional) delay each flight imposes on other subsequent flights (Hall, 1991; Ghali and Smith, 1995). For such purposes, it includes development of three component-models: i) a model for congestion, ii) a model for the system delays and their costs, and iii) a model for the assessment of the feasibility of realisation of each flight during the congestion period after a charge has been imposed.

The model for congestion   Up to date, congestion at transport facilities is usually modelled by using queuing and simulation models. Newell (1982); Odoni and Bowman (1997), and Hall (1989) all use one of the models based on a diffusion approximation to deal with severe congestion. In this paper, this model is modified in order to appropriately handle the problem of quantifying the congestion to be charged at a given airport.
   Figure 4.8.1 is a visualisation of this model as a graphical representation of a typical congestion process at a given airport during congestion period *T*. For the 'point-to-point' airport, period *T* may last for several hours or even for the whole day. For the 'hub-and-spoke' airport, period *T* may relate to the duration of one 'wave' of incoming and outgoing flights of one or a few airlines. As can be seen, the lines $A(t)$ and $D(t)$ represent the cumulative count of those flights requesting service and those being served, respectively, by time (t). Since the cumulative number of flights is assumed to be large ($\gg 1.0$) both types of counts, actually the step

functions of time, can be considered as their continuous (smooth) counterparts. The curves $A(t)$ and $D(t)$ can be derived either as a single realisation or as the average of many realisations in period T. Since both curves change in time, their relationship also changes. By deriving functions $A(t)$ and $D(t)$ with respect to the variable (t) the intensity of demand and the capacity can be obtained, respectively, as follows: $\lambda(t) = dA(t)/dt$ and $\mu(t) = dD(t)/dt$. Consequently, the demand capacity ratio $\rho(t)$ as the time-dependent function can be as obtained as $\rho(t) = \lambda(t)/\mu(t)$.

In Figure 4.8.1, dependent on the values of the function $\rho(t)$ compared with the 'critical' value 1.0, three sub-periods can be observed. In the first one $(0, t_1)$, $\rho(t)$ is less than *1.0*, that is $\lambda(t) < \mu(t)$ implying that only 'random effects' cause congestion. The curve $A(t)$ lies below curve $D(t)$. During the second sub-period $\tau \equiv (t_1, t_2)$, $\lambda(t)$ becomes first equal to and after that greater than $\mu(t)$, the function $\rho(t)$ becomes first equal to and then greater than 1.0, respectively. Consequently the curve $A(t)$ equalises and goes above curve $D(t)$. In this case, 'deterministic effects' dominate as causes of congestion completely over-shadowing the previously dominant 'random effects'. Finally, during the sub-period (t_2, T), again $\lambda(t) < \mu(t)$ and the ratio $\rho(t)$ drops below *1.0*. The curve $A(t)$ needs additional time to equalise and then drop below curve $D(t)$. As soon as this happens, it is an indication that the 'random effects' dominate again as a cause of congestion, similar to sub-period $(0, t_1)$. Obviously, only congestion during the period T should be considered for charging since it will certainly produce delays longer than the threshold of 15 minutes (Newell, 1982; Hall, 1991). To estimate this congestion and associated delays dependent on time, the period T is divided into K increments Δt (that is $K\Delta t \approx T$). Increment Δt has two features: i) first, it is sufficiently short in comparison with sub-period T to enable registering of changes of congestion and delays, and ii) it is sufficiently long to guarantee independency between the cumulative flight arrival and the cumulative flight departure process, including their independency during the successive increments. For example, if (τ) is a period of several hours during the day, Δt will certainly be a quarter, a half or an hour. Under such circumstances, the functions $A(t)$ and $D(t)$ can be treated as the processes of independent increments or the diffusion processes (Newell, 1982).

If the differences between the cumulative flight demand and the cumulative airport capacity in (k)th and $(k+1)$st time increment Δt, $A(k + 1) - A(k) \equiv A_{k+1} - A_k$ and $D(k + 1) - D(k) \equiv D_{k+1} - D_k$, respectively, are the stochastic variables with a normal probability distribution, the difference $Q_{k+1} = A_{k+1} - D_{k+1}$, which represents the queue in the $(k+1)$st increment Δt, will also be the stochastic variable with a normal probability distribution $(k \in K)$ (Newell, 1982). Consequently, the queue of flights in the $(k+1)$st increment Δt can be approximated as follows:

$$Q_{k+1} = Q_k + \bar{Q}_{k+1} + B_{k+1} = Q_k + (\lambda_{k+1} - \mu_{k+1})\Delta t + B_{k+1} \text{ for } k = 0, 1, 2, , , , K-1 \quad (4.8.1)$$

where

Q_k is the actual queue in (k)th increment Δ_t;

$\bar{Q}_{k+1}$ is the average queue in $(k+1)$st increment Δ_t;

λ_{k+1} is the intensity of flight demand in $(k+1)st$ increment Δt;

μ_{k+1} is the airport capacity, that is the flight service rate, in $(k+1)st$ increment Δt;

B_{k+1} is the anticipated deviation of the flight queue from its average value in $(k+1)st$ increment Δt.

Equation 4.8.1 indicates that the average queue of flights either increases or decreases in accordance to the behaviour of $\lambda_{k+1} > \mu_{k+1}$ or $\lambda_{k+1} < \mu_{k+1}$. The anticipated deviation B_{k+1} can be estimated as follows (Newell, 1982):

$$B_{k+1} \cong \sqrt{\Delta t(\sigma_{a,k+1}^2 / \overline{t}_{a,k+1}^3 + \sigma_{d,k+1}^2 / \overline{t}_{d,k+1}^3)} * C \quad \text{for } k = 0, 1, 2, ..., K-1 \qquad (4.8.2)$$

where

$\overline{t}_{a,k+1}; \overline{t}_{d,k+1}$ is the average inter-arrival and service time, respectively, of a flight in $(k+1)$st increment Δt;

$\sigma_{a,k+1}; \sigma_{d,k+1}$ is the standard deviation of inter-arrival and service time, respectively, of a flight in $(k+1)$st increment Δt;

C is a constant ($C = \Phi^{-1}(1 - p)$, where Φ^{-1} is the inverse laplace function and p is the probability that the queue in the $(k+1)$st increment Δt will spill out of the confidence interval ($\overline{Q}_{k+1} \pm B_{k+1}$).

In Equation 4.8.2, the variances of distributions of the flight inter-arrival and flight service time are assumed to be independent in the successive (k)th and $(k+1)$st increment Δ_t (Newell, 1982).

In Equation 4.8.1 and Figure 4.8.1, at the beginning of period T, the function $\rho(t)$ takes the value one for the first time, and the deterministic queue begins to build up. However, this queue continues to the queue built up due to the previously dominating 'random effects'. The already existing queue $\overline{Q}_0$ can be approximated as follows (Newell, 1982):

$$\overline{Q}_0 \equiv Q_{m/(\lambda_m = \mu_m)} = \left\{ \left(\frac{1}{\left[(\sigma_{a,m}/\overline{t}_{a,m})^2 + (\sigma_{d,m}/\overline{t}_{d,m})^2 \right]^2} \right) * (1/\mu_m) * (d\rho_m / dt) \right\} \quad (4.8.3)$$

where

m is the time increment Δt in which the intensity of flight demand becomes for the first time equal to the flight service rate, that is to capacity ($m \in K$).

Other symbols are analogous to those in the previous expressions.

Similar expressions (1)–(3), if reasonable, can be applied to determine the flight queue for the frequent congestion periods in which the demand/capacity ratio $\rho(t)$ is close, equal to or greater than one, which is the characterisation of the hub-and-spoke operations.

The model for the system delays and their costs   From the Equations 4.8.1–4.8.3, delay of a flight joining the queue in the $(k+1)$st increment Δ_t can be approximated as follows:

$$d_{k+1} = Q_{k+1} * (\bar{t}_{d,k+1} + B_{d,k+1}) = Q_{k+1} * \left[\bar{t}_{d,k+1} + \sigma_{d,k+1} * \Phi^{-1}(1-p)\right] \qquad (4.8.4)$$

where the symbols are as in the previous expressions.

Equation 4.8.4 assumes that the flight service rate (that is the airport capacity) does not change during the service of queue Q_{k+1}. In addition, from Figure 4.8.1, the marginal delay, a flight arriving during the $(k+1)$st increment Δt imposes on all subsequent flights until the end of the congested period T can be determined as:

$$w_{k+1} \cong T - \left[(k+1)\Delta t + d_{k+1}\right] \equiv (\bar{t}_{d,k+1} + B_{d,k+1}) * \sum_{l=k+1}^{K} \left[1/(\bar{t}_{a,l} + B_{a,l}\right] * \Delta t =$$

$$= \left[\bar{t}_{d,k+1} + \sigma_{d,k+1} * \Phi^{-1}(1-p)\right] * \sum_{l=k+1}^{K} \left\{1/\left[\bar{t}_{a,l} + \sigma_{a,k} * \Phi^{-1}(1-p)\right]\right\} * \Delta t \qquad (4.8.5)$$

where all symbols are as in the previous expressions.

Let M airlines schedule their flights during the congestion period τ. The flights of each airline are uniformly distributed over this period. Consequently, the number of flights scheduled by these airlines from the $(k+1)$st increment Δt to the end of the congestion period τ ($\tau \equiv K\Delta t$) is equal to: $N(\tau - k\Delta t) = \sum_{i=1}^{M} N^i(\tau - k\Delta t)$, where $N^i(\tau - k\Delta t)$ is the number of flights of an airline (i). If an additional flight is scheduled by airline (i), the total marginal cost this flight will impose on the succeeding flights during the period ($\tau - k\Delta t$) can be determined as follows:

$$C^i_{m,k+1} = \left[1 - N_i(\tau - k\Delta t)/N(\tau - k\Delta t)\right] * \left[\bar{t}_{d,k+1} + \sigma_{d,k+1}\Phi^{-1}(1-p)\right] *$$

$$\sum_{l=k+1}^{K} c_l(n_l) * \left\{1/\left[\bar{t}_{a,l} + \sigma_{a,l}\Phi^{-1}(1-p)\right]\right\} * \Delta t \qquad (4.8.6)$$

where

$c_l(n_l)$ is the average cost per unit of delay of a flight with the capacity of (n_l) seats scheduled in the (l)th increment Δt (in monetary units per unit of time); this cost may include the cost of aircraft and passenger delay.

Other symbols are as in previous expressions.

The model for assessment of feasibility of a given flight   The congestion charge is efficient if it compromises the expected profitability of a new flight. Thus, the charge deters access of a new flight or removes a flight from the airport during the congested period. If the charge is $C^i_{m,k+1}$ and if the average cost per unit of time of a given flight of capacity (n) of airline (i) in the $(k+1)$-th increment Δt is $c^i_{k+1}(n)$, the total social cost imposed by this flight will be as follows:

$$C^i_{f,k+1} = c^i_{k+1}(n) * \left[t^i_{f,k+1} + d_{k+1}\right] + C^i_{m,k+1} \qquad (4.8.7)$$

where

$t^i_{f,k+1}$ is the duration of a given flight of airline (i) scheduled in the (k+1)st increment Δt.

Other symbols are as in the previous expressions.

If the airfare includes the cost of the congestion charge, the expected revenue from a flight can be estimated as follows:

$$R^i_{f,k+1} = p^i_{k+1}(L,C^i_{m,k+1}) * \lambda^i_{k+1}\left[p^i_{k+1}(L,C^i_{m,k+1})\right] * n^i_{k+1} \qquad (4.8.8)$$

where

$p^i_{k+1}(L,C^i_{m,k+1})$ is the average airfare 'corrected' by the congestion charge $C^i_{m,k+1}$ imposed on a given flight scheduled by airline (i) on route (L) in the (k+1)st increment Δt;

$\lambda^i_{k+1}[p^i_{k+1}(L,C^i_{m,k+1})]$ is the expected load factor influenced by the 'corrected' airfare of a given flight of airline (i) in the (k+1)st increment Δt;

n^i_{k+1} is the seat capacity of a given flight of airline (i) in the (k+1)st increment Δt.

The 'corrected' airfare $p^i_{k+1}(L,C^i_{m,k+1})$ in expression 4.8.8 can be determined as follows:

$$p^i_{k+1}(L,C^i_{m,k+1}) = p^i_{k+1}(L) + C^i_{m,k+1} / n^i_{k+1} \qquad (4.8.9)$$

where

$p^i_{k+1}(L)$ is the basic average airfare of a given flight of airline (i) on route (L) in the (k+1)st increment Δt.

From Equations 4.8.7 and 4.8.8 it follows that a given flight will be unprofitable under the following condition:

$$R^i_{f,k+1} - C^i_{f,k+1} = p^i_{k+1}(L,C^i_{m,k+1}) * \lambda^i_{k+1}\left[p^i_{k+1}(L,C^i_{m,k+1})\right] * n^i_{k+1} - c^i_{k+1}(n) * \left[t^i_{f,k+1} + d_{k+1}\right] - \\ - C^i_{m,k+1} \leq 0 \qquad (4.8.10)$$

where all symbols are as in previous expressions.

If charge $C^i_{m,k+1}$ determined by the Equation 4.8.6 does not fulfil the condition 4.8.10, it should be increased to at least make the expected profit per flight negative, other factors remaining constant.

4.8.5 Application of Models for Charging Congestion

4.8.5.1 Setting up the case

The proposed models for charging for congestion at an airport are applied to the case of New York's (NY) LaGuardia Airport. This is an example of a heavily congested airport with the most advanced ideas about introducing this concept (Odoni and Fan, 2002). As one of the three largest airports in the New York area LaGuardia serves the US domestic short- and medium-distance 'point-to-point' traffic. It is the origin and destination of about 92 per cent of its total flights, carrying about 45–55 per cent business passengers. One of the main driving forces of this type of traffic is the proximity of the airport to Manhattan, the centre of New York City – about 18 km away. After the September 11/2001 terrorist attack there was an immediate sharp decline in traffic, but this has gradually recovered and reached, by the end of 2002, the annual number of about 22 million passengers and 358 thousand flights. The average number of passengers per flight has always been relatively stable, from 58 to 62, during the past five years (58–62) (PANYNJ, 2003).

Currently, 20 airlines operate at the airport. Three have the greatest market share in terms of the number of flights and the number of passengers, respectively: US Airways (38 per cent; 14.2 per cent), Delta (18 per cent; 17.2 per cent), and American (17 per cent; 18.5 per cent). Two right angle-crossing runways, each 7,000 ft (2,135 m) long, mostly influence the type of the fleet and the length of the routes-markets served to/from the airport. The fleet mostly consists of aircraft types B737/717 and A320 with a capacity of 100–150 seats, and of the smaller regional jets and turboprops with a capacity of 70–110 seats. The average route length to/from LaGuardia Airport is about 1,200 km (Backer, 2000; PANYNJ, 2003).

The airport runway capacity is about 80 (40/40) aircraft movements – flights – per hour under VMC and 64 (32/32) aircraft movements per hour under IMC rules. The aircraft are accommodated at 60 apron parking stands.

The intensity of hourly flight demand frequently exceeds the airport capacity of both the runway system and the apron and causes severe congestion. Since there is no available land, the options for relieving such congestion by physical airport expansion, under conditions of predicted growth of about 19 per cent by the year 2010 in comparison with the year 2002, appear to be very limited. Possible options actually consist of increasing the average aircraft size and increasing the runway capacity by innovative operational procedures and technologies. The former option already took place in the year 2001 after flights by the aircraft B767-400ER, with a seat capacity of about 280 seats, were introduced (AN, 2001). The latter option is still waiting to take place and is expected to raise the runway capacity by about 10 per cent under VMC and by about 3 per cent under IMC rules (FAA, 2003a). None of the options includes demand management by modifying the current service charging system based on aircraft weight. The unit charge is set at $6.55 for each 500 kilograms or thousand pounds of the aircraft's maximum take-off weight. Each flight is additionally charged a fixed amount of USD100 if it arrives between 8 a.m. and 9 p.m. (PANYNJ, 2003a). Nevertheless, neither of the above-mentioned options seems to efficiently cope with the prospective long-term growth of demand beyond the year 2010. This may again instigate a search for economic measures of demand management based on past experience. For example, the initial trials of slot auction, that is 'slottery', substantially mitigated congestion in the year 2000. For the future, charging for congestion might be reconsidered. The following numerical example illustrates what its effects might be.

4.8.5.2 Inputs

Two groups of inputs are used for the application of the proposed models: i) inputs on the demand and capacity for estimating the congestion and delays under given

**Table 4.8.1   Parameters of distributions of the flight inter-arrival and
inter-departure time in a given example: New York LaGuardia
Airport (US)**

Time of the day	Demand		Capacity	
	Flight inter-arrival time		Flight service time	
Hour (k)	Mean ($t_{a,k}$) (s/flight)	St Dev.($\sigma_{a,k}$) (s/flight)	Mean ($t_{d,k}$) (s/flight)	St Dev ($\sigma_{d,k}$). (s/flight)
1	–	–	–	–
2	–	–	–	–
3	–	–	–	–
4	–	–	–	–
5	–	–	–	–
6	50.72	9.972	52.56	7.488
7	52.20	3.942	52.92	7.524
8	50.76	3.123	52.20	4.608
9	49.68	4.716	52.56	7.776
10	50.04	4.860	52.20	7.164
11	50.40	1.764	51.12	6.912
12	50.76	2.376	51.12	6.984
13	48.96	3.096	50.76	6.912
14	51.84	3.744	50.76	6.336
15	50.04	3.312	50.40	7.056
16	48.24	2.916	50.40	7.020
17	48.60	5.148	50.04	7.022
18	51.48	8.640	50.04	7.020
19	50.76	5.292	50.40	7.704
20	51.84	7.992	49.68	6.624
21	59.67	5.220	49.32	6.012
22	78.12	16.236	49.32	5.976
23	23.84	36.468	50.40	7.308
24	–	–	–	–

s – seconds
Source: FAA (2003); Janić (2005)

circumstances; ii) inputs on the aircraft operating costs and airfares for assessing
profitability of particular flights.

Inputs for estimating congestion and delays   The distribution of the hourly number
of flights and corresponding capacity at NY LaGuardia airport for every day in
July 2001 are used as inputs for the proposed queuing model, based on a diffusion
approximation to estimate congestion and delays. These distributions are determined
by using 31 daily traffic realisations at NY LaGuardia (FAA, 2003, 2003a). Each
distribution for each hour is assumed to be normal or nearly normal and independent
of the others. Table 4.8.1 gives the main parameters of these distributions for an
average day. In addition, constant C is set to be 1.96 in all experiments, which
implies that the queues stay within given confidence boundaries with a prescribed
probability of 0.95 (Newell, 1982).

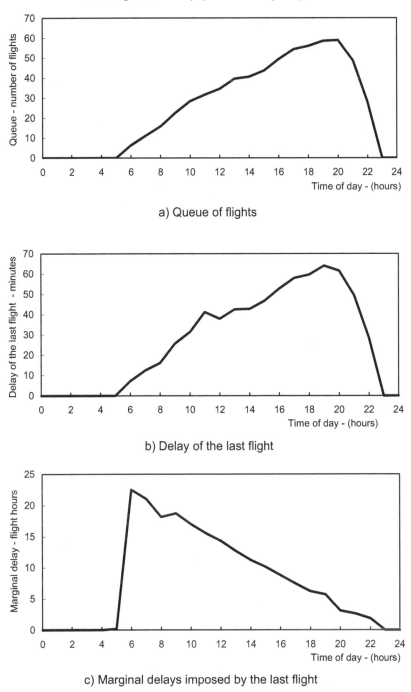

a) Queue of flights

b) Delay of the last flight

c) Marginal delays imposed by the last flight

Figure 4.8.2 The system congestion and delays in a given example: a) Queue of flights; b) Delay of the last flight; c) Marginal delays imposed by the last flight

Aircraft/flight operating costs   Aircraft/flight operating costs are standardised according to a seat capacity metric and expressed in monetary units per block hour. The cost relevant data related to the US airlines are obtained by the convenient linear regression technique as follows: $C = 21.97S + 11.99; R^2 = 0.934; N = 150$, where the independent variable is the cost per flight C and the dependent variable is the number of seats per flight S. The size of the sample is N (FAA, 1998). It indicates that the cost per flight almost linearly increases with the increasing aircraft seat capacity – size. For example, as the regression equation shows, the average cost of an aircraft/flight of 100–150 seats such as B737 or B717 at NY LaGuardia Airport varies between USD2,209 and USD3,307/h or from USD37 to USD55/min, respectively. The average cost of an aircraft of 280 seats such as for example B767-400 ER is USD6,162/h or USD103/min. This cost does not include the cost of passenger time.

Airfares   The average airfare per passenger at NY LaGuardia Airport is determined by the regression technique applied to the 1998 US data, modified for changes in the value of the US dollar in the year 2002 as follows: $p(L) = 9.56L^{0.3903}$; $R^2 = 0.940$; $N = 28$ (Cheng-Sheen, 2000; Mendoza, 2002). As can be seen, the average airfare increases at a decreasing rate with increasing route length, which reflects a decrease in the average aircraft unit cost with an increasing in the non-stop flying distance (Janić, 2001). At NY LaGuardia Airport, for an average flight of about 1,200 km the corresponding airfare should be USD152 (Mendoza, 2002). These airfares do not include the congestion charge.

4.8.5.3 Results
The results from the experiments with the model are shown in Figures 4.8.2, 4.8.3, and 4.8.4. Figure 4.8.2 shows congestion and delay created by a flight joining a queue at different times during the day.

Figure 4.8.2a shows that during an average day the queue of flights starts to develop early in the day, immediately after opening the airport at 6 o'clock in the morning. If the airport operates at its declared capacity, the queue grows gradually during the day and reaches a maximum in the evening around 20:00 hours. After that time the intensity of demand significantly diminishes, and so the long queue disappears relatively quickly, up to 23 hours. In this case, an average of 35 flights and a maximum of 59 flights wait in the queue.

Figure 4.8.2b shows the flight delays during the day. As can be seen, the delay of the last flight in the queue changes in line with the change of queue length. The airport's declared capacity makes the average and maximum delay per flight about 35–40 and 65 minutes, respectively.

Figure 4.8.2c illustrates variations of the marginal delay due to changing the scheduling time of the last flight in the queue. As can be seen, if scheduled early in the morning, this flight will impose a longer marginal delay than otherwise. In the given example, scheduled at 6 o'clock in the morning, the flight will impose an additional delay of about 22 flight-hours on the subsequent flights by the end of the congestion period – in this case the day. If scheduled later during the day, the flight will impose smaller additional delays, as intuitively expected, since it will affect a

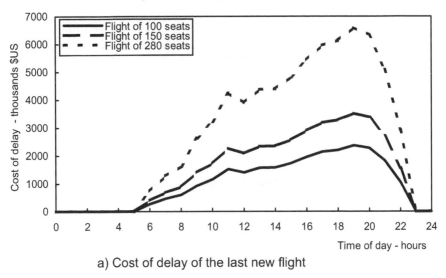

a) Cost of delay of the last new flight

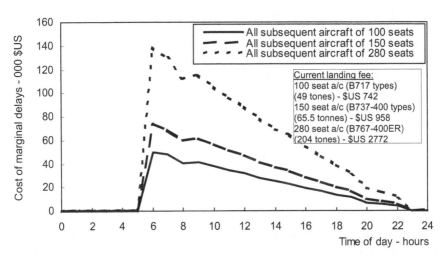

b) Cost of the marginal delays imposed by the last-new flight

Figure 4.8.3 The system cost of the last new flight in a given example: a) Cost of delay of the last new flight; b) Cost of the marginal delays imposed by the last new flight

smaller number of subsequent flights. The average marginal delay imposed by a flight scheduled at any time during the day on other flights is about 10–12 flight-hours.

Figure 4.8.3 shows the social cost generated by the last flight joining the queue at NY LaGuardia Airport under given traffic conditions.

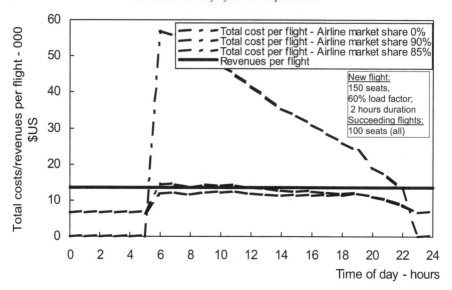

Figure 4.8.4 Conditions of profitability of an additional flight in a given example

In particular if a greater number of more expensive – higher – capacity flights arrive behind a given flight. For example, a flight entering the queue around 6 o'clock in the morning will impose a marginal cost of delays of about USD50, USD75 and USD150 thousands on all subsequent flights with a capacity of 100, 150 and 280 seats, respectively. Obviously, the difference between the cost of marginal delays and the current cost of landing based on the aircraft weight is considerable.

By summing up the cost of delay in Figure 4.8.3 (a, b), the total social costs of a given flight can be estimated.

Figure 4.8.4 shows the conditions of profitability of a flight joining the queue at NY LaGuardia under given conditions from the airline point of view. A congestion charge is imposed on such a flight of 2 hours and capacity of 150 seats.

The flight operational cost, not including the cost of passenger time, is USD3300. If the average load factor is 60 per cent and the average airfare is USD152, the revenue of this flight will be USD13680 (heavy line in Figure 4.8.4). If the airline does not have any market share at the airport, this flight, fully burdened by the congestion charge based on the cost of marginal delays imposed on all the succeeding flights with a capacity of 100 seats, will be unprofitable if it arrives at any time during the day before 22:00 hours. However, if the given airline already has a significant market share, for example, about or higher than 85–90 per cent, the given flight might be profitable under the same conditions.

This result confirms some doubts that the congestion charging may disfavour competition at airports by imposing unacceptably high burdens on new entrants and only a modest burden on the airlines that already hold a high market share at the airport. For the given traffic scenario, the charge particularly discourages the flights of new entrants carried out by smaller aircraft earlier during the congested period. This is particularly true if they arrive before the larger aircraft, and vice versa. If these flights are by the airlines that already have a high market share, such flights will further strengthen the position of these airlines at the given airport.

4.8.6 Concluding Remarks

This section has presented the models for charging for congestion at an airport. Currently, congestion charging is not practised at airports despite introducing the different charging mechanisms for peaks and off-peaks. Charging for congestion would actually mean internalising one of the externalities of the air transport system – time losses of aircraft/flights and passengers due to interference and competition for the scarce resources – constrained airport infrastructure capacity. The models developed have been: based on the queuing diffusion approximation theory used to quantify the relevant queues and delays during the congestion period. The models have been applied to New York's (NY) LaGuardia Airport (US).

The application indicates that the proposed models could be efficiently used to deal with the problem of charging for congestion at airports like NY LaGuardia. In particular, the queuing model based on diffusion approximation has enabled realistic quantification of flight queues and delays. Two other models have estimated the congestion charge and profitability of a particular flight for the given traffic (congestion) scenario and charging conditions.

The results obtained from the particular models have shown that congestion charging could be used as a demand-management economic instrument at airports under the following conditions:

Congestion is exclusively the consequence of the relationships between demand and airport capacity causing delays longer than the threshold of 15 minutes, that is when the demand/capacity ratio approaches or exceeds the value one.

Congestion should be created by many flights of different competing airlines in order to raise the need for internalising the marginal costs of delays these flights impose on each other since a single airline with many flights has already internalised the costs of the marginal delays of these flights. The airports with many competing airlines performing point-to-point operations are therefore more likely candidates for a congestion-charging scheme than the airports with a few airlines operating hub-and-spoke networks even though in both cases the congestion can be significant and relevant for charging.

The congestion charge appears to be effective in preventing access earlier during the congestion period of flights carried out by smaller aircraft before flights carried out by larger aircraft, and vice versa. This implies that the congestion charging stimulates exclusion of the earlier arrivals by smaller-regional aircraft and instead favours the use of larger aircraft that are less sensitive to the time of arrival.

Congestion charging appears to encourage additional flights of airlines which already have a significant airport market share, that is those with more flights and therefore already internalised cost of marginal delays. In this sense it contributes to consolidation of the market position of the incumbents, discourages new entries and thus compromises competition at the airport during the congested period.

Despite being theoretically clear regarding the distribution of benefits and losses, congestion charging looks pretty fuzzy if it was to be applied in practice, particularly if the potentially charged flights were not realised and tools not collected.

4.9 Comparison of the Environmental Performances of High-Speed Systems (HSS)

4.9.1 Background

The obsession of many people with increasing travel speed has extended the travel distance within a given time budget. For example, over the past two centuries, the travel distance has extended from about 50 km by horseback to about 650 km by car per day, and 200–650 km/h by High-Speed Rail (HSR), Air Passenger Transport (APT) and eventually TRANSRAPID MAGLEV (Ausubel and Marchetti, 1996).

In addition to the increasing travel speed and improvement in efficiency and effectiveness of transport services, HSR and Apt, particularly in the European Union (EU), have been expected to significantly contribute to the micro- and macro-spatial, socio-economic and political development of particular regions through their cohesiveness, integration, internationalisation, and globalisation (CEC, 1993, 1995). However, in addition to the obvious benefits to the travelling public, both systems have created impacts on people's health and the environment (flora, fauna, ecosystems) at the local, regional, and global scale, which in general have imposed both preventive and remedial costs (externalities) on society (CEC, 1993; AEAT, 2001; EEC, 2001b). The most common direct impacts include energy consumption, air pollution, land-take (use), noise, safety, congestion and delays. As mentioned throughout this book, the last three have also often been categorised as the socio-economic burdens (CEC, 1993; EEC, 2001b). In many cases, particular impacts and their externalities have been used for the assessment of the effects of the substitution of operations by alternative transport modes aiming to mitigate their current and prospective impacts. HSR, for example, has always been considered as a potentially viable High-Speed (HS) alternative to Apt (Levison et al., 1996; EC, 1998).

This section presents an overview of the environmental performances of High Speed Systems – HSR and Apt – in the EU based on the marginal values of particular impacts. The marginal values are those generated by a single HS vehicle – train and/or aircraft – under the most common operating conditions. These values are expressed by the 'quantity of impact per unit of output' usually adopted to be p-km (passenger-kilometre). In addition, some figures for the marginal costs of potential environmental damages caused by particular impacts are provided for comparative purposes.

4.9.2 Development of High-Speed Rail and Air Passenger Transport

The High-Speed Rail (HSR) and Air Passenger Transport (APT) have developed in the European Union (EU) during the last half of the twentieth century.

APT emerged as the HS option in the 1930s. Over time, the system has continuously modernised through development of aircraft capabilities, airline strategy and governmental regulation (Boeing, 1998).

The aircraft capabilities have included increasing speed, payload, and take-off-weight, which have all contributed to increasing the vehicle technical productivity, efficiency, effectiveness, and safety (Horonjeff and McKelvey, 1994; Boeing, 1998).

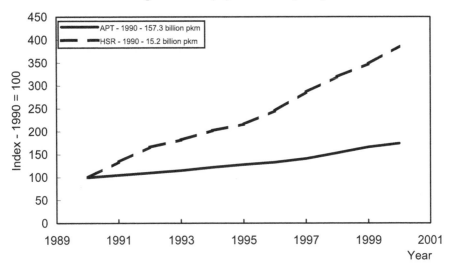

Figure 4.9.1 Development of APT and HSR traffic in the EU (1990–2000)
Compiled from EC (2002; 2004); UIC (2004)

The airline strategy has dealt with improvement of Apt efficiency, effectiveness and safety under conditions of relatively stable long-term growth in demand by both passengers and freight. Most airlines have introduced bigger, faster, safer, quieter, more fuel-efficient and less air-polluting noisy aircraft. The airports have increased their capacity to accommodate the growing demand through spatial expansion on the one hand and by using modern passenger, freight and aircraft service facilities and equipment on the other. Air traffic control has followed the above efforts by implementing innovative ground and airspace navigational aids and surveillance equipment (Horonjeff and McKelvey, 1994).

Governmental regulation has consisted of the gradual deregulation/liberalisation of the air transport market and creating conditions for privatisation of air transport operators (airlines, airports, and air traffic control). The objectives have been to support the system efficiency and effectiveness in the broadest sense, both at the national and international scale. Deregulation of the US air transport market in 1978 and gradual liberalisation of the European Union (EU) internal air transport market in 1997 appear to be the most important global regulatory achievements (Janić, 1997).

In general, over the past decade (1990–2003), APT demand in the EU has grown at an average annual rate of about 5 per cent. The heavy line in Figure 4.9.1 shows the general trend (EC, 2004).

The most recent forecasts have indicated that similar growth rates will be sustained and that the total volume of APT demand will double by the year 2015 (ATAG, 2000). The growth of demand has put pressure on the airport capacity and air traffic control capability, thus causing frequent congestion and delays, and increasing the impact of APT on the environment in terms of noise, air pollution, and land take.

HSR certainly represents a culmination in the increase in travel speed on the railways. Over the past century and a half, the railway achievable technical speed has increased by more than 10 times from about 50 to 500 km/h (Janić, 1999). In the

EU, certainly the most significant railways' attainment consists of the planning of a Trans-European HSR network with a length of about 29,000 km and an expected investment cost of about €240 billion (infrastructure plus rolling stock). The building started at the beginning of the 1980s with the prioritising of the national projects in France and is going to be completed by the year 2010, thus connecting cohesive intra-EU international links (CEC, 1995).

HSR passenger demand has grown during the past decade (1990–2000) at an average annual rate of 16 per cent (about three times faster than that of APT) (see dotted line in Figure 4.9.1). The growth has been strongly dependent, in addition to the other demand driving forces, on the progress in the building of the new HSR infrastructure. In addition, the rapid growth has enabled HSR to account for about 15 per cent of the total share in competitive passenger markets (CEC, 1995; UIC, 2002).

The third High-Speed system – TRANSRAPID MAGLEV – based on the idea of magnetic levitation and currently not operating in Europe is not considered (Kertzschmar, 1995).

4.9.3 Interaction between High-Speed Rail and Air Passenger Transport

Interactions between HSR and APT may materialise in the form of competition and complementarity. Both can result in the substitution of operations between the two modes, driven by their commercial viability, market strength, and different institutional and environmental constraints. The modal substitution might improve the internal efficiency of each particular system (substitution through competition), and reduce, in the broadest sense, their cumulative impact on the environment (substitution through complementarity) (EC, 1998).

4.9.3.1 Competition
Competition between HSR and APT in the European Union (EU) takes place in passenger transport market-corridors with sufficient demand. For such competition, airports at the ends of these corridors do not necessarily have to be included in the HSR network. The passengers make a choice of transport alternative based on the perceived generalised cost (out of pocket travel cost), dependent on factors such as the system's service accessibility costs, departure frequency, fares, and the overall quality of service (CEC, 1995). The alternative with the lower generalised cost is expected to be chosen. However, since other factors may also influence the final modal choice (type of travel – business/leisure, age – old/young, safety, and so on.), each HS alternative has always retained some market share. To date, in the short-haul, densely inhabited corridors of length between 300 and 500 km in which both systems operate, HSR has taken over a significant market share from APT. The most important factor influencing such a development seems to be the HSR travel time, which has been reduced to between two and four hours. Other factors have been competitive fares and departure frequencies. Figure 4.9.2 shows that the market share of HSR decreases almost more than proportionally with as travel time increases (ITA, 1991; Janić, 1993; EC, 1998, 2002; Janić, 1997; EEC, 2001b).

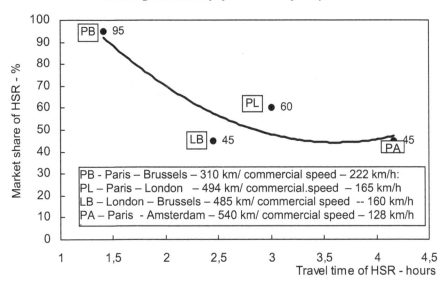

Figure 4.9.2 Market share of HSR in the corridors of lengths between 300 and 500 km

Compiled from EC (2002)

4.9.3.2 Complementarity

The HSR and APT may complement each other through offering so-called complementary services in particular markets. From the viewpoint of users – passengers, they are complementary if the combination, instead of a single transport mode, is preferred for travelling between given origins and destinations (ITA, 1991; EC, 1998). From the viewpoint of transport operators, HSR is a complement to APT if it replaces the short-haul 'feeder' flights connecting into and out of the long-haul flights by using 'feeder' trains according to a compatible (balanced) timetable (EC, 1998). Generally, three types of complementary networks, if commercially viable, can exist (Janić, 2003): i) HSR may partially replace APT in collecting and distributing passenger flows between a hub airport and particular spokes. One such example is Frankfurt Main airport (Germany), where many short haul domestic APT services have been replaced by equivalent HSR services; ii) HSR may completely substitute APT by providing 'feeder' services between a hub and spokes while APT exclusively connects hub airports to each other. One of the examples, which still has to take place is the APT connection between Paris (France) and Rome (Italy), which would be 'fed' by the HSR instead of short haul APT services (ITA, 1991; Janić, 2003); and iii) Apt may connect hub airports with spokes while HSR provides exclusive surface connections between hub airports themselves. One example is an HSR line connecting Paris CDG airport and Lyon-Satolas Airport in France, which is partially 'fed' by APT (air passengers) (ITA, 1991; EC, 1998; Janić, 2003).

Several conditions should be fulfilled for setting up the above networks:

- The airports should be connected to the HSR network;
- The timetables of HSR and APT should be coordinated; and

- Ticketing including convenient checking and transferring of passengers and their baggage between two the modes should be provided.

The first condition has been partially fulfilled by inclusion of some European hub airports into HSR network such as Frankfurt-Main, Lyon-Satolas, Paris CDG, Stockholm-Arlanda, and Copenhagen-Kastrup (EC, 2004). The latter two conditions are fulfilled through the so-called 'inter-modal code sharing agreements' between airlines and HSR operators, when airlines have acquired trains and operated them within their own timetables (EEC, 2001b).

4.9.4 The Environmental Performances – Quantification of Impacts

Operation of the HSR and APT create direct impacts, which may damage the environment (flora, fauna, ecosystems) and people's health. Direct impacts include energy consumption, air pollution, noise, land-take (use), safety and congestion. The first four have a physical dimension. The last two also have a social dimension. All have an economic dimension in the form of external costs of the perceived or real environmental damage. The level of direct impacts is primarily dependent on the duration of operations and the activities causing them. However, due to their depositing character, the short-term impacts cumulate and cause damage to the environment in the long term. Energy consumption and air pollution may cause both local and global damage as pointed out throughout this book. Other impacts may generally cause damage at a local level (EC, 1996; Janić, 1999).

4.9.4.1 Energy consumption
Energy is needed to power the vehicles used for transport operations. The HSR and APT vehicles, the HS trains and aircraft, use different types of energy. This, together with the diversity of technologies (vehicles capacity) and operating conditions (different route lengths and load factors), makes their comparison complex in terms of energy consumption. In order to overcome such complexity, certain HS trains can be compared with the aircraft types used on short- and medium-haul routes, where both systems may interact. In addition, it is assumed that 100 passengers are on board each vehicle, the HS train and an aircraft. This provides a unique basis for comparison of the two technologies and partially eliminates the diversity of operating conditions.

The HS trains are mostly electrically powered. Their marginal energy consumption (quantity of energy per unit of output – kWh/p-km) is mainly proportional to the cruising speed. It is lower during the accelerating/decelerating phase of a trip and higher but reasonably constant during cruising at constant speed (of about 250 km/h). Some recent calculations and measurements (AEAT, 2001) have shown that French TGV trains (Sud-Est, Atlantique, Reseau and Duplex) consume about 19 kWh/km, which, divided by 100 passengers onboard, gives an average energy consumption rate of 0.19 kWh/p-km (The average train capacity is 430 seats/train). A German ICE train consumes about 22 KWh/km, which, divided by 100 passengers gives an average energy consumption rate of 0.22 kWh/p-km (AEAT, 2001; ITA 199; UIC, 2002) (The average train capacity is 380 seats/train).

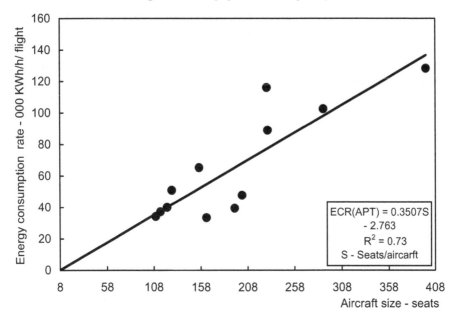

Figure 4.9.3 Dependence of the energy consumption rate on aircraft size while cruising for one hour at optimal altitude
Compiled from Archer (1993)

The APT vehicles – aircraft – are powered by kerosene and/or aviation gasoline. Generally, each burned kilogram of kerosene generates 12.03 kWh of energy (Chevron, 2002). The energy consumption rate (energy-fuel consumption per unit of time) differs for different aircraft types (capacity) and stage of flight. In general, they are higher for larger aircraft, and for any aircraft the consumption rate is higher during take-off and climbing, lower during cruising at optimal altitudes and the lowest while approaching, landing, and taxiing on the ground. Figure 4.9.3 illustrates an example of the above dependability (cross-sectional data for different aircraft types) (Archer, 1993). As can be seen, the energy consumption rate increases more or less proportionally with the increase in the aircraft size. To create a comparable measure to the energy consumption rates of HSR, the rates in Figure 4.9.3 are divided by the aircraft average cruising speed, capacity (seats), and average load factor as follows:

$$ECR_{avg}(ATP) = 1000/(vS\lambda)*[0.3507*S - 2.763] \text{ in } (kWh/pkm) \qquad (4.9.1)$$

where

$ECR_{avg}(APT)$ is the energy consumption rate of an aircraft (KWh/p-km);
S is the aircraft capacity (seats);
v is the aircraft cruising speed (km/h);
λ is the average vehicle load factor ($0 \leq \lambda \leq 1$).

From Equation 4.9.1, for example, the average energy consumption rate of an aircraft with 100 seats and 100 passengers on board (short haul jet) is 0.380, of 150 seats

Table 4.9.1   Air pollution rates of CO_2 dependent on the type of fuel used for generation of electricity for the HSR

HS train/Source for gross electricity generation (k)	Natural gas[1]	Coal (lignite)[2]	Fuel oil[3]	Nuclear/ Water/ Wind[4]	Averages $\sum_{k=1}^{4} p_k APR_k$
TGV – France					
Proportion of source in gross electricity generation – p_k^{1}	0.020	0.055	0.019	0.890	
IOE (kgCO$_2$/kWh)[1]	0.147	0.211	0.345	0	–
ECR (kWh/p-km)[2]	0.190	0.190	0.190	0.190	–
APR = IOE × ECR (gCO$_2$/p-km)[3]	27.93	40.09	65.66	0	4.011
ICE (Germany)					
Proportion of source in gross electricity generation – p_k^{1}	0.111	0.501	0.010	0.353	
IOE (kgCO$_2$/kWh)[2]	0.147	0.211	0.345	0	–
ECR (kWh/p-km)[3]	0.220	0.220	0.220	0.220	–
APR = IOE × ECR (gCO$_2$/p-km)[4]	32.34	46.24	75.90	0	27.515

[1] Compiled from EC (2001)
[2] IOE – Intensity of Emission: Converted from (EIA, 2002)
[3] ECR – Energy Consumption Rate: Estimation from subsection 4.1 (per 100 passengers onboard)
[4] Air Pollution Rate (per 100 passengers onboard)

(medium haul) 0.586, and of 400 seats (long haul) 1.618 KWh/p-km. The aircraft are assumed to cruise at their fuel-optimal altitudes at a speed of 850 km/h. Based on the above figures and evidence from some other studies, the advantage of HSR over ATP in terms of the energy consumption rates is evident (Levison et al., 1996; AEAT, 2001).

4.9.4.2 Air pollution
Air pollution consists of gases and particulates, the products of burning the fuels needed to power the transport vehicles while carrying out services. The HSR and APT use different energy sources and therefore create different forms of air pollution. HS trains mostly use electric energy. Aircraft (APT) use kerosene and gasoline, which are derivatives of crude oil. The important air pollution gases are CO, NO$_x$, SO$_2$, VOCs, CO$_2$ and PM$_{10}$. Since CO$_2$ prevails in the total emitted quantities, it is analysed in more detail, although NO$_x$ and SO$_2$, as has been mentioned earlier in this book, are also now being recognised as important greenhouse gases (ICAO, 1993a; EIA, 2002).

The quantity of polluting gasses generated to power an HS train for a given trip depends on the amount of energy consumed and the air pollution from the electricity plant generated to produce that amount of energy. The electricity may be obtained from several sources, according to the 'average generating mix,' which is specific for each country (ITA, 1991; AEAT, 2001; EC, 2004). The mix may include the use of non-renewable (coal, crude oil, natural gas, nuclear) and renewable (wind, solar, hydro and biomass) sources of energy. Due to the potentially high diversity of sources used, it appears to be relatively complex to generalise the quantity of air pollution emissions by the HSR. However, some generalisation can be made if, for example, is assumed that the composition of sources for generating electricity for HSR is the same as that for the country as the whole. Table 4.9.1 provides an example of such a generalisation.

As can be seen, due to their slightly higher energy consumption rates and different composition of sources used for generation electricity in Germany and France, respectively, ICE trains have a much higher average air pollution rate in terms of CO_2 than TGV trains. The HSR air pollution spreads around the electricity production plants and has, dependent on the type of pollutant, both a local and global depositing character.

The marginal quantity of polluting gases generated by aircraft (APT) is much higher than that from HS trains due to the higher energy consumption rates and the types of fuel used (kerosene or gasoline). The dominant pollutants during a flight in terms of quantity are CO_2 and water vapour (The quantity is proportional to the amount of kerosene consumed multiplied by constant factors of 3.16 and 1.18, respectively) (Archer, 1993; ICAO, 1993a; Levison et al., 1996). As an example, the air pollution rate of CO_2 is calculated as follows. The quantity of kerosene of 83 g needed to generate 1 kWh of energy is multiplied by the constant emission rate of kerosene of 3.16 $kgCO_2/kg$ in order to obtain the intensity of emission of CO_2 per unit of energy of 262.6 g/kWh (EIA, 2002). By multiplying the intensity of emission by the energy consumption rate, the average air pollution rate of CO_2 can be obtained.

For particular aircraft categories, each with 100 passengers onboard, this looks as follows: *0.380 kWh/p-km * 262.6 gCO₂/KWh = 99.8gCO₂/p-km* for an aircraft of 100 seats; *0.598 kWh/p-km * 262.6 gCO₂/kWh = 153.9gCO₂/p-km* for an aircraft of 150 seats; and *1.618 kWh/p-km * 262.6 gCO₂/kWh = 424.9gCO₂/p-km* for an aircraft of 400 seats.

Air pollutants generated by APT spread and deposit locally, around the airports, and globally, in the airspace around the aircraft at cruising altitudes (troposphere) (Janić, 1999).

The above discussion shows that HSR affects environment less than APT in terms of the air pollution rates and spatial deposition of pollutants (Janić, 1999).

4.9.4.3 Noise

Noise is defined as unwanted sound. For HSR, noise mostly depends on the technology in use. In general, HS trains generate noise as the wheel-rail noise, pantograph/overhead noise, and aerodynamic noise (Levison et al., 1996). It is a short time event, which impacts during the time when an HS train passes by. This noise is usually measured in dB (A) scale (decibels). However, traditionally, the noise in the vicinity of airports and railway lines has been expressed by the Equivalent Continuous Sound Level (L_{eq}). This is defined as the steady noise level over a defined period (usually 18 h for APT and 24 h for HSR), which contains the acoustic energy equivalent to the sum of the noise energy of the individual events over that period (AEAT, 2001). The measurements of the individual noise events have shown that the noise is different for different types of HS trains and is positively correlated with their cruising speed. This is shown in Figure 4.9.4. In addition, this noise decreases

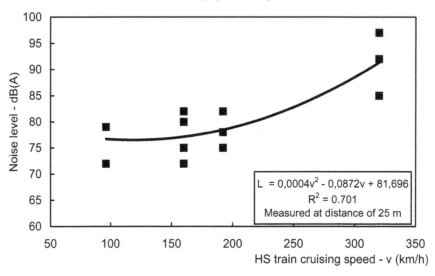

Figure 4.9.4 Dependence of the HS train noise on its cruising speed: TGV, ICE, MAGLEV, Shinkansen, Amtrak, Turbotrain
Compiled from Levison et al. (1996); Kertschmar (1995)

with the increasing distance of the observer from the HS train, but in many cases it is not a linear decline (Kertzschmar, 1995; Levison et al., 1996; Geerlings, 1998).

The APT noise is generated during aircraft landing, taking-off and sideline flight at low altitudes. Consequently, it is very high in the vicinity of airports, and negligible while the aircraft are at cruising altitudes (Smith, 1989). This noise mostly originates from the aircraft engines and airframe configuration. For phases of flight carried out at and around airports, the noise is usually estimated (measured) at the noise 'reference locations,' which are either the aircraft noise certification points or some other selected locations in the vicinity of the airport. The aircraft are certificated to maximum noise levels at the certification points in terms of so-called noise Category 3 and 4 (ICAO, 1993b). The noise emitted by aircraft mostly depends on the aircraft size (that is take-off weight). Usually, the bigger (heavier) aircraft are noisier than the smaller ones (ICAO, 1993b; Janić, 1999; AEAT, 2001). For example, by using cross-sectional data for 13 aircraft types, a causal relationship between the aircraft size and level of noise in dB(A) has been estimated:

$$L_A = 61.094 \; W^{0.094} \; T^{-0.103}$$

$$(22.608)(3.679)(27.455)$$ (4.9.2)

$$R^2 = 0.982 \text{ and } N = 13$$

where

L_A is the aircraft noise in dB (A) measured at the noise certification points;

W is the aircraft take-off weight (tonnes) (from 42 tonnes of aircraft BAE146-200 to 395 tonnes of aircraft B747-400SR);

T is the time when a particular aircraft entered service (years, 1,2, .., 30).

The numbers below particular coefficients of the regression equation are *t*-statistics reflecting the relative importance of corresponding independent variables.

Due to different factors influencing noise and a quite different spatial pattern, HSR may have an advantage compared with APT as regards the noise at the beginning (departure) and end (arrival) of the trip since, during these phases, HS trains tend to run at lower speeds, thus generating a lower level of noise. The APT aircraft is more advantageous during the cruising phase of the trip, particularly when compared with HS trains cruising close to heavily populated areas. However, since the harmonised indicators for noise from HSR and APT are not available, it is necessary for each comparative study to agree about the methodology to enable the two modes to be compared fairly in terms of noise burden. This agreement should inevitably include the size and density of population around airports and along HSR lines.

4.9.4.4 Land-take (use)

Land-take means using a piece of land for building transport infrastructure. HSR and APT can be compared with respect to the size of land needed to build tracks and runways, total route area and airfields, and total airport area and supporting rail infrastructure, such as stations and depots (ATAG, 1996; AEAT, 2001).

The land taken by HSR is dependent on the length of the line and is not influenced by the volume of traffic. HSR typically uses a land surface increment of 3.2 ha/km for tracks although some recent research has shown that this surface increment can be smaller, about 2.0 ha/km (CEC, 1993, 1993a: 1995; AEAT, 2001). The total land taken for an HSR line is then roughly proportional to the product of length of line and the above constant factor.

For APT, land-take is primarily used for building airports. Generally, the land taken for building and operating an airport increases with traffic volume, which in turn determines the number, length and configuration of the runways. The typical land increment is about 30 ha/km for an airport airside infrastructure intended to accommodate the aircraft movements (runway, taxiway, and apron) (Horonjeff and McKelvey, 1994).

Land use is a common term related to the intensity of use of taken land. In general, the intensity of land use can be determined by dividing the total land taken for building some infrastructure of a given system by the volume of traffic carried out during a given period of time. Some research has shown that the intensity of land use tends to be comparable for APT and HSR, 3.23 and 2.86 million p-km/year/ha of land taken, respectively (CEC, 1993, 1993a). Generally, in the EU, APT is in a favourable position compared with HSR with respect to land-take, since most airports already exist and only usually require some carefully planned expansion, while the new land needs to be taken to build completely new HSR lines.

4.9.4.5 Safety

Generally, safety can be defined as the acceptable level of risk or as the chance of injury, damage and loss of life and/or property. It is measured by the probability of occurrence of an event, which may cause the above undesirable outcome (Janić, 1999). Accidents and incidents on HSR and APT represent the burdens on society and the environment. The accidents result in considerable damage or complete demolition of vehicles, and many lead to loss of life, severe injuries, and loss of property of those directly involved and of third parties. Incidents usually result in repairable damage of vehicles, properties and light injuries.

Statistical data on incidents and accidents for both HSR and APT show that these are relatively very rare events. For example, in the EU, the HSR has had only one serious accident (German ICE crash at Eschede in 1998, when about 100 people died and 150 were injured). Incidents have been more frequent. During the period 1983–2001, the French TGV has had 12 incidents (train derailment, collision with objects on the tracks, and terrorist attacks) in which seven people died and 173 were injured (TGVweb, 2002). During the same period, the system carried 339.5 billion p-km. The annual death rate has been $7/(339.5 \times 18) = 0.001,14$ deaths per billion p-km/year.

The accidents and incidents related to Apt system have also been rare compared with the number of departures and volume of p-km (passenger kilometres) carried during a given period of time (Boeing, 2002). The analysis has shown that the safety of air transport at a global scale has constantly improved over time. For example, during the period 1970–1993 the annual death rate has fallen from 0.018 to 0.004 (deaths per billion p-km/year) and has stabilised around the latter value despite further growth in air traffic. There have been 105 deaths resulting from flight operations of European air transport operators over EU territory during the period 1990–1999. Divided by the total cumulative output over the same period of 2012.84 bn p-km, this gives the average annual death rate of $105 / (2012.84 \times 9) = 0.005,80$ deaths per billion p-km/year (ATAG, 2000).

These very low figures compared with the volume of output illustrate that both HSR and ATP have been safe systems in terms of the annual death rates. According to the above figures, it seems that HSR could be prioritised over Apt.

4.9.4.6 Congestion

Congestion occurs whenever demand exceeds the capacity of a service facility.

In the transport sector, it causes delays for vehicles and users between origins and destinations. In addition to travel time, congestion creates a direct burden on society in terms of lost time. Indirectly, congestion contributes to the increasing energy consumption and air pollution, and generates the need for building additional infrastructure, which may require acquisition of further land. In addition, the perceived risk of incidents and accidents, and thus the safety burden may rise due to the increased complexity of the on-line and off-line management and control of operations.

HSR is usually considered to be free of congestion, since its capacity and control system are designed to prevent congestion at terminals, stops and lines. Data on HSR congestion and delays are mostly lacking, but a 'freedom' from congestion means that punctuality rate should be close to 100 per cent and eventual delays always less than 15 minutes (in comparison with Apt) (EC, 1998; ITA, 1998).

Unlike at the HSR, congestion in Apt has become an inherent operational characteristic. Some figures indicate that the proportion of flights in European airspace delayed for more than 15 minutes has increased in the period 1985–2001. It was minimal (10 per cent) during the 1990s and maximal (25–30 per cent) in

2000/01. The most common value has been 15–20 per cent, which means that on average every fifth flight is delayed by more than 15 minutes (AEA, 2004).

Like punctuality, the average delay per flight has also increased both in absolute and relative terms. For example, at five major European hubs, Paris CDG, London Heathrow, Amsterdam Schiphol, and Frankfurt Main this delay was 4.67 min/flight in 2001, which is far above a target of 3.5 min/flight set up by EUROCONTROL for all flights (EEC, 2004).

The above figures show that HSR, compared with Apt, inherently offers less of a burden in terms of lost time due to congestion, and consequently less congestion-related additional burdens/emissions such as air pollution and risk of incidents and accidents. However, due to their lower cruising speeds, the total travel time of HSR may be longer than that of Apt over comparative distances, which may compromise its above advantages.

4.9.5 Evaluation of Particular Impacts

The impacts from HSR and Apt operations may cause damage to the environment at a local and global scale in both the short and long-term (EC, 1996b). The scale and scope of damages is usually proportional to the volume of operations and marginal rates of impacts on one side, and the size and relative closeness of the affected environment and people to the sources of contamination on the other. Table 4.9.2 provides a summary of the previously elaborated marginal impacts generated by the HSR and Apt (EC, 1996b). The purpose is to compare them and assess the potential for cumulative mitigation, which could be achieved by substitution between the two modes.

As can be seen, by replacing Apt with HSR, the most important cumulative impacts would be reduced. In general, the damages caused by the impacts of particular burdens/emissions from transport operations have been evaluated in monetary terms as the external costs (externalities). The purpose has been to design a fair pricing policy for paying the costs of external damage by those who actually cause them (EC, 1996b; COFAR, 2000; EEC, 2001b). In some earlier studies, the

Table 4.9.2   The marginal impacts from HSR and APT: Summary

Type of impact/HS mode	HSR		APT[1]
	TGV	*ICE*	
Energy consumption (kWh/p-km)	0.190	0.220	0.380–0.586[1]
Average air pollution (CO_2/p-km)	4.011	27.515	99.8–153.9[1]
Noise (dB(A))	$0.000,4v^2 - 0.087,2v + 81.986$		$61.094\ W^{0.094}\ T^{-0.1032}$
Land-take (ha/km)	3.2		30.0
Safety (deaths/billion p-km/year)	0.00114		0.00580
Congestion (% of punctuality) Average delay (min/flight)	–/–		25–20/4.75

[1] Aircraft of 100 and 150 seats, which may interact with HS trains (100 passengers are onboard for both types of vehicles); v – HS train cruising speed; W – an aircraft take-off weight; T – time (years); '–' lack of data

Table 4.9.3   The total external costs of HSR and APT (euro/p-km): Summary

Cases	HS mode *HSR*	HS mode *APT*
Case 1	0.00201	0.07530
Case 2	0.02000	0.04800
Case 3	0.01000	0.04800
Case 4	0.01200	0.02900
Case 5	0.01000	0.01780
Case 6	0.0051–0.0055	–

[1] Corridor Los Angeles – San Francisco (USA) (Levison et al., 1996)
[2] Total rail and air traffic in EU (1995) without congestion (EEC, 2001; COFAR, 2000);
[3] Corridor Paris-Brussels with HSR Thalys (UIC, 2004);
[4] Corridor Paris-Vienna (UIC, 2004);
[5] Air/rail – general (EC, 1996a);
[6] (EC, 1996b)

average costs of particular externalities have usually been estimated by using the top-down approach. Recent studies have dealt with quantification of the marginal costs of particular externalities by using the bottom-up approach. The basis for calculating these latter costs has been the individual vehicle trip or traffic flow on a particular route. The cost of damage by one additional vehicle or trip has then been estimated (EC, 1996b).

Calculation of the external costs of HSR and Apt has been carried out in order to provide an argument in favour of the implementation of HSR in the corridors where Apt already operates. Both of these approaches have been used.

Table 4.9.3 provides some aggregate figures on the cost of particular burdens/emissions such as air pollution, noise, safety, congestion, climate change, urban effects, and up/downstream processes.

As can be seen, despite the obvious differences caused by the differences and specificity in particular cases, the marginal external costs are generally lower for HSR than Apt, which indicates that this system has better environmental performances. However, for the full-scale comparison and assessment of the effects of substitution, the systems' operational and passenger time costs as well as the real load factors on particular corridors should also be taken into account (ITA, 1991; CEC, 1993a; Levison et al., 1996). There has been some academic research, which has taken into account the above-mentioned factors. The multi-criteria analysis reflecting the preferences of particular actors involved has been applied. In all cases the HSR has been shown to be the preferable HS alternative to Apt under the given circumstances (Janić, 2001).

4.9.6 Concluding Remarks

This section has presented an overview of the environmental performance of the HSR and APT in the European Union (EU). It aimed to show the potential for mitigating the environmental impacts by substitution of two modes in particularly densely used corridors. The direct impacts such as energy consumption, air pollution, noise, land-take (use), safety, and congestion have been considered as the systems' environmental performances. The marginal costs (externalities) of these impacts calculated for different conditions have also been presented. The comparison has shown that HSR has a better environmental performance than APT, which means that any kind of substitution, either through competition or complementarity, currently driven mostly by the commercial viability rather than by the environmental constraints would contribute to the mitigation of the cumulative environmental damages.

References

AB (1999), 'Airports-top 100 Rankings', *Airline Business*, June, 58–70.

AB (2000), 'The Airline Ranking', *Airline Business*, July, 86–94.

ACI Europe (1998), *The Social and Economic Impact of Airports: Creating Employment and Prosperity* (Brussels: Airport Council International Europe).

ACI Europe (2002), *Airport Charges in Europe*, Report of Airport Council International (Brussels: Airport Council International Europe).

Adler, N. (2002), *'Barriers and Implementation Paths to Marginal Cost Based Pricing: Rail, Air and Water Transport'*, *3rd MC-ICAM Seminar, 2–3 September*, 5.

AEA (2001–2005), *Yearbook 2000/5* (Brussels: Association of European Airlines).

AEAT (2001), *A Comparative Study of the Environmental Effects of Rail and Short-Haul Air Travel, Report for Commission for Integrated Transport* (Abingdon, UK: AEA Technology).

Airwise News (2001), 'Delta Launches Boeing 767-400ER At LaGuardia', http://news.aiwise.com.

Allan, S.S., Beesley, A.J., Evans, E.J., Gaddy, G.S. (2001), *'Analysis of Delay Causality at Newark International Airport,'* *4th USA/Europe Air Traffic Management R&D Seminar, Santa Fe, USA*, 11.

Andreatta, G. and Romanin-Jacur, G. (1987), 'Aircraft Flow Management under Congestion', *Transportation Science*, **21**(4), 249–253.

Archer, L.J. (1993), *Aircraft Emissions and Environment* (Oxford: Oxford Institute for Energy Studies).

ATA (2002), *System Capacity, Part I: Airline Schedule, Airport Capacity, and Weather*, http://www.air-transport.org/public/industry.

ATAG (1996), *The Economic Benefits of Air Transport* (Geneva: Air Transport Action Group).

ATAG (1996a), *Air Transport and the Environment* (Geneva: Air Transport Action Group).

ATAG (2000), *European Air Traffic Forecast 1980–2010* (Geneva: Air Transport Action Group).

Ausubel, J.H. and Marchetti, C. (1996), 'Elektron: Electrical Systems in Retrospect and Prospect', *Daedalus*, **123**(summer), 139–170.

Aykin, T. (1994), 'On the Location of Hub Facilities', *Transportation Science*, **22**, 155–157.

Aykin, T. (1995), 'The Hub Location and Routing Problem', *European Journal of Operational Research*, **83**, 200–219.

Aykin, T. (1995a), 'Networking Policies for Hub-and-Spoke Systems with Applications to the Air Transport System', *Transportation Science*, **29**, 201–221.

Backer, C. (2000), 'Airports: Top 1000 Ranking', *Airline Business*, 55–86.

Beatty, R., Hsu, R., Berry, L. and Rome, J. (1998), *'Preliminary Evaluation of Flight Delay Propagation Through an Airline Schedule'*, 2nd USA/Europe Air Traffic Management R&D Seminar, Orlando, 1–4 December, 9.

Bianco, L. and Bielli, M. (1992), 'Air Traffic Management: Optimization Models and Algorithms', *Journal of Advanced Transportation*, **26**, 131–167.

Blumstein, A. (1960), *An Analytical Investigation of Airport Capacity, Report TA-1356-8-1* (Cornell: Cornell Aeronautical Laboratory).

Boeing (1998), *Evolution of the World Fleet: Time Line*, http://www.boeing.com.

Boeing (2002), *Statistical Summary of Commercial Jet Aeroplane Accidents: Worldwide Operations 1959–2001*, http://www.boeing.com.

Bontekoning, Y. and Janić, M. (2003), *'Modelling Intermodal Hub-And-Spoke Freight Transport Systems: Rail-Rail Terminals Versus Shunting Yards'*, 2003 Meeting of the Association of American Geographers, 4–8 March, New Orleans, Louisiana, USA.

Brueckner, J.K. (2002), 'Internalisation of Airport Congestion', *Journal of Air Transport Management*, **8**, 141–147.

BTS (2001), *Airline Service Quality Performance Data* (Washington, DC: US Department of Transportation, Bureau of Transportation Statistics).

Button, J.K. and Stough, R. (1998), *The Benefits of Being a Hub Airport City: Convenient Travel and High-Tech Job Growth, Report, Aviation Policy Program* (Fairfax, VA: George Mason University).

Button, K. (2002), 'Debunking Some Common Myths About Airport Hubs', *Journal of Air Transport Management*, **8**, 177–188.

Campbell, K.C., Cooper, W.W., Jr, Green Baum, D.P. and Wick, L.A. (2000), *'Modelling Distributed Human Decision Making in Traffic Flow Management Operations,'* 3rd USA/Europe Air Traffic Management R&D Seminar, Napoli, Italy, 9.

Carlson, F., *Airport Marginal Cost Pricing: Discussion and an Application to Swedish Airports, Discussion Paper* (Gutenberg, Sweden: Department of Economics, Gutenberg University), 20.

Caves, R.E. and Gosling, G.D. (1999), *Strategic Airport Planning* (Amsterdam: Elsevier Science).

CEC (1993), *European High-Speed Rail Network: Socio-economic Impact Study, Final Report* (Brussels: Commission of the European Communities).

CEC (1993a), *The European High-Speed Train Network: Environmental Impact Assessment, Executive Summary* (Brussels: Commission of the European Communities).

CEC (1995), *High-Speed Europe,* High Level Group 'The European High Speed Train Network' (Brussels: Commission of the European Communities).

CEC (2001), *A Sustainable Europe for a Better World: A European Union Strategy for Sustainable Development, COM* (2001), 264 (Luxembourg: Commission of the European Communities).

Cheng-Sheen, A.H. (2000), *An Analysis of Air Passenger Average Trip Lengths and Fare Levels in USA Domestic Markets, Working Paper UCB-ITS-WP-2000-1, NEXTOR Aviation Operations Research* (Berkeley, CA: Institute of Transport Studies, University of California, Berkeley).

Cheung, W., Leung, L.C. and Wong, Y.M. (2001), 'Strategic Service Network Design for DHL Hong Kong', *Interfaces*, **31**, 1–14.

Chevron (2002), 'Aviation Fuels', Chevron Product Corporation. http://www. chevron.com/prodserv/fuels.

COFAR (2000), *Landside Accessibility and Ground Transport*, Report on Research Phase II, Theme 2, Common Options for Airport Regions (Delft, the Netherlands: COFAR).

Coleman, R.J. (1999), 'Environmentally Sustainable Capacity', *Proceedings of the ECAC/EU Dialogue with the European Air Transport Industry: Airport Capacity – Challenges for the Future, Salzburg, Austria.*

Corbett, J.J. (2002), 'Small Communities are Concerned about Congestion Pricing', *The Air and Space Lawyer*, **17**(summer), 17–21.

Daniel, J.I. (1995), 'Congestion Pricing and Capacity of Large-Hub Airports: A Bottleneck Model with Stochastic Queues', *Econometrical*, **63**, 327–370.

Daniel, J.I. (2001), 'Distributional Consequences of Airport Congestion Pricing', *Journal of Urban Economics*, **50**, 230–258.

Daniel, J.I. and Pahoa, M. (2000), 'Comparison of Three Empirical Models of Airport Congestion Pricing', *Journal of Urban Economics*, **47**, 1–38.

DeCota, W. (2001), 'Matching Capacity and Demand at LaGuardia Airport', *NEXTOR, Airline and National Strategies for Dealing with Airport and Airspace Congestion*, (University of California, Berkeley, USA), 26.

Dell'Olmo, P. and Lulli, G.M. (2001), *A Dynamic Programming Approach for the Airport Capacity Allocation Problem, Report by Department of Statistics, Probability and Applied Statistics* (Rome: University of Rome 'La Sapienza').

DERA (1999), *Aircraft Engine Exhaust Emission Databank* (London: Defence Evaluation and Research Agency).

DETR (1999), *Oxford Economic Forecasting: The Contribution of the Aviation Industry to the UK Economy* (London: Department of the Environment, Transport and Regions).

DETR (2000), *Noise from Arriving Aircraft: Final Report of the ANMAC Technical Working Group* (London: Department of the Environment, Transport and Regions).

DETR (2000), *The Future of Aviation, The Government's Consulting Document on Air Transport Policy* (London: Department of the Environment, Transport and Regions).

Doganis, R. (1992), *The Airport Business* (London: Routledge).

Doganis, R. (2002), *Flying-Off Course: The Economics of International Aviation* (London: Routledge).

Donohue, G.L. (1999), 'A Simplified Air Transportation Capacity Model', *Journal of ATC*, **41**(April/June), 8–15.

Donohue, G.L. (2000), *'United States and European Airport Capacity Assessment Using the GMU Macroscopic Capacity Model (MCM)'*, *3rd USA/Europe Air Traffic Management R&D Seminar, Napoli, 13–16 June*, 10.

EC (1996a), *External Costs of Transport in ExternE, Report* (Brussels: European Commission).

EC (1996b), 'Towards Fair and Efficient Pricing in Transport: Policy Options for Internalising the External Cost of Transport in European Union, Green Paper', *Bulletin of the European Union*, Supplement 2/96.

EC (1997), *External Cost of Transport in ExternE, European Commission, Non Nuclear Energy Programme* (Cologne, Germany: IER Germany).

EC (1998), *Interaction between High-Speed Rail and Air Passenger Transport, Final Report, Action COST 318* (Luxembourg: European Commission).

EC (2001), *Concerted Action on Transport Pricing Research Integration – CAPRI, Final Report, European Commission, Transport RTD of the 4th Framework Programme, ST-97-CA-2064* (Brussels: European Commission).

EC (2002), *Developing EU (International) Rail Passenger Transport, OGM-Final Report European Commission* (Brussels: European Commission).

EC (2004), *Statistical Yearbook: Transport – 2004* (Luxembourg: European Commission).

ECMT (1998), *Efficient Transport for Europe: Policies for Internalization of External Costs* (Paris: European Conference of Ministers of Transport).

Economides, N. (1996), 'The Economics of Networks', *International Journal of Industrial Organization*, **14**, 673−699.

EEC (1997), *Testing Operational Scenarios for Concepts in ATM – TOSCA II, Interim Report, EEC Task FSO-11* (Brussels: EUROCONTROL).

EEC (1998), *Data on European Airports* (Brussels: EUROCONTROL).

EEC (1999), *Cost of the En-Route Air Navigational Services in Europe, EEC Note 8/99* (Bretigny-sur-Orge, France: EUROCONTROL).

EEC (2001a), *Review of Research Relevant to Rail Competition for Short Haul Air Routes* (Paris: EUROCONTROL Experimental Centre).

EEC (2001b), *Delays to Air Transport in Europe − Annual 2000*, CODA-Central Office for Delay Analysis (Brussels: EUROCONTROL).

EEC (2002), *An Assessment of Air Traffic Management in Europe during Calendar Year 2001: Performance Review Report* (Brussels: EUROCONTROL).

EEC (2002), *CODA − Delays to Air Transport in Europe-Annual Report 2000* (Paris: EUROCONTROL/ECAC).

EEC (2004), *Enhanced Flight Efficiency Indicators, EUROCONTROL Experimental Centre, EEC / SEE/2004/011* (Bretigny-sur-Orge, France: EUROCONTROL).

EEC/ECAC (2001), *Study on Constraints to Grow, Report* (Brussels: EUROCONTROL Agency).

EEC/ECAC (2001a), *ATFM Delays to Air Transport in Europe, Annual Report 2000* (Brussels: EUROCONTROL).

EEC/ECAC (2002), *ATFM Delays to Air Transport in Europe: Annual Report 2001* (Brussels: EUROCONTROL).

EIA (2002), *Voluntary Reporting on Greenhouse Gases: Form – 1605* (Washington, DC: US Department of Energy).

EPA (1999), *Evaluation of Air Pollution Emissions from Subsonic Commercial Jet Aircraft, EPA–420R–99–013* (Washington, DC: ICF Consulting Group, Environmental Protection Agency).

FAA (1993), *Aviation System Capacity – Annual Report* (Washington, DC: Federal Aviation Administration).

FAA (1995), *Total Cost for Air Carrier Delay for the Years 1987–1994, Report – APO-130* (Washington, DC: Federal Aviation Administration).

FAA (1998), *Economic Values for Evaluation of Federal Aviation Administration Investment and Regulatory Decisions, Report FAA-APQ-98-8* (Washington, DC: Federal Aviation Administration).

FAA (2000), *FFP1 Performance Metrics: Results to Date, Version 1.0* (Washington, DC: Federal Aviation Administration).

FAA (2002), *Airport Capacity Benchmark Report 2001* (Washington, DC: Federal Aviation Administration).

FAA (2003), *Airport Capacity Benchmarking Report 2001* (Washington, DC: Federal Aviation Administration).

FAA (2003a), *Aviation Policy and Plans (APO) – FAA OPSNET and ASPM* (Washington, DC: Federal Aviation Administration).

Ferrar, T.A. (1974), 'The Allocation of Airport Capacity with Emphasis On Environmental Quality', *Transportation Research*, **8**, 163–169.

FI (1999a), *'Commercial Airline Directory'*, *Flight International*, September, 62–67.

FI (1999b), 'Commercial Aircraft/Engine Directory', *Flight International*, September, 62–74/December, 52–61.

Fraport (2004), *EDDF-SOP – Standard Operating Manual – Frankfurt Main Airport, V9.6*. http://www.vacc-sag.org.

Fron, X. (2001), 'Dealing with Airport and Airspace Congestion in Europe', in *ATM Performance, the Paper Presentation, EUROCONTROL Meeting, Brussels, March*, 25.

GAO (2004), *Commercial Aviation: Legacy Airlines Must Further Reduce Costs to Restore Profitability, Report GAO-04-836 to Congressional Committees* (Washington, DC: United States Government Accountability Office).

Geerlings, H. (1998), 'The Rise and Fall of New Technologies: MAGLEV as Technological Substitution?', *Transportation Planning and Technology*, **21**, 263–286.

Ghali, M.O. and Smith, M.J. (1995), 'A Model for the Dynamic System Optimum Traffic Assignment Problem', *Transportation Research B*, **29B**, 155–170.

Gilbo, E.P. (1993), 'Airport Capacity: Representation Estimation, Optimization', *IEEE Transactions on Control Systems Technology*, **1**, 144–153.

Gilbo, E.P. (1997), 'Optimizing Airport Capacity Utilization in Air Traffic Flow Management Subject to Constraints at Arrival and Departure Fixes', *IEEE Transactions on Control Systems Technology*, **5**, 490–503.

Gilbo, E.P. (2003), *'Arrival/Departure Capacity Trade-Off Optimisation: A Case Study at the St. Louis Lambert International Airport (STL)'*, *5th USA/Europe Air Traffic Management R&D Seminar, Budapest, 23–27 June, Hungary*, 10.

Gilbo, E.P. and Howard, K.W. (2000), *'Collaborative Optimization of Airport Arrival and Departure Traffic Flow Management Strategies for CDM'*, *3rd USA/Europe Air Traffic Management R&D Seminar, Naples, 13–16 June, Italy*, 10.

Groenewege, A.D. (1999), 'Air Distances between Major World Airports' in *Compendium of International Civil Aviation Organisation* (Montreal: International Press/Aviation Development Corporation), Appendix 5.

Hall, R.W. (1989), 'Configuration of an Overnight Package Air Network', *Transportation Research*, **23A**, 139–149.

Hall, R.W. (1991), *Queuing Methods for Services and Manufacturing* (London: Prentice-Hall).

Hall, R.W. (2001), 'Truck Scheduling for Ground to Air Connectivity', *Journal of Air Transport Management*, **7**, 331–338.

Hall, R.W. and Chong, C. (1993), *'Scheduling Timed Transfers at Hub Terminals'*, in *Transportation and Traffic Theory*. Daganzo, C.F. (ed.) (Amsterdam: Elsevier Science), 217–236.

Hammer, J.A. (2000), *'Case Study of Paired Approach Procedure to Closely Spaced Parallel Runways'*, *Air Traffic Control Quarterly*, **8**, 223–252.

Harris, R.M. (1972), *Models for Runway Capacity Analysis*, The MITRE Corporation Technical Report, MTR-4102 (The MITRE Corporation).

Helme, M.P. and Lindsay, K. (1992), 'Optimization of Traffic Flow to Minimize Delay in the National Airspace System', *IEE Proceedings on Control Theory and Applications*, **1**, 435–437.

Hockaday, S.L.M. and Kanafani, A. (1974), 'Development in Airport Capacity Analysis', *Transportation Research*, **8**, 171–180.

Horonjeff, R. and McKelvey, F.X. (1994), *Planning and Design of Airports*, 4th edn (New York: McGraw-Hill).

Huang, S.C.A. (2000), *An Analysis of Air Passenger Average Trip Lengths and Fare Levels in USA Domestic Markets*, NEXTOR Working paper WP–UCB-ITS-WP-2000-1 (Berkeley, CA: Institute of Transportation Studies, University of California, Berkeley).

ICAO (1993a), *'Aircraft Engine Emissions'*, *Environmental Protection*, **2**, Annex 16.

ICAO (1993b), *'Aircraft Noise'*, *Environmental Protection*, **1**, Annex 16.

ICAO (1998), *Rules of the Air and Air Traffic Services: Procedures for Air Navigation Services*, Doc. 4444–RAC/501, 13th edn (Montreal: International Civil Aviation Organization).

ICAO (2004), *Aerodromes: Volume I: Aerodrome Design and Operations*, 4th edn (Montreal: International Civil Aviation Organisation).

Ignaccolo, M.A. (2003), 'Simulation Model for Airport Capacity and Delay Analysis', *Transportation Planning and Technology*, **26**, 135–170.

IPCC (1999), *Aviation and the Global Atmosphere, Special Report–Summary for Policy Makers, WMO – VNEP* (Geneva: Intergovernmental Panel on Climate Change).

ITA (1991), *Rail/Air Complementarity in Europe: The Impact of High-Speed Rail Services* (Paris: Institute du Transport Aèrien).

ITA (2000), *Cost of Air Transport Delay in Europe, Final Report* (Paris: Institut de Transport Aèrien).

Janić, M. (1993), 'A Model of Competition Between High-Speed Rail and Air Transport', *Transportation Planning and Technology*, **17**, 1–23.

Janić, M. (1994), 'Modelling Extra Aircraft Fuel Consumption in En-Route Environment', *Transportation Planning and Technology*, **18**, 163–186.

Janić, M. (1997), 'Liberalisation of the European Aviation: Analysis and Modelling of the Airline Behaviour', *Journal of Air Transport Management*, **3**, 167–180.

Janić, M. (1999), 'Aviation and Externalities: The Accomplishments and Problems', *Transportation Research D*, **4**, 159–180.

Janić, M. (2001), *Analysis and Modelling of Air Transport System: Capacity, Quality of Services and Economics* (Amsterdam: Gordon and Breach).

Janić, M. (2001), 'Development of High-Speed Systems in Europe: Multi Attribute Ranking of Alternatives', in *Analytical Advances in Transport Systems and Spatial Dynamics: New Perspectives*. Gastaldi, M. and Reggiani, A. (eds) (Aldershot: Ashgate Publishing), 169–188.

Janić, M. (2003), 'Modelling Operational, Economic and Environmental Performance of an Air Transport Network', *Transportation Research D*, **8**, 415–432.

Janić, M. (2003), 'The Potential to Modal Substitution', in *Towards Sustainable Aviation: Trends and Issues*. Upham, P., Maughan, J., Raper, D. and Thomas, C. (eds) (London: Earthscan).

Janić, M. (2004), *'Air Traffic Flow Management: A Model for Tactical Allocation of Airport Capacity'*, ATRS (Air Transport Research Society) 2004 Conference, 1–3 July, Istanbul, Turkey, 31.

Janić, M. (2004a), 'Expansion of Airport Capacity: Case of London Heathrow Airport', *Transportation Research Record*, **1888**, 7–14.

Janić, M. (2005), *'An Evaluation of Sustainability and Reliability of (Air) Transport Networks'*, VII NECTAR Conference, Las Palmas, Grand Canaria, Spain, 2-4 June, 20.

Janić, M. (2005), 'Modelling Charges for Congestion at an Airport', *Transportation Planning and Technology*, **28**, 1–27.

Janić, M. (2005a), 'Modelling Consequences of Large Scale Disruptions of an Airline Network', *Journal of Transportation Engineering*, **131**, 249–260.

Janić, M. (2006), *'Model of Ultimate Capacity of Dual-Dependent Parallel Runways'*, 85th TRB (Transportation Research Board) Annual Conference, Washington DC, USA, 20.

Janić, M. and Reggiani, A. (2002), 'An Application of the Multiple Criteria Decision Making (MCMD) Analysis to the Selection of a New Hub Airport', *European Journal of Transport and Infrastructure Research*, **2**, 113–142.

Janić, M. and Stough, R. (2003), *'Congestion Pricing at Airports: Dealing with an Inherent Complexity'* ERSA (European Regional Science Association) 2003 Congress, Jyvakyla, Finland, 23–26 August, 35.

Janić, M. and Tosic, V. (1982), 'Terminal Airspace Capacity Model', *Transportation Research*, **16A**, 253–260.

Jeng, C.Y. (1987), 'Routing Strategies for an Idealized Airline Network'. PhD dissertation, Institute of Transport Studies, University of California, Berkeley, USA.

Johnsonbaugh, R. and Schaefer, M. (2003), Algorithms, (New York, USA: McMillan Professional).

Johnston, E.L. (2000), *'Airline Integrated Recovery and Simair'*, The AGIFORS Conference, Budapest, Hungary, 28.

Kanafani, A. and Ghobrial, A.A. (1985), 'Airline Hubbing – Some Implications for Airport Economics', *Transportation Research A*, **19A**, 15–27.

Kara, B.Y. and Tansel, B.C. (2001), 'The Latest Arrival Hub Location Problem', *Management Science*, **47**, 1408–1420.

Kertzschmar, R. (1995), *'Transrapid Maglev: Prospects for Fast Regional Transportation Service'*, 1st European Workshop on High Speed Maglev Transport Systems: European Prospects, University of Padua, Padua, Italy.

Kim, D., Barnhart, C., Ware, K. and Reinhardt, G. (1999), 'Multimodal Express Package Delivery: A Service Network Design', *Transportation Science*, **33**, 391–407.

Lettovsky, L., Clarke, M.D.D., Johnson, E.L. and Smith, B.C. (1999), *'Real-Time Recovery: Aircraft, Crew and Passengers'*, the AGIFORS Conference, 18–21 April, Istanbul, 22.

Levison, D., Gillen, D., Kanafani, A. and Mathieu, J.-M. (1996), *The Full Cost of Intercity Transportation – A Comparison of High Speed Rail, Air and Highway Transportation in California, UCB–ITS–RR–96–3* (Berkeley, CA: Institute of Transportation Studies, University of California, Berkeley).

Liang, D., Marnane, W. and Bradford, S. (2000), *'Comparison of the U.S. And European Airports and Airspace to Support Concept Validation'*, 3rd USA/Europe Air Traffic Management R&D Seminar, Napoli, 13–16 June, 15.

Lufthansa (2003), *Annual Report 2002* (Cologne: Deutsche Lufthansa AG CGN IR).

Lyle, C. (1999), 'Maturing Industry Must Continue to Cope with Challenges of Growth in 21st Century', *ICAO Journal*, **54**, 20–24.

Marianov, V., Serra, D. and Revelle, C. (1999), 'The Location of Hubs in a Competitive Environment', *European Journal of Operations Research*, **114**, 363–371.

Mayer, E., Rice, C., Jaillet, P. and McNerney, M. (1999), *Evaluating the Feasibility of Reliever and Floating Hub Concepts when a Primary Hub Experiences Excessive Delays, Report* (Austin, Texas: University of Austin).

Mendoza, G. (2002), *New York State Airport Air Fare Analysis* (Washington, DC: Aviation Service Bureau, NYSDOT).

NASA (1998), *Air Traffic and Operational Data on Selected US Airports with Parallel Runways*, NASA/CR-1998-207675 (Langley, VA: National Aeronautic and Space Administration).

NASA (1999), *Speed Profiles for Deceleration Guidance During Rollout and Turnoff (ROTO)*, NASA/TM-209829-1999 (Langley, VA: National Aeronautic and Space Administration).

Nash, C. and Sansom, T. (2001), 'Pricing European Transport System: Recent Developments and Evidence from Case Studies', *Journal of Transport Economics and Policy*, **35**, 363−380.

Nero, G. (1999), '*A Note on the Competitive Advantage of Large Hub-and-Spoke Networks*', *Transportation Research E*, **35**, 225–239.

Newell, G.F. (1979), 'Airport Capacity and Delays', *Transportation Science*, **13**, (London: UK), 201−241.

Newell, G.F. (1982), *Application of Queuing Theory* (UK: Chapman & Hall).

OAG (2002), *Executive Flight Guide – Europe* (Frankfurt: OAG Office).

Odoni, A.R. and Bowman, J. (1997), *Existing and Required Modeling Capabilities for Evaluating ATM Systems and Concepts*, Final Report No. NAG2-997 (Boston, MA: International Center for Air Transportation, Massachusetts Institute of Technology).

Odoni, A.R. and Fan, T.C.P. (2002), '*The Potential of Demand Management As a Short-Term Means of Relieving Airport Congestion*', *4th USA/Europe Air Traffic Management R&D Seminar, Santa Fe, USA, 3–7 December*, 11.

Offerman, H. and Bakker, M. (1998), '*Growing Pains of Major European Airports*', *2nd USA/Europe Air Traffic Management R&D Seminar, Orlando, USA*, 10.

O'Kelly, M.E. (1998), 'A Geographer's Analysis of Hub-and-Spoke Networks', *Journal of Transport Geography*, **6**, 171−186.

O'Kelly, M.E. and Bryan, D.L. (1998), 'Hub Location with Flow Economies of Scale', *Transportation Research B*, **32**, 605−616.

O'Kelly, M.E., Bryan, D., Skorin-Kapov, D. and Skorin-Kapov, J. (1997), 'Hub Network Design with Single and Multiple Allocation: A Comparative Study', *Location Science*, **4**, 125−138.

Ott, J. (2003), '"De-Peaking" American Hubs Provides Network Benefits', *Aviation Week and Space Technology*, February, 13−15.

PANYNJ (2003), *La Guardia Airport: Traffic Statistics* (New York: The Port Authority of New York and Port of New Jersey).

PANYNJ (2003a), *Schedule of Charges for Air Terminals* (New York: The Port Authority of New York and Port of New Jersey).

Pels, E., Nijkamp, P. and Rietveld, P. (2000), 'A Note on Optimality of Airline Networks', *Economic Letters*, **69**, 429−434.

RC (1999), *Integrated Information Systems (I²S)* (Cedar Rapids, IA: Rockwell Collins).

Richetta, O. (1995) 'Optimal Algorithms and a Remarkable Efficient Heuristic for the Ground Holding Problem in Air Traffic Control', *Operations Research*, **43**, 758−770.

Richetta, O. and Odoni, A.R. (1993), 'Solving Optimally the Static Ground Holding Policy Problem in Air Traffic Control', *Transportation Science*, **27**, 228−238.

Richetta, O. and Odoni, A.R. (1994), 'Dynamic Solution to the Ground Holding Problem in Air Traffic Control', *Transportation Research A*, **28A**, 167−185.

Rosenberger, J.M., Schaefer, A.J., Goldsman, D., Johnson, E.L., Kleyweght, A.J. and Nemhauser, L.G. (2000), *A Stochastic Model of Airline Operations. Working Paper* (Atlanta: Georgia Institute of Technology).

Ruijgrok, C.J.J., (2000), *Elements of Aviation Acoustics*, (Amsterdam, The Netherlands: Het Spinhuis Publishers).

Schaefer, L. and Millner, D. (2001), *'Flight Delay Propagation Analysis with the Detailed Policy Assessment Tool'*, *Proceedings of the 2001 IEEE Systems, Man and Cybernetics Conference, Tucson, USA*, 5.

Schavell, A.Z. (2000), *'The Effects of Schedule Disruptions on the Economics of Airline Operations'*, *3rd USA/Europe Air Traffic Management R&D Seminar, Napoli, Italy. 13–16 June*, 11.

SG (2002), *Annual Report-2001* (Amsterdam: Amsterdam Schiphol Airport).

Shangyao, Y. and Chung-Gee, L. (1997), 'Airline Scheduling for the Temporary Closure of Airports', *Transportation Science*, **31**, 72–83.

Smith, M.J.T. (1989), *Aircraft Noise* (Cambridge, Massachusetts, USA: Cambridge University Press).

Spangenberg, J.H. (2002), 'Institutional Sustainability Indicators: An Analysis in Agenda 21 and a Draft Set of Indicators for Monitoring Their Effectivity', *Sustainable Development*, **10**, 103–115.

Stamatopoulos, M.A., Zografos, K.G. and Odoni, A.R. (2004), 'A Decision Support System for Airport Strategic Planning', *Transportation Research C*, **12**, 91–117.

Swedish, W.J. (1981), *Upgraded Airfield Capacity Model*, The MITRE Corporation Technical Report, MTR-81W16 (Washington, Virginia, USA: The MITRE Corporation).

Terrab, M. and Odoni, A.R. (1993), 'Strategic Flow Management in Air Traffic Control', *Operations Research*, **41**, 138–152.

TGVweb (2002), *High-Speed Rail Safety: Tgv Accidents*, http://www.mercurio.iet. unipi.it/tgv.

Tosic, V. and Horonjeff, R. (1976), 'Effects of Multiple Path Approach Procedures on Runway Landing Capacity', *Transportation Research*, **10**, 319–329.

UIC (2002), *'Shaping the Railway of the 21st Century: AIR/HSR from Competition to Complementarity'*, *the Paper Presentation* (Paris: International Union of Railways), http://www.uic.asso.fr/d.

Velazco, E.E. (1995), 'Air Traffic Management: High-Low Traffic Intensity Analysis', *European Journal of Operational Research*, **80**, 45–58.

Venkatakrishnan, C.S., Barnett, A. and Odoni, A.R. (1993), 'Landings at Logan Airport: Describing and Increasing Airport Capacity', *Transportation Science*, **27**, 211–227.

Vickery, W. (1969), 'Congestion Theory and Transport Investment', *American Economic Review*, **59**, 251–260.

Vranas, P., Bertsimas, D.J. and Odoni, A.R. (1994), 'Dynamic Ground Holding Policies for a Network of Airports', *Transportation Science*, **28**(3), 275–291.

Welch, D.J. and Lloyd, T.R. (2001), *'Estimating Airport System Delay Performance'*, *4th USA/Europe Air Traffic Management R&D Seminar, Santa Fe, USA*, 11.

Winston, W.L. (1994), *Operations Research: Applications and Algorithms*, 3rd edn (Belmont: Duxbury Press).

Wojhan, O.W. (2000), *Airline Network Structure and the Gravity Model*, *Working Paper* (Hamburg: University of Hamburg, Department of Economics).

Wojhan, O.W. (2001), *Airline Network Structure if Passengers Value Time and Service Quality, Working Paper* (Hamburg: University of Hamburg, Department of Economics).

Wu, C.-L. and Caves, R. (2002), 'Research Review of Air Traffic Management', *Transport Reviews*, **22**, 115–132.

Yu, G. (1998), *Operational Research in the Airline Industry* (Amsterdam: Kluwer).

Chapter 5

Epilogue

This book has elaborated the concept of the sustainability of the air transport system and its components – airlines, airports and Air Traffic Control/Management (ATC/ATM). Unlikely other studies and documents dealing with sustainability, the book has explicitly included six instead of the commonly considered three dimensions of the system's performance, such as technical/technological, operational, economic, social, environmental, and institutional. In particular, the inherent dependability of particular performances has been emphasised, implying the need for an integral approach to sustainability of the air transport system.

Specifically, Chapter 1 introduced the concept of a sustainable entity, sustainable transport, and a sustainable air transport system.

Regarding geographical scale, dealing with the sustainability of the air transport system could be carried out on a local (airport, local community), regional (the system of given country or a continent) and global (the entire world) scale. This could also be carried out regarding the individual system components such as airlines, airports and ATC/ATM. In this context, sustainability implies increasing the absolute positive difference between the system's full social benefits and the costs of the impacts on society and the environment (that is the externalities). This approach differs from the well-established approach considering sustainability only as a relative and/or absolute reduction or stagnation of consumption of the non-renewable resources and related impacts with a simultaneous increase in the system output. Consequently, the present air transport system would never be considered as sustainable by either approach, mainly due to the continuous increase in the consumption of (non-renewable) fuel and related emissions of air pollutants, and noise around the expanding airports. Development of technology has been considered as a relatively efficient remedy.

Chapter 2 examined the different performances of airlines, airports, and ATC/ATM. The technical/technological performances of airlines relate mainly to the techno characteristics of aircraft such as speed, carrying capacity, and technical productivity, and the airline fleet composition. The operational performances embrace the fleet capacity, and the air-route network layout. Two different network layouts were considered: hub-and-spoke and point-to-point network. The economic characteristics embrace revenues, costs, and profits. In this context, three economic/business models were analysed: full cost (legacy), low-cost, and charter airlines. The social performances mainly relate to the direct and indirect employment dependent on the airline fleet size, characteristics of the network, and economic/business model. The environmental performances include fuel consumption, air pollution, noise and waste. The airline institutional performances embrace the structure of the airline industry and market regulation, including airline ownership. In particular, it was pointed out that only the economically strong and stable airlines have been able to include sustainability issues in their long-term development plans.

The technical/technological performances of airports consist of their size (land taken), and the infrastructure, facilities and equipment in the airside and landside areas used to handle the expected demand such as aircraft, passengers, and freight (cargo). The operational performances relate to the airport demand, capacity, quality of services and outcome of their dynamic interaction. The economic performances embrace the revenues from aeronautical and non-aeronautical services, costs, and profits. As in the case of airlines, the airport's social performances consist of the airport-related direct and indirect employment, as well as contributions to the local economies. The environmental performances refer to noise, local air pollution, land taken for expansion, and waste. The airport institutional performances include airport ownership, recognised as an important factor influencing its sustainable development.

The ATC/ATM technical/technological performances consist of the radio-navigational facilities and equipment on the one side, and the regulation of movements of the aircraft/flights by using the separation rules on the other. In addition to the unavoidable safety requirements, the operational performances include the demand in terms of the number of aircraft/flights, the airspace and air traffic controllers' capacity, and quality of service. This last is expressed by the efficiency and effectiveness of realised flights, that is deviations of the actual from the optimal four-dimensional trajectories. The economic performances relate to the revenues obtained from charging the airlines for the air navigational services and the costs of providing these services. The social performances mainly refer to direct employment. The environmental performances include safe aircraft guidance along the four dimensional fuel-optimal trajectories and management of the air traffic flows, which have contributed to reducing fuel consumption and the related emissions of air pollutants. In addition, this includes guiding the air traffic around airports aiming to reduce the noise burden. The *ATC/ATM* institutional performances have mainly referred to commercialisation of the air traffic services aiming speed up the implementation of new technologies through more intensive investments with simultaneous protection of the required level of safety, efficiency and effectiveness of services.

Chapter 3 examined the concept of sustainable development of the air transport system. This included assessment, evaluation and comparison of its full social benefits and costs. The costs concerned are either the expenses for mitigating or remedying the damage to people's health and the environment or the expenses to prevent or avoid the damage. In addition, the need to monitor development of the air transport system aiming to appropriately locate the prospective sources of benefits and costs were recognised. As a result, indicator systems were designed reflecting particular dimensions of the system's performances regarding their relevance for particular actors involved. The main actors are users-passengers and freight shippers, airlines, airports, ATC/ATM, aerospace manufacturers, local and central governments (policy makers), national and international air transport organisations and lobby groups, and community members around airports. As well, some real-life examples of sustainable development in the air transport system illustrating the general principles have been discussed.

Chapter 4 dealt with modelling sustainability in the air transport system. It included: an analysis and modelling of sustainability and vulnerability of the airline hub-and-spoke networks to disruptions; quantification of the extra fuel consumption due to flying at the fuel non-optimal altitudes; influence of new technologies and operational procedures on the runway's ultimate capacity and consequently land use at airports; optimal utilisation of the available airport capacity; principles of demand management by charging for congestion at airports; and a comparison of the environmental performances of the air and other High-Speed surface transport systems regarding their substitutive potential.

The presented material might give rise to the question of whether the air transport system is already sustainable, and if it will be able to move along a sustainable trajectory in the long term under current and prospective circumstances. If the full system benefits and costs are considered, the answer is certainly 'Yes! It is and will continue to be sustainable'. Reasoning behind such optimism is the continuous widening of the positive differences between the benefits and costs expressed per unit of the system output, mainly thanks to the permanent improvements and innovations in technology, operations, economic/business model(s), and safety in an increasingly liberalised institutional environment. In particular, the unit benefits have increased nearly or just slightly less than proportionally and the unit costs have decreased more than proportionally with increasing (that is growth) system output (the number of aircraft departures, the volume of p.-km or t-km).

The further technological and operational improvements directed to reducing the aircraft fuel consumption and noise and increasing safety of the ATC/ATM are being developed. Specifically, they include consideration of using new fuels such as liquid hydrogen, improving the technical and operational reliability of the system's components by using new materials, and implementation of innovative procedures for more efficient and effective utilisation of the available capacity of the airport and airspace infrastructure. Nevertheless, there are still some important matters of concern. Firstly, despite the fact that the share of air transport in the total transport fuel consumption and associated emissions of CO_2 is expected to stay at the present level of about 3–3.5 per cent, pressure on depletion of the increasingly scarce non-renewable fuel (crude oil) reserves will certainly continue. Secondly, depositing air pollutants such as CO_2, NO_x and H_2O globally, that is directly in the high atmosphere (tropopause) where they act as 'greenhouse' gases contributing to global warming and thus climate change will also continue. Even with hydrogen-based fuels the increased emissions of NO_x and H_2O from their burning will become a new matter of concern despite removing the most important air pollutant – CO_2. Nevertheless, the real contribution of air transport to the phenomenon of global warming remains to be investigated.

Dealing with the sustainability of the air transport system further could include development in two directions: i) consistent implementation of the concept and principles of sustainability throughout the system respecting the relevance for the particular actors involved; ii) designing a convenient tool for the monitoring and assessment of the system's sustainability at different levels – local, national and international. The tool would be based on the indicator systems and their measures, which would be regularly updatable in terms of their values and influencing factors.

The author does not doubt that sooner or later, the actors within the air transport system will manage to achieve consensus on the future trajectory of the system's sustainable development. New technologies, operational procedures, and economic-business models of the air transport enterprises, applied to a more liberalised (market) environment are guarantees for success, despite the further strengthening of the constraints and targets on particular system impacts on society and the environment.

Index